Studies in Systems, Decision and Control

Volume 498

The series "Studies in Systems, Decision and Control" (SSDC) covers both new developments and advances, as well as the state of the art, in the various areas of broadly perceived systems, decision making and control–quickly, up to date and with a high quality. The intent is to cover the theory, applications, and perspectives on the state of the art and future developments relevant to systems, decision making, control, complex processes and related areas, as embedded in the fields of engineering, computer science, physics, economics, social and life sciences, as well as the paradigms and methodologies behind them. The series contains monographs, textbooks, lecture notes and edited volumes in systems, decision making and control spanning the areas of Cyber-Physical Systems, Autonomous Systems, Sensor Networks, Control Systems, Energy Systems, Automotive Systems, Biological Systems, Vehicular Networking and Connected Vehicles, Aerospace Systems, Automation, Manufacturing, Smart Grids, Nonlinear Systems, Power Systems, Robotics, Social Systems, Economic Systems and other. Of particular value to both the contributors and the readership are the short publication timeframe and the world-wide distribution and exposure which enable both a wide and rapid dissemination of research output.

Indexed by SCOPUS, DBLP, WTI Frankfurt eG, zbMATH, SCImago.

All books published in the series are submitted for consideration in Web of Science.

Kofi Kissi Dompere

The Theory of Epistemic Fields

Fuzzy-Semantic Foundations of Intellectual Categories and Knowledge Factories

Springer

Kofi Kissi Dompere
Department of Economics
Howard University
Washington, DC, USA

ISSN 2198-4182 ISSN 2198-4190 (electronic)
Studies in Systems, Decision and Control
ISBN 978-3-031-42469-4 ISBN 978-3-031-42470-0 (eBook)
https://doi.org/10.1007/978-3-031-42470-0

This Springer imprint is published by the registered company Springer Nature Switzerland AG
The registered company address is: Gewerbestrasse 11, 6330 Cham, Switzerland

DEDICATION

To all scholars, researchers, students and supporters working on the epistemic foundations of Information-Knowledge connectivity to the system of neuro-decision-choice spaces of human behaviors over the spaces of problem-solution, question-answer and success-failure dualities, and their connections to the spaces of natural and non-natural transformations under relational continuum and unity for the continual construction of methodological approaches with methods and techniques to the understanding of the development of our universal knowledge systems and total intellectual heritage of humanity to the benefit of organization of societies in setting peace against war and nonviolence against violence peace against war as well as working very hard for the living and praying hard for the dead.

To Patrice Lumumba, Thomas_Sankara,_Samora Machel
Robert Mugabe, Ahmed Sékou Touré, Modibo Keita, Ahmed Ben Bella,
who had the courage and vision,
to work to help to decolonize Africa,
and to set the new face of Africa's complete emancipation
under information-knowledge and problem-solution disequilibrium states.
Finally, to Fella Anikulapo Kuti,
The Great African musician and liberation activist as a true African
With music is a weapon and water has no enemy

.

ACKNOWLEDGEMENTS

The theory of rational process, also characterized as the theory of rationality, depending on collectively acceptable rules over the space of thinking-reasoning dualities, irrespective of how it is conceived and interpreted, affects all areas of human thought and the derived laws of thought that may guide decision-choice behavior toward actions in managing our social set-ups in time and over time. There are many varieties of problem-solution dualities that constitute the space of problem-solution dualities for the construct of the theory of the general rational process over the space of ignorance-knowledge dualities to define the epistemic transversality conditions between opposites, where the theory of general rationality is defined under the conditions of unbreakable relations between information-knowledge and problem-solution dualities over the space of identification-transformation dualities. The space of identification-transformation dualities establishes the telescopic past in relation to past universal existence of *what there was* , telescopic present in relation to present universal existence of *what there is* and telescopic future in relation to *what there would be,* where *what there was* defines the structure of *what there is* which defines the conditions of variety and categorial identities for initializing the conditions of (info-static [R17.17], categorial conversion and necessity [R17.15]) for *what there would be* (info-dynamics [R17.18], philosophical consciencism, freedom and sufficiency [R17.16]) in Sankofa-anoma conceptual system of forecasting from the past, discounting from the future and deciding in the present to set the conditions of variety future. Thanks to those working on category theory in philosophy and mathematics and similarly those working on conflicts, tensions, and principle of opposites with dualities and polarities for the understanding of general dynamics.

In all the neuro-decision-choice actions, the Sankofa-anoma conceptual systems under the principle of opposites with relational continua and unity establish a universal principle of past-preset-future connectivity in an unbreakable chain of universal information-time interdependencies of variety give-and-take sharing modes over the space of identification-transformation dualities, where the fuzzy paradigm finds its methodological and analytical residence and power. In this respect, I thank all those involve in the development of paradigms of thought to help our understanding of our agreements and disagreements on collective thinking and reasoning as well as the understanding of the meaning of information-knowledge conditions in relations to uncertainty, identity and change and their relational connectivity to the spaces of destruction-construction and input-output dualities as well as to recognize the limitations of paradigm of thought. Important thanks to all working to prevent the greatest danger to integrity of the process of individual and collective knowing and information-knowledge development is the process of destruction of the conditions of thinking process which forms the foundation of reasoning over the space of problem-solution dualities to evade the social constraints of ideological faith and scientific credulity on thinking-reasoning dualities.

Special thanks go to friends, critics and those who have taken interest and time to read any of my monographs, whose positive and negative encouragements respectively have made this work enjoyable to the finish. My thanks also go to all my graduate students at Howard University in my courses in economic theory, mathematical economics and cost-benefit analysis for allowing me to introduce them to new techniques and logic of thinking and reasoning in the space of relativity with inexact information structure and the use of cost-benefit methods and techniques with attempts to free them from the rigid logical structure of the classical system over the space of absoluteness with exact information structure. Once again, I express grate appreciation to Ms. Jasmin Blackman for prove-reading the manuscript from beginning to the end. I accept all responsibilities for all errors.

EPISTEMIC FIELDS

PREFACE

I: REFLECTIONS ON CONCEPTS AND DEFINITIONS

This monograph is an epistemic continuity of my examination and development of the theories of information and knowledge as economic production and not as communication. It is special in the sense that the examinations and discussions combine conditions of neuro-decision-choice actions for knowing and conditions of information-knowledge production as interdependent social production, the results of which are inputs to all areas of socio-economic production of the social system of varieties in the space of input-output dualities. As a general neuro-decision-choice over the socio-economic space, the transformation processes and their inputs must be specified, and similarly, the outputs of the input transformation processes must be specified. The distinguishing conditions of the inputs and outputs of the transforming processes must be explicitly specified and stated for categories to establish a framework in examining conditions of categorial conversions, where varieties of categories lose their identities and enter new categories. Here, two organic conditions must be developed: one is category formation, and the other is categorial conversion.

The category formation leads to the theory of categorial conversion under either the classical paradigm of thought, the fuzzy paradigm of thought or the hybrid, where necessity and initial conditions are established through a system of neuro-decision-choice actions over the identification problem-solution dualities [R17.15] [R17.17]. The categorial conversion leads to the development of the theory of transformation processes that connects categories, where the developments of factors show the arrow dynamics among categories and the development of the theory of philosophical consciencism to establish sufficient and freedom conditions of intra-categorial and inter-categorial conversions over the transformation problem-solution dualities [R17.16] [R17.18]. The phenomena of variety identifications and variety transformation are treated as the general problem-solution dualities. In other words, the general knowing process and information-knowledge accumulation are about *variety identification* and *variety transformation*.

The variety identification problem-solution dualities are related to what there was and what there is over the space of telescopic past-present dualities. The variety transformation problem-solution dualities are related to *what there is* and *what there would be* over the space of telescopic present-future dualities, or, in general, are related to *what there was*, passing through *what there is* to what *there would be* over the Sankofa-anoma information-knowledge trinity of telescopic past-present-future structure, where the present resides in the past and the future resides in the present, and the past-present conditions shape the future structures over the space of imagination-reality dualities. The human intellectual efforts are about the neuro-decision-choice actions over the space of variety identification-transformation dualities to develop information-knowledge stocks for their uses against surprise, increase outcomes in the space of anticipation-expectation dualities, and the development and management of organizations and their institutions, where the institutions are specific in their problem-solution dualistic-polar structures in the epistemic field contained in the space of relativity-absoluteness dualities under a given paradigm of thought, where such paradigm of thought must be explicitly stated. The development of the paradigm of thought and its laws of thought and reasoning must be made explicit in relation to categories as departments in the epistemic field.

The neuro-decision-choice actions acquire increasing efficiency under the principle of sequential actions with relational connectedness under the give-and-take sharing principle in a state of oneness over the space of problem-solution dualities. The sequential actions require creating sets of problem-solutions with similar characteristic dispositions by an appropriate method of indicator function into either fuzzy decomposition or classical decomposition into categories, where the collection of these categories constitute the epistemic field and the epistemic field embedding with intra-categorial and inter-categorial relations and transformations with dynamic relations and information. The intra-categorial and inter-categorial give-and-take sharing relations are dynamics that allow categories to be combined as well as to be divided to serve the knowing and information-knowledge development through neuro-decision-choice actions over the space of problem-solution dualities.

Let us keep in mind that the correct and incorrect identifications of variety problems and variety solutions belong to the space of identification problem-solution dualities, where the variety identifications reveal variety identities and behaviors in static states, and the results of knowing through the neuro-decision-choice actions generate information-knowledge conditions of telescopic past (*what there was*) and telescopic present (*what there is*). Similarly, the correct and incorrect transformations from variety problems to variety solutions belong to the space of variety transformation problem-solution dualities, where the variety transformations reveal changes in variety characteristic dispositions in identities and behaviors in dynamic states, and the results of knowing through the neuro-decision-choice actions generate information-knowledge conditions of telescopic future (*what would be*), where new epistemic fields are induced. The space of variety identification problem-solution dualities and the space of variety transformation problem-solution dualities are contained in the space of imagination-reality dualities, where variety problems belong to the space of realities contained in the space of actuals, and the space of variety solutions belongs to the space of imaginations which is contained in the space of potentials, all the varieties of which are defined by their characteristic dispositions.

II: THE SPACE OF ACTUAL-POTENTIAL DUALITIES AND THE SPACE OF IMAGINATION-REALITY DUALITIES

The most important epistemic structure is the understanding of the nature, structure, differences, and similarities of all the spaces as defined and their relationships to one another and to the space of ontological-epistemological dualities, and how the ontological and epistemic spaces are abstracted from the space of ontological-epistemological dualities under the conditions of the general principle of opposites with relational continuum and unity. The ontological space contains the space of actual-potential dualities. The epistemological space contains the space of imagination-reality dualities. All the space of imaginations is contained in and abstracted from the space potentials to create a relational continuum and unity with the space of realities. All-natural activities directly take place over the ontological space. All neuro-decision-choice actions and social behaviors take place over the epistemological space that is relationally connected to the ontological space for knowing and information-knowledge accumulation. Unlike the natural processes over the ontological space, complications arise for neuro-decision-choice actions over the epistemological space. These complications are generated by indirect processes for the neuro-decision-choice actions to connect the epistemological space to the ontological space through the space of imagination-reality dualities to which we now reflect on.

III: THE SPACE OF IMAGINATION-REALITY DUALITIES, POSSIBILITY PROBABILITY, POSSIBLE WORLD, AND IMPOSSIBLE WORLD

The relational connectivity of possibility, probability and the space of imagination is presented to resolve certain misunderstanding and confusion about the concepts and measurements of possibility, impossibility, probability and improbability, and how the concepts of possibility and probability may be extended to the concepts of possible world, impossible world and the space of possible-world-impossible-world dualities under the principle of opposites and a system of thinking and reasoning. Some of the elements in the set of terms with confusion in the spaces of definitions and explications are probability, possibility, uncertainty, information and knowledge. The destination of a zero probability of an event as an impossible outcome is analytically troubling in the sense of neglecting the notion that an event may be possible and yet improbable. An improbable event is not necessarily an impossible event under dualistic-polar conditions within the principle of opposites in relation to the space of possibility-impossibility dualities and the space of probability-improbability dualities. Every variety belonging to the probability space also belongs to the possibility space, while every variety belonging to the improbability space belongs to the space of possibility-impossibility dualities in the sense that it may belong to either the possibility space or impossibility space. In this respect, the work done by Dubois and Prade is not only helpful but instructive in understanding possibilistic thinking and necessity relative to probabilistic thinking and freedom [R5.17] [R5.20]. It helps in the understanding of relational separation, continuum and unity of the space of possibility and the space of probability, as well as contributing to the development of the possibility-probability or the fuzzy stochastic hybrid conceptual view on the general uncertainty over the space of quantity-quality dualities and the use of possibilistic-probabilistic reasoning in understanding the behavior of uncertain outcomes in the space of varieties as is also reflected in [R5.19] [R21.5].

The concepts of possibility and probability are mutually creating in terms of likelihood of outcomes under cognitive capacity limitations in relation to an information-knowledge system and neuro-decision-choice actions of human behaviors over all spaces contained in the epistemological space. The cognitive capacity limitations over the information-knowledge system generate uncertainty of qualitative and quantitative dispositions of varieties over the space of imagination-reality dualities as a sub-space of actual-potential dualities. The uncertainties are contained in the space of imagination, where there is the qualitative uncertainty and quantitative uncertainty. The qualitative uncertainty resides in possibility defined as the likelihood of variety anticipation and the collection of these variety anticipations is the possibility space for the construct of relevant fuzzy numbers and possibilistic reasoning. The concept of qualitative uncertainty is the difference of what would and what would not be of a variety in terms of characteristic disposition. The quantitative uncertainty resides in probability defined as the likelihood of variety expectation given the anticipation and the collection of all variety expectations constitutes the probability space for the construct of probability numbers and probabilistic reasoning. The hybrid of qualitative-quantitative uncertainty resides in the space of possibility-probability variety defined as the double likelihood of anticipation-expectation varieties, and the collection of all the simultaneous anticipation-expectation varieties constitutes the space of possibility-probability dualities for the construct of hybrid numbers and possibilistic-probabilistic reasoning in the space of fuzzy-classical paradigms of thought in knowing and information-knowledge production. The concepts of the double likelihood, the possibility-probability hybrid or the fuzzy-stochastic hybrid has given rise to the concept of probability of probability or imprecise probability or inexact probability over the space of subjective-objective dualities. These added qualifications on the concept and quantification of probability are due to cognitive capacity limitations from the space of acquaintance as well as the efficiency of the classical paradigm of thought which are linked to the space of imaginations containing the possibility as a primary category with probability as a derived category in the space of parent-offspring dualities.

The probability space, improbability space and the space of probability-improbability dualities are contained in the possibility space. This translates into the idea that the space of varieties with zero probabilities as improbabilities is contained in either the possibility space or the impossibility space, and similarly, the space with positive probabilities is contained in the possibility space and hence by dualistic-polar conditions the space of probability-improbability dualities is contained in the space of possibility-impossibility dualities. In this respect, we may ask the question as to what role does the impossibility and impossibility space play in examining the neuro-decision-choice activities in the space of imagination-reality dualities as a sub-space of the space of actual-potential dualities. In defining the concepts of probability and possibility the concept of information is used without adherence to laws of definition and explication. For example, let us look at the following definition: *Uncertainty refers to epistemic situations involving imperfect or unknown information.* The definitions of corresponding synonyms and antonyms are similarly troubling. Uncertainty as involving imperfect information is meaningless unless we know what information is. This is the definition-explication problem of information in all areas of knowing, knowledge accumulation and neuro-decision-choice actions.

In all the discussions and theories on concepts and phenomena of information, probability, uncertainty, doubt, belief, decision, choice and many such elements in the space of production-consumption dualities, the nature and meaning of information are assumed as known and given where the phenomenon and the concept of information are continually set in equality with the concepts of knowledge, fact, data, and evidence. In this respect, the relational differences and similarities of these interrelated phenomena and concepts are neglected in epistemic processes in explanatory and prescriptive processes for amplification of uncertainties in all neuro-decision-choice processes increasing the zone of riskiness. In the development, understanding and communication of theories and applications, it is useful to follow the structures of the *phenomena*, concepts, definitions and measurements in all areas of knowing and information-knowledge accumulation. In this respect, what is the phenomenon of the concept of information? What is the definition of the concept of information? What are the relationships among information and all the varieties, spaces, zones, categories, sets, fields, and neuro-decision-choice actions over the space of individual-collective dualities? Substantial clarification of what constitutes inputs and outputs into decision choice processes must be made explicit since these input-output processes are related to the spaces of successes-failure and risky-riskless dualities. Generally, knowledge is an input into neuro-decision-choice processes and information is an output of neuro-decision-choice actions which then becomes input into knowing and knowledge processes to produce knowledge as input into neuro-decision-choice actions over the problem-solution transitional dynamics of the interchangeabilities of problem-solution outcomes. The phenomenon, concept and the meaning of information and their relationships to the phenomena, concepts and meanings of fact, data, knowledge and evidence are discussed in [R17.17] [R17.18]. From these phenomena, concepts, definitions and the constructed spaces, uncertainties are generated from the differences and comparisons of the same variety with differences of characteristic dispositions in the space of actual-potential dualities and the space of imagination-reality duality, where the variety transitional transformations generate different uncertainties in the space of possibility-probability dualities which result from the conditions of the space of ontic-epistemic dualities. The organizational relations among the spaces are presented in Figure IIIA.

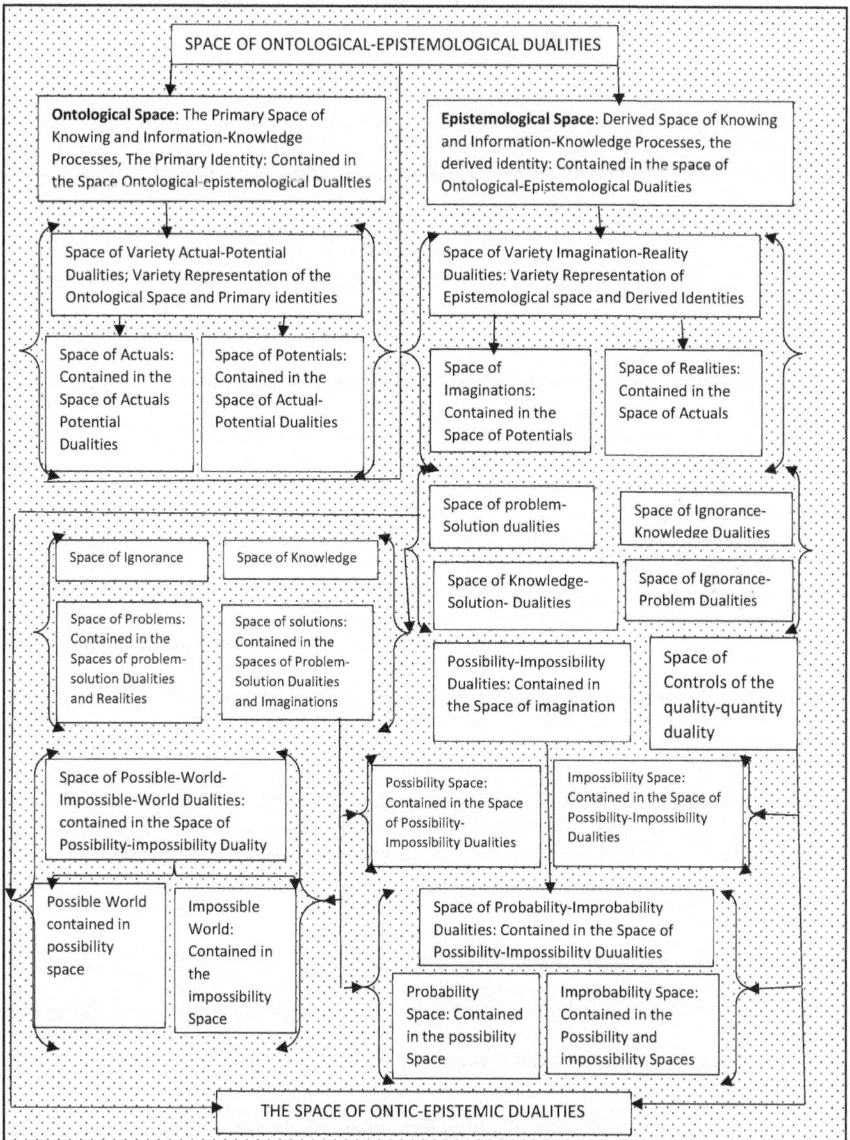

Fig IIIA ORGANIZATIONAL PAREMIOLOGY OF THE GEOMETRY OF SPACES AND SUB-SPACES FOR KNOWING, NEURO-DECISION-CHOICE ACTIONS AND INFORMATION-KNOWLEDGE PROCESSES UNDER THE GENERAL PRINCIPLE OF OPPOSITES

IV ANALYTICAL CONFUSIONS ON THE CONCEPTS AND PHENOMENA OF POSSIBILITY AND PROBABILITY

The confusions over possibility and probability, impossibility and improbability with corresponding spaces and possible world and impossible world center around the confusions on their phenomena and concepts. The differences and similarities of the concepts and phenomena concerning these varieties must be established for epistemic clarities in definitions and measurements with a clear understanding that the identity of a phenomenon proceeds concept, definition, explication and measurement and the roles they play in neuro-decision-choice actions over the space of imagination-reality dualities. The analytical system connecting all these ideas is made explicitly by the development of the theory of epistemic fields and epistemic field embeddings that will allow us to discuss phenomena and concepts, composition-decomposition aggregations-disaggregation over the space of relativity-absoluteness dualities with lexical elasticity and conceptual flexibility. This lexical elasticity and conceptual flexibility may be related to the Brouwer fixpoint theorem seen in the theory of definitions and mathematical representation relative to synonyms, antonyms, fuzzy set covering and linguistic flexibility in use and application in the family of ordinary languages (FOL) and not in the family of abstract languages (FAL). A variety in the space of imagination is said to be possible if it has a transformation function that converts variety imagination to variety reality. It is said to be impossible if such transformation function cannot be found. This transformation function is functor for each variety which is also defined in the space of problem-solution dualities in the epistemic field supported by relational and information fields the theory of which this monograph is about.

The theory of the epistemic fields with the supporting relational and information fields provides us with a self-contained framework to trace the epistemic journey over the space of ignorance-knowledge dualities. It, further, allows a connectivity to how mathematical theories of category, functor and conformal mapping may help in the understanding of their applications to the intra-categorial and inter-categorial transformations as well as neuro-decision-choice actions with transitional uncertainties linking the space of problem-solution dualities to the space of cost-benefit dualities under the psycho-economic principle of non-satiation, where more is preferred to less for variety benefits and less is preferred to more for variety costs as analytical structures are imposed on the neuro-decision-choice behavior with cognitive capacity limitations over the space of identification-transformation dualities as we journey over the space of ignorance-knowledge dualities. In this respect, it may be useful to offer another definition with a corresponding explication of the possibility space through the conceptual system of both functor and categorial conversion, where a variety possibility may be defined in relation to either the space of functors or the space of categorial convertors in the sense that a variety is said to be possible if there is either a functor or categorial converter that can transform it from the space of imaginations to the space of realities in the sense that every variety possibility is made up of the variety in the imagination space with a corresponding functor from the space of functors or a categorial converter from the space of categorial convertors. Under the principle of opposites, a variety impossibility is composed of a variety without a corresponding functor from the space of functors or corresponding categorial converter from the space of categorial convertors in the space of imaginations to transform it to the space of realities.

Given the functor and categorial approach to the definition and explication of the possibility space, we may say that the presence of either functor or categorial convertor generates anticipation where the corresponding spaces generate the space of anticipations which correspond to the possibility space and the space of non-anticipations corresponding to the impossibility space.

In this respect, the probability space contains expectation in addition to anticipation as generated by either a functor or a convertor in the possibility space. The improbability space is a sub-space of the space of possibility-impossibility dualities, in that, not all varieties with anticipation have expected outcomes, but some varieties in the space of anticipation and all the space of non-anticipation have no expectations. These are the differences between possibility and probability and the possibility space and the probability space. The similarity is that the possibility, impossibility, possibility-impossibility duality, the probability, improbability, probability-improbability dualities and all the corresponding spaces are contained in the space of imaginations.

V: THE SPACE OF PRODUCTION AND REPRODUCTION OF REAL LIFE, CONFORMAL MAPPING AND THE SPACE OF PRODUCTION AND REPRODUCTION OF PROBLEM-SOLUTION DUALITIES

The epistemic structure of each problem-solution duality over the space of static-dynamic dualities is a production and reproduction of problem-solution dualities over the space of parent-offspring dualities. The space of problem-solution dualities is a conformal mapping onto the space of the space of production and reproduction of real life through neuro-decision-choice actions over the space of cost-benefit dualities. The important epistemic point is that every variety problem is also a variety solution where every variety solution generates a new problem requiring a new solution. The conformal mapping shows that every cost-benefit duality is mapped onto a problem-solution duality with further mapping onto the space of production and reproduction of real life, preserving important key characteristics of genealogy and heredity that allow the present to be connected to the past in historically telescopic processes to establish conditions of the present to be connected to the telescopic future under Sankofa-anoma tradition in information continuum and unity for the historical process in organicity. Such a conceptual system sees the meaning of life as defined by a system of problems and the essence of life is defined as the way problems are solved in never-ending problem-solution processes. Cost-benefit conditions are used to define problems, solutions and problem-solution dualities which are used to define the space of production-consumption dualities which is mapped onto the space of production and reproduction of real life with a past-present-future telescopic structure.

The epistemic field with information and relational fields containing categories and sub-categories create a framework of categorial conversions which is then related to conformal mapping which helps to preserve the essential properties over the space of identification-transformation dualities. In human neuro-decision-choice actions over the space of life-death duality, the theory of categorial conversion in addition to the theory of info-statics provide us with initial conditions and necessity for change, while the theory of philosophical consciencism in addition to the theory of info-dynamics provide us with sufficient conditions and freedom for transformation. Every variety identification presents conditions for variety transformation in either intra-categorial dynamics, intra-categorial dynamics or both over the space of incentive-disincentive dualities as established over the space of input-output dualities.

VI: THE EPISTEMIC FIELDS AND INTERNAL MULTIVATIONS

In the development of an approach to knowing and information knowledge development, it is useful to account for some important internal forces of qualitative motivation and encouragements such as courage, hope, despair, curiosity, creativity, critical thinking, critical reasoning, perseverance and other cognitive characteristics that increase motivations and efficiencies of efforts in the journeys over the space of ignorance-knowledge dualities, where the journeys are intra-categorial and inter-categorial transformations of variety ignorance to variety knowledge through the individual and collective neuro-decision-choice actions over the space of problem-solution dualities. The theory of the epistemic fields must provide a framework to integrate these cognitive characteristics into the knowing processes and information-knowledge accumulation as neuro-decision-choice actions over the spaces of production-consumption dualities, and input-output dualities over the action space.

In these processes, all the phenomena and concepts must be defined in terms of characteristic dispositions over the relevant spaces which must be defined as collections of varieties for identifications and transformations and all-knowing varieties must be established through variety signal dispositions from variety characteristic dispositions under cognitive capacity limitations. The conceptual ideas contained in the theory of knowing as production of intellectual flows and the theory of information-knowledge accumulation as intellectual capital stocks are that the theory of knowing and the theory of the information-knowledge system are completely interdependent and integrated as foundations of general systems of variety production-consumption dualities, input-output dualities, destruction-replacement dualities over the space of variety identification-transformation dualities relative to the space of imagination-reality dualities containing the space of possible-impossible dualities, and where the space of possibility contains the space of probability-improbability dualities. The outcomes of neuro-decision-choice actions in knowing must be identified and subjected to quality control before they become available as inputs into information-knowledge processes under a paradigm of thought to be accepted into the intellectual capital stocks. It is here, that critical needs of analytical definitions and explications become necessary for category formations, categorial identities, categorial conversions, variety identities and variety transformations relative to the complex system of neuro-decision-choice actions over all spaces of human endeavors.

It is the need for analytical definitions and explications relevant for critical understanding of knowing and information-knowledge accumulation as encompassing a complete system of spaces and sub-spaces of human productions why I criticized the classical theory of knowledge and the definition of knowledge as *justified true belief* as confusing, insufficient, and unhelpful. How is *justified true belief* related to information, how is information related to knowledge that is in a *justified true belief*, how is knowledge as *justified true belief* related to data, how is data related to fact and how is fact related to evidence and evidential things and how are paradigms of thought related to the production of these concepts and ideas? Within the fuzzy paradigm of thought, these concepts are seen as fuzzy categories and the directions of their relations as obtained by fuzzy conformal mappings.

The theoretical and applied difficulties in the general understanding of knowing and information-knowing accumulation over the space of fundamental-applied dualities through the classical approach to knowledge are found in the lack of explicit definition and explication of information and the role of information as an input into the production of knowledge and the roles information and knowledge as input-output outcomes in neuro-decision-choice behavior, where is the primary category of knowing and knowledge is a derived category in the space of input-output dualities. In this respect every neuro-decision-choice is a productive activity over the space of identification-transformation dualities that contains the spaces of ignorance-knowledge and problem-solution dualities. These difficulties lead to the interchanging use of information, knowledge, fact, data, and evidence in thought without any attention to their lexical elasticities, defined by membership characteristic functions in the fuzzy epistemic space, in such a way that their interchanging uses become a menace to creative thinking, reasoning, and knowing progress over the space of imagination-reality dualities. The approach in the development of the theories of information-knowledge processes, info-statics, info-dynamics, entropy and problem-solution dualities supported by the fuzzy paradigm of thought helps to resolve some of the theoretical and applied difficulties, where explicit definitions and explications are given to information, knowledge, fact, data and evidence as a sequential conceptual system on the basis of characteristic-signal dispositions and to show their roles in communication over the space of source-destination dualities and how they affect neuro-decision-choice actions over the space of human behavior. The phenomenon and the concept of information are not only input-output phenomena but encompass everything in similarity, commonness, differences, and distinction that must be made explicit without which human behavior will be impossible in our socio-natural environment. It is useful to note that the fuzzy paradigm of thought provides a way to deal with lexical elasticities in degrees of belonging and substitutions within synonyms and antonyms in communication under the principle of opposites.

VII. THE MONOGRAPH

The monograph is about the development of the theory epistemic fields with relational and information fields. The construct of the epistemic field starts with category formation of the space of problem-solution dualities with the use of appropriate indicator functions for knowing and information-knowledge accumulation, where there are complex intra-categorial and inter-categorial relations making explicit the understanding of give-and-take sharing modes in categorial sub-spaces of problem-solution dualities, where the resulting conditions in one category of problem-solution dualities may provide some keys to unlock the conditions of problem-solution dualities of other categories locked in relational continuum and unity. The process presents a framework to understand institutional creations with relevant departments in relation to neuro-decision-choice actions for knowing through research, teaching, and learning over the space of socio-natural problem-solution dualities, the results of which generate experiences towards increasing information-knowledge stocks and perfection. Given the need for a segmental system of neuro-decision-choice actions over the general space of socio-natural problem-solution dualities, the monograph is set to develop a theory of epistemic fields with relational and information fields through category formation in a dynamic system of give-and-take sharing modes. The categories are used to partition the general space of socio-natural problem-solution dualities into categories of problem-solution dualities that contain problem-solution varieties with similar characteristic dispositions. The theory presents a framework for analyzing variety identities, variety relations and the role of information that provides a system of continual interactions over the space of positive-negative dualities in knowing through neuro-decision-choice actions over the identification problem-solution dualities and transformation problem-solution dualities in relational continuum and unity over the space of cost-benefit dualities.

This approach will allow us to discuss the conditions of a number of question-answer dualities regarding the role that cognitive qualitative variety characteristic dispositions play in critical thinking and reasoning relative to creativity, courage, innovation, curiosity problem-solution identifications and transformations under knowing as the production of intellectual investment flows and information-knowledge accumulation as the production of intellectual capital stocks. . These questions come to us always as primary and derived relative to neuro-decision-choice actions in all areas of human actions. We must keep in mind that the categorial partition of the space of socio-natural problem-solution dualities are a strategy and tactic to maximize the effectiveness of neuro-decision-choice actions over the general space of problem-solution dualities by simultaneously taking advantage of intra-categorial and inter-categorial results within the space of acquaintance-description dualities under a paradigm of thought.

The theory of epistemic fields with the supporting relational and information fields must be constructed to maintain the phenomena of the expansion of the space of socio-natural problem-solution dualities as well as the expansions of the relational and information fields, where these expansions have no upper bounds as old varieties exist and new varieties enter the space of varieties through the results of the natural and neuro-decision-choice actions over the space of socio-natural varieties as connected to the space of destruction-construction dualities that relate to new categories of problem-solutions dualities and natural restructuring of the ontological environment. In other words, the knowing processes and information-knowledge processes are not only interdependent but are mutually creating from neuro-decision-choice actions irrespective of the category of knowing. Every category of knowing is a category of problem-solution dualities under decision-choice actions passing through the space of imagination-reality dualities as a sub-space of actual-potential dualities, where the space of imaginations contains the space of complex interactions of the spaces of possibility-impossibility dualities, probability-improbability dualities, possibility-probability dualities, possible world, impossible world, and possible-world-impossible-world dualities.

The *space of imaginations* contains the spaces of uncertainty-certainty dualities, doubt-surety dualities and decidability-undesirability dualities, visions, goals and objectives demanding courage, hope creativity and tenacity where the outcomes of neuro-decision-choice actions are nothing more than experiences, and problem-solution dualities and information-knowledge accumulation is either justified or unjustified experience that provides conditions for corrections and revisions of forces operating within opposites towards a zone of perfection under the principle of non-satiation. Alternatively, the *space of realities* contains the complex interactions of the space of problems, individual-collective dualities, cognitive characteristic dispositions, cognitive limitations relative to courage, creativity, innovations and other such epistemic qualities in thinking, reasoning, and action. The other relevant questions relating to the need for the development of the theory of epistemic fields are stated in the preface contained in [R8.16] and how all these monographs are relationally interconnected to knowing as a production of intellectual investment flows and information-knowledge accumulation as intellectual capital accumulation to define the meaning of life and establish the essence of life within the space of problem-solution dualities.

The theory of the epistemic fields is developed to deal with: 1) the understanding of how the space of problem-solution dualities is partitioned into categories ; 2) the understanding of the relationships among intra-categorial transformations and conversions, inter-categorial transformations and conversions, and relations to the *mathematical theory of conformal mapping, category and functors*; 3) the understanding of incentive structures in categorial movements under the principles of non-satiation in the infinite search for perfection; 4) the understanding of the relations among, thinking, reasoning neuro-decision-choice actions, vision, goals, objectives, knowing, knowledge and ignorance as categorial conversions for cognitive journeys; 5) the understanding of the relevant spaces for knowing as the production of intellectual investment flows and information-knowledge accumulation as the production of intellectual capital stocks; 6) the understanding of knowing, and information-knowledge accumulation as the foundational and primary system of production-consumption dualities from which derived systems of production-consumption dualities emerge over the spaces of input-output dualities and destruction-production dualities; 7) the understanding of categorial diversities, relational continuum and unity in the space of problem-solution dualities; 8) the understanding of the establishment of foundations of categories and departments of knowing, learning, research and teaching over the space of problem-solution dualities in the development of information-knowledge stocks; 9) the understanding of relational connectivity of information-knowledge stocks and social philosophical consciencism as guides to neuro-decision-choice actions of elements in the space of individual-collective dualities within the space of thinking-reasoning dualities; 10) the understanding of the rise of the space of systemic good-evil dualities such as racism, risk, crime, violence, injustice, justice, kindness, wickedness and many such moral varieties in the space of acceptable-unacceptable dualities of human behavior under the general principle of some codes of conduct in the social set-up that despises iniquities, lies and related baleful behaviors; and 11) how together these ten conditions provide an integrated framework for the development of institutional arrangements and their organizations over the space of problem-solution dualities.

VIII: ORGANIZATION OF THE MONOGRAPH AND THE CHAPTER SUMMARIES.

The monograph is organized with a preamble, prologue, preface, acknowledgement and nine chapters which are concluded with an epilogue. Each chapter deals with a specific set of analytical difficulties within the development of the theory of the epistemic fields with relational and information fields in support of the understanding of the development of the theory of problem-solution dualities and polarities, info-statics, info-dynamics and entropy under fuzzy information and a paradigm of thought for the general understanding of the principles of opposites, tension, conflicts, diversity and unity of knowing, science and general information-knowledge accumulation as the primary category of production-consumption dualities.

There is a *preamble to the monograph*. The preamble is composed of a series of quotations from important works by different authors in philosophy, science, mathematics, economics, decision-choice theories, and other relevant areas to the information-knowledge questions about the general science and art of knowing under cognitive capacity limitations that generate questions over the spaces of certainty-uncertainty dualities and projected to the space of doubt-surety dualities. The preamble is used to initialize the idea that all information-knowledge systems are generated by knowing actions irrespective of the area of human endeavor through neuro-information-decision-choice processes over the space of problem-solution dualities within the ontological-epistemological duality. Again, the objective of the preamble is to introduce the reader to the conflict zones of knowing and to point out the epistemic zones of irreducible disagreements in knowing as intellectual investment and an information-knowledge system as intellectual capital accumulation enterprise.

The *prologue* is analytically structured to follow the contents of the preamble. The prologue introduces a set of essential epistemic propositions in the theoretical system on categories, the epistemic field, epistemic-field embeddings, relational fields and information fields and their relationships to categories of intellectual investment flows and information-knowledge accumulation in intellectual capital stocks. The prologue is used to set the stage of the development of the theory. Based on these essential propositions, the concepts of methodology, techniques and methods for discussing the propositions are defined, explicated and analyzed to provide conditions of their differences and commonness in theoretical and applied analytics. The techniques and methods are toolbox collections in any methodology of interest for knowing, thinking, and reasoning over the space of the problem-solution dualities. The prologue is also used to introduce the concepts of exact and inexact information structures and how they relate to the space of absoluteness and the space of relativity to affect the methodology of thinking and reasoning over the space of decidability-undecidability dualities. The prologue points to the path of the theoretical journey and the parameters of containments and deviations regarding the applicable areas over the space of question-answer dualities and polarities to establish diversity and unity principles in knowing and science.

The *preface*: the prologue is followed by a preface which introduces the motivation and conditions for writing the monograph, what it is about and how the conceptual framework of the monograph is connected to other previous monographs by the same author on information-knowledge processes and general conditions of categories of knowing. The general conceptual frame makes explicit the structure of the theory of problem-solution duality and polarity and how it forms the foundations of knowing and information-knowledge accumulation in all areas of cognitive activities over the space of fundamental-applied dualities in knowing. The analytical connectedness of these monographs is found in the consistent use of the principle of opposites, inexact information, fuzzy paradigm of thought with its mathematics, logic and corresponding toolbox of techniques and methods of analysis, where knowing is seen as an intellectual investment flow in a discovery of knowledge and information-knowledge is seen as an intellectual capital accumulation of discoveries, providing us with factories of production and consumption in the general input-output space and decision-choice actions over the space of problem-solution dualities. Together, these monographs present the meaning and essence of my works, research agenda and possible contribution to organization of knowing, research, and teaching, where information-knowledge stock-flow disequilibrium dynamics is seen as an economic production and related to continual neuro-decision-choice dynamics over the space of problem-solution dualities that resides in temporary equilibrium states and permanent disequilibrium dynamics. In a sense, the interrelated works provide foundations for explaining the continual process of transformation of varieties in terms of parent-offspring replacement processes at the levels of both fundamental and applied knowledge on the dynamics of emergence of the new and the disappearance of the old, where the end is in the beginning and the beginning is in the end in continuum and connectivity, as well as take a critical look at the logic of the economic theory of production in which the production of intellectual investment flows and intellectual capital stocks are based.

Chapter 1 of this monograph is devoted to the development of the theory of the *epistemic fields* and corresponding *relational fields* based on *information fields* relevant in understanding the creations of specific and multidisciplinary areas of study, foundations of the creation of departmental systems of research, learning and teaching, and the creation of different institutions of research for information-knowledge and problem-solution transformation strategies to achieve desired social goals and objectives. It further provides the logical foundations for multidisciplinary approaches to neuro-decision-choice activities over the space of problem-solution dualities which are viewed as variety elements in the ontological space in categories containing institutions and departments. It is about the epistemic understanding of *diversity and unity* and *diversity and disunity* of knowing through neuro-decision-choice actions over the space of problem-solution dualities, where nothing is knowable without neuro-decision-choice actions and nothing is transformable without the interactive processes of the space of neuro-decision-choice actions and the space of problem-solution dualities that form the foundation of the development of information-knowledge systems as stock-flow disequilibrium dynamics over the space of simplicity-complexity dualities through knowing, where given the system of neuro-decision-choice actions, nothing is knowable without information, defined as characteristic-signal disposition that establishes varieties in similarities, commonness, differences, diversities, categories and relations. It also establishes the required logical conditions for specific and interdisciplinary approaches in developing educational institutions and departments of research, learning and teaching through the understanding of category formation over the space of problem-solution dualities and categorial conversions for interdisciplinary collaboration to also offer a framework for the development of social institutions and departments of policy analytics over intra-categorial and inter-categorial problem-solution dualities for individual cooperation and collective efficiencies in dynamic conditions of problem-solution dualities.

Chapter 2 provides the understanding of the interactive relations among the phenomena of categorial dividedness, a general analytical space, a characteristic-signal field and its connectivity to the epistemic fields supported by the corresponding relational and information fields. The objective is to show how thought and practice are integrated to relate variety imaginations to variety realities with further understanding of how the possibility space and probability space are contained in the space of imaginations to present the space of necessity-freedom dualities and how they are related to knowing and information-knowledge processes as stock-flow disequilibrium dynamics. The chapter also provides relevant definitions and characteristic dispositions of the information-knowledge system from the space of problem-solution dualities and how the indicator functions are used to construct categories of areas of knowing with interdisciplinary dynamics as well as epistemic spaces. From the indicator functions three epistemic fields are presented as $((\sigma)$-field or (σ)-algebra$)$ induced by the classical indicator function, $((\alpha$-field$)$ or $(\alpha$-algebra$))$ induced by the fuzzy indicator function and $((\gamma$-field$))$ or $(\gamma$-algebra$)$ induced from the characteristic-signal disposition. The chapter provide a framework for the understanding of diversity with relational continua and unity in knowing and information-knowledge development as socio-natural production-consumption processes in all areas within the space of imagination-reality dualities contained in the space of actual-potential dualities under the general principle of opposites. The $((\gamma$-field$))$ or $(\gamma$-algebra$)$ provides us with an analytical space of the development of hybrid systems in knowing and information-knowledge systems under variety characteristic-signal dispositions.

Chapter 3 deals with the conceptual and analytical foundations of the phenomena and conditions of diversity and unity in knowing and science through the space of problem-solution dualities and polarities, where the primary principle of diversity and unity of knowing and the primary principle of diversity and unity of science and non-science are established for identity and difference. This is followed by the development of supporting derived principles of knowing and science. Categorial connectivity analytics are undertaken to show how all the knowing and science are in relation to neuro-decision-choice actions under social philosophical consciencism at generation points, where the dynamics of the social philosophical consciencism shape the dynamics of generational preferences over the space of problem-solution dualities and polarities and where the information-knowledge intellectual stocks shape the social philosophical consciencism to further control the problem-solution dualities and the knowing processes. From the space of problem-solution dualities, dynamics of social philosophical consciencism, methods and techniques of thinking and reasoning are introduced as meta-paradigms and distinguished from the organic paradigm of thought. All neuro-decision-choice actions and the structure of variety problem-solution balances are defined over the space of cost-benefit dualities under the socio-psychological principle of non-satiation. Under these conditions, the knowledge by acquaintance as the primary category of knowing and the knowledge by description as the derived category are introduced and their relationships with organic paradigms of thought and meta-paradigms of thought are established. The chapter is ended with discussions on differences and similarities of information. knowledge, data, fact, and evidence.

Chapter 4 of this monograph presents two essential elements for the construct of the principles of diversity and unity of knowing as well as the principles of diversity and unity of science, where the latter is seen in relation to the former which is the parent in a parent-offspring process in the epistemic field and characteristic-signal field, where the meaning of problem-solution duality is defined by goals and objectives over the space of cost-benefit dualities. Knowing is a primary foundation in all human actions over the space of production-consumption dualities, where the space of consumption-production dualities is intimately connected to the space of neuro-decision-choice actions in all areas of human existence, where any category of knowledge as intellectual stock is a derivative from the category of knowing as an intellectual investment flow. The construct of the diversity and unity principles of knowing and science is based on the fundamental notion that all human activities are neuro-decision-choice actions to solve *problems* in relation to *things,* analytically called varieties with characteristic dispositions that establish variety identities, commonness and differences called *information-knowledge conditions*. This foundational framework allows for the development of an epistemic field and epistemic field embedding as mathematical and philosophical entities to examine the thinking-reasoning path to knowing, the diversity and unity of knowing and the partition of the space of knowing into categories of knowing in terms of disciplines and inter-disciplines of intellectual investment flows and information-knowledge accumulations as intellectual capital stocks. This chapter examines extra factors in support of the primary and derived principles of diversity and unity in knowing and then questions the idea of disunity of intellectual investments and accumulation of information-knowledge stocks. It sets the stage for examination of the organization of knowing as the production of *intellectual investment flow* and conditions of variety information knowledge claims as production of *intellectual capital stocks* over the space of ignorance-knowledge dualities.

Chapter 5 is devoted to the analysis of the fuzzy paradigm of thought as an organic paradigm for thinking and reasoning and in the development of fuzzy meta-paradigms for the analysis of the conditions in the space of ignorance-knowledge dualities. From the cognitive framework of the fuzzy paradigm of thought the technique and method of fuzzy optimization are used to analyze the optimal conditions of varieties in the space of ignorance-knowledge dualities, where variety ignorance and variety knowledge are mutually constraining for abstracting the optimality which is then used to establish and define the transversality conditions in the journey from the zone of ignorance to the zone of knowledge in specifying the variety identity for knowing. The conditions of optimality are used to decompose the zones of knowledge and ignorance for specifying the optimal knowledge and optimality fuzzy decomposition for neuro-decision-choice actions over the space of the problem-solution dualities. Discussions are made with examples on the use of fuzzy numbers for optimizing ignorance-knowledge relationality and the use of the conditions of optimization for decomposition of the space of ignorance-knowledge dualities into the zone of ignorance and the zone of knowledge. The cumulative results of the chapter is a development of a theory of ignorance-knowledge dualities that allows the specification for the conditions of zones of ignorance and knowledge for a confidence claim of variety identities with fuzzy conditionality as known where the fuzzy paradigm of thought is coordinated with the principle of opposites with relational continuum and opposites.

In chapter 6, we explore certain ideas of conceptual and epistemic anomalies such as paradox, phantom, possible world, impossible world, imagination, fantasy, curiosity, and creativity in relation to reality, actuality, potentiality, and other such analytical structures that emerge in knowing and information-knowledge accumulation. These concepts are placed in a system of decision-choice actions in relation to elements in the spaces of decidability-undecidability dualities and input-output dualities to provide foundations for further explorations into these concepts and phenomena and their effects in social progress. The epistemic concern here, is how these anomalies relate to the space of possibility-probability dualities that houses the spaces of necessity-freedom dualities and risk-benefit dualities as one journeys between variety ignorance and variety knowledge in ignorance-knowledge dualities in the space of intellectual investments and information-capital accumulations over the space of individual-collective dualities. The discussions on the anomalies are then linked to the discussions of categorial conversion in [R17.15] and to the discussions of social philosophical consciencism in [R17.16], where the contents in philosophical consciencism include the anomalies, conceptual weapons that generate structural conditions in the spaces of justice-injustice, war-peace, and other dualistic-polar structures. The space of imagination-reality dualities is introduced as an analytical space and shown to be a sub-space of the space of actual-potential dualities. The space of imaginations is shown as a derived space from the space of the potentials and then shown to contain the possibility and probability spaces, while the space of realities is shown to be contained by the space of the actuals. The space of imagination-reality dualities is argued to generate all the intellectual anomalies the structures of which are discussed, and to examine the effective integration of hope, courage, curiosity into creativity and innovations in knowing and information-knowledge processes.

Chapter 7 follows the works in chapter 6 to reflect on the concepts and phenomena of possibility with possibility space, impossibility with impossibility space and their relationships to possible and impossible worlds relative to the telescopic conditions over the space of identification-transformation dualities with further relations to the space of static-dynamic dualities providing a framework to the logical debates on the true-false conditions of varieties contained in the possible worlds relative to the space of imagination-reality dualities. The space of imaginations is shown as a derived space from the space of the potential and them shown to contain the possibility and probability spaces, which then give rise to the concepts and philosophical reflections of possible and impossible worlds. The concepts and phenomena of possibility and possibility are defined and explicated with comparative differences and similarities differently from the existing ones. Their similarities and differences of the relationships to the concepts and phenomena of possible and impossible worlds are discussed relative to the spaces of true-false and information-knowledge dualities. The chapter is concluded with examples of mathematical representations of dualistic-polar membership functions that may be used under dualistic-polar conditions with the use of the fuzzy paradigm of thought.

The *epilogue* of this monograph provides a connectivity with the prologue and the prologue with the epistemic development of theories on knowing as social production of intellectual investment flows and information-knowledge accumulations as social production of intellectual capital stocks as telescopic projections from the space of past-present dualities to the space of present-future dualities relative to a space of development of theories on organizational analytics as a central role of simultaneity of oneness, diversity, unity and disunity, where knowing and information-knowledge accumulation are seen in terms of intellectual investment-capital stock-flow dynamics. The fuzzy rationality, principles of opposites with relational continua and unity, the theory of info-statics, identification problem-solution dualities, the theory of info-dynamics, transformation problem-solution dualities, the theory of problem-solution dualities as production-consumption dynamics over the spaces of input-output dualities and destruction-construction dualities, are all connected to the space of varieties as contained in the space of imagination-reality dualities as a derived category by neuro-decision-choice actions from the space of actual-potential dualities. The connectivity systems of actions for variety identifications and transformations are undertaken through the development of organizations and sub-organization containing institutions and departments which are relevant for neuro-decision-choice actions and policies relative to dynamic categories of problem-solution dualities. The system of ideas and theories as outlined so far points to the development of the theory of organizational analytics and the contents institutions within the processes of explanatory and prescriptive sciences, to give us pathways of understanding of the conditions of past-present dualities and parent-offspring dynamics of institutions and problem-solution dualities to connect the space of imaginations as a sub-space of the space of potentials to the space of realities as a sub-space of the space of actual which allows us to provide explanations to cognitive anomalies and systemic outcomes in the spaces of true-false dualities and good-evil dualities with relational continuum and unity under the principle of opposites. Foundations are, therefore, prepared to analyze the concepts and phenomena of functor and categorial converter and how they are affected by curiosity, creativity, perseverance, fear, and other such qualities in the space of knowing and information-knowledge accumulation relative to the thinking-reasoning processes.

PREAMBLE

I: QUOTATIONS

(QI)*Many problems in the social and administrative sciences concern the division of some given total into a number of components. The question may then arise: how large is the degree of "dividedness"? The entropy provides an answer to this question,* [R20.13, p.1].

(QII)*Modern systems theory has been developed almost exclusively on what is commonly called "the state space approach" The problem we encounter here is the fact that the state space approach is not easily explained or defended in a few; nor is it easy to explain the relation between the state space point of view and the notion of "an input-output relation parametrized by (evolving in) time." Indeed, it was Zadeh [38,40] who pointed out the size of the problem by observing the state space of a processor to be an even more complex parametrization of it than time itself. It suffices to say there is no simple phrase to describe the state space point of view of processor unless we accept that a processor is an input-output relation parametrized (receiving in) state. In other words, one has to take the "state" of a processor as a primitive concept* [R17.67, p.140]

(QIII)*Neither acquiescence in skepticism nor acquiescence in dogma is what education should produce. What it should produce is a belief that knowledge is attainable in a measure though with difficulty; that much of what passes for knowledge at any given time is likely to be more or less mistaken, but that the mistakes can be rectified by care and industry* [R3.25, p. 24].

(QIV)*Education in credulity leads by quick stages to mental decay; it is only by keeping alive of free inquiry that the indispensable minimum of progress can be achieved..* [R3.25, p. 26].

(QV)*One of the chief obstacles to intelligence is credulity, and credulity could be commonly diminished by instructions as to the prevalent form of mendacity* [R3.25, p.27].

(QVI)*It is a progressive development, in which the bounds of knowledge are stage by stage enlarged, in which the agreement of ideas and theories with objective reality is stage by stage increased, and in which stage by stage what was provisional and hypothetical gives place to what is assumed and verified....*

The progress of knowledge always comes up against barriers which arise from the limitations of existing knowledge add existing practice. But while the progress of knowledge always faces barriers to further advances, knowledge progresses by finding how to get over them. [R17.11b, p. 179].

(QVII)*From the Submission of our desires springs the virtue of resignation; from the freedom of our thought springs the whole world art and philosophy, and the vision of beauty by which, at least, we half reconquer the reluctant world. But the vision of beauty is possible only to unfettered contemplation, to thoughts not weighted by the load of eager wishes; and thus Freedom comes only to those who no longer ask of life that it shall yield them any of those personal goods that are subject to the mutation of Times.* [R3.25, P15].

(QVIII)*Today, however, philosophy has little need to resort to crypticisms. Speaking in general terms, I may say that philosophy has fashioned two branches of study which enable it to solve the problem of categorial conversion in a satisfying way. These tools are Logic and Science,* *both of which owe their origin and early development to the demands of philosophy. The conceptual tools which philosophy has fashioned in Logic, and by means of which it can cope with the formal problems of categorial conversions, are contained in nominalism, constructionism and reductionism. For philosophy's model of categorial conversion. It turns to science. Matter and energy are two distinct, but, as science has shown, not unconnected or irreducible, categories. The inter-reducibility of matter and energy offers a model for categorial conversion. And another model is given in the distinction between physical change and chemical change, for in chemical change physical quantities give rise to emergent qualities*. [R17.44, P21]

(QIX)*Everything in the universe maintains its continuity (reproduces itself) through polarization in the image of the mystery of the original polarization and the alternation of polarities makes for existence (apparent life) growth, maturity, and aging.* [R17.59, p12]

(QX)*The title of this note was selected so as to provide an excuse for discussing some rather general matters. Let us first consider the question of which of probability and statistics came first historically. This question is like the one about eggs and chickens. The question whether eggs or chickens came first could in principle be given a meaning by using precise definitions of eggs and chickens, and even then, probably somebody would be to answer the question* [R24.14a, p. 59]

II: REFLECTIONS ON THE QUOTATIONS IN THE PREAMBLE

The contents of the quotations require some reflections to connect them to the development of the monograph and its connectivity to the theory of knowing, information-knowledge development, and neuro-decision-choice actions over the general space of human behaviors, where problem-solution, question-answer and success-failure dualities are defined with relational continuum and unity. The important interest, here, is not only on the questions on the connotations of information, knowledge, and their relationships but the whole activities on information-knowledge processes, their effects on cognition, human life, its quality, progress and how they constitute the core of socio-economic production in universal existence. Information-knowledge items are input-output varieties as well as constitute the core of the production of socio-economic varieties and the information-knowledge items, where every variety production is information, and every information item is a production.

All variety productions take place through neuro-decision-choice actions over the space of problem-solution dualities under preferences, where problems are replaced by solutions and solutions generate new problems. The complexity of the space of problem-solution dualities is such that the varieties of the problem-solution dualities are placed in cohorts through a fuzzy partition to create sub-spaces of problem solution dualities in relational continuum and unity of existence, where the relational continuum and unity of existence define the relational continuum and unity of knowing as intellectual investment flows and information-knowledge accumulation as intellectual capital stocks. Here comes the conceptual relevance of decomposition of unity into sub-categories as contained in quotation one **(QI)** and two **(QII)** in terms of oneness and dividedness as strategies in neuro-decision-choice activities over the space of problem-solution dualities.

The results of neuro-decision-choice actions over the sub-spaces of the space of problem-solution dualities must not be taken as dogma, settled knowledge and conclusions. Knowing as investment intellectual investment flows and information-knowledge accumulation as intellectual capital stocks rest on belief where the neuro-decision-choice results are problem-solution processes that are accomplished over the spaces of decidability-undecidability, success-failure and error-correction dualities under the Sankofa-anoma telescopic time trinity over the space of learning-non-learning dualities. It is here, that the practice of the principle of doubt acquires its utility in the processes of knowing and information-knowledge production, while an important obstacle to neuro-decision-choice efficiency over the space of problem-solution dualities is principle of credulity that prevents the mind from entering the house of free inquiry over the space of thinking-reasoning dualities, where freedom of thought liberates visions encapsulated in the space of imagination that houses possibility containing necessity, and probability containing freedom. The knowing and information-knowledge as neuro-decision-choice activities are always constrained by cognitive capacity limitations that enhance the ignorance conditions in the space of problem-solution dualities requiring increasing industry as pointed in the quotations **(QIII-QVII).**

The knowing process as the production of intellectual investment flows and the information-knowledge accumulation as the production of intellectual capital stocks proceed in terms of movement from one category to another over the space of ignorance-knowledge duality induced by categorial conversions or functorial transformations given the structure of category formations in the space of problem-solution dualities, where institutions arise in response to categorial problem-solution needs, where old institutions may give way to new ones or the old and new institutions may be consolidated into oneness in response to the new conditions of the problem-solution dualities to facilitate categorial conversions over the space of ignorance-knowledge dualities. In the knowing and information-knowledge processes, we are confronted with variety categorial conversions which are achieved by logic and science in multiplicities of directions as pointed out in the quotation **(QVIII),** where the relational continuum and unity of the universal existence under the principle of mater-energy equilibrium complemented by the principle of information-knowledge disequilibrium are such that every variety functions over the space of identification-transformation dualities under parent-offspring dynamics over the space of input-output dualities. The relational continuum and unity reveal the actions of dualities through the poles of the polarities to give rise to static-dynamic conditions over beginning-aging dualities in parent-offspring structures as pointed out in quotation **(QIX).**

In this respect, the space of variety parent-offspring dualities is always like the space of egg-chicken dualities as reflected in the quotation **(QX).** Both are reflections of universal beginning-end duality that evades cognitive understanding due to cognitive capacity limitations of cognitive agents over the past-present-future telescopic structure in continuum and unity, producing paradoxes in mathematical representations over the spaces of thinking-reasoning, computability-uncomputability, decidability-undesirability, success-failure dualities under paradigms of thought. It is the conditions of cognitive capacity limitations that generate energy for increasing information-knowledge stocks and progress over the social space, where progress is obtained by neuro-decision-choice actions over the space of problem-solution dualities by setting solutions against problems and hence knowledge against ignorance in a never-ending process over intergenerational space as is reflected in quotation **(QVI).** The conditions required for correctness of mathematical representation and computation are pointed out in this monograph where the theory of epistemic field is presented and discussed.

Finally, the quotations in the preamble will provide us with some foundations in understanding the source and phenomena of uncertainty, space of certainty-uncertainty duality, the rise of imagination and the space of imaginations and how imagination gives rise to the phenomena of possibility, impossibility, possibility-impossibility duality, and the corresponding spaces which contain necessity, initial conditions, and anticipations for the logical experiment of possible worlds in cognition. Similarly, how the same space of imagination gives rise to the phenomena of probability, improbability, probability-improbability dualities, and the corresponding spaces which contain conditions of freedom, expectations and sufficiency over the space of identification-transformation dualities. The analytical framework allows one to examine how expectations relate to anticipations, to present anticipation-expectation dualities, necessity-freedom dualities, and possibility-probability dualities to give rise to possibility and probability theories. The framework also provides us with the relationship between necessity and freedom and their roles in the spaces of neuro-decision-choice actions, problem-solution dualities, social progress, and the evolving civilizations.

The way probability has been introduced to us in mathematics, philosophy, economics and other areas of knowing presents some epistemic difficulties and confusions such that analytical difficulties arise over the spaces of quality-quantity dualities and subjective-objective dualities with unnecessary disagreements without any meaningful discussions and references to the role of cognitive capacity limitations that relate to the whole space of human actions over the spaces of input-output and cost-benefit dualities that generate the spaces of goals, objective, visions and problem-solution dualities where these spaces are affected by and determined from the conditions of social philosophical consciencism in relation to the differences of inter-generational preferences over the space of problem-solution dualities.

From the space of ontic-epistemic dualities arises the *general uncertainty principle*. This general uncertainty principle characterizes all areas of knowing and knowledge production over the spaces of quality-quantity dualities, exact-inexact dualities, vagueness-clarity dualities. Even though the uncertainty principle is general, it has sub-categories such as uncertainty principle in physics, economics, all areas of decision-choice theory, all areas of measurements and over the space of production-consumption dualities and quality controls. It is here that the general theory of entropy requires increasing development and refinement. Again, all anomalies, ill-posed problems, incomputable problems, Gödel's completeness, and incompleteness condition are explainable form the conditions from ontic-epistemic dualities as derived category and the set of human cognitive capacity limitations is the primary category which gives rise to the space of knowability-unknowability dualities.

A PROPOSAL

1. Statement of Aims and Rationale behind the Monograph

a) The ***aim*** of this monograph is to provide a pure and applicational framework in support of the rise of institutions and organization of these institutions for problem-solving over the space of human actions in thought and practice over the universal existence of space of social problem-solution relative to the space of matter-energy dualities. The framework is seen in relation to the conception of human existence as defined by the space of problem-solution dualities in that the meaning of life is problem solving and the essence of life is the nature of solutions and the method of obtaining. The framework is analytically self-contained over the space of imagination-reality dualities as derived from the space of actual-potential dualities which establishes the ontological apace while the space of imagination-reality dualities is established the epistemological space. The resulting framework is the monograph on the theory of epistemic fields containing a multiplicity of categorial institutions and departments of problem-solving dualities with decompositions and integrations to reveal the variety identities, the understanding of the dynamics of a number of current and future problem-solution phenomena as they relate to human existence over the space of socio-natural dualities relative to the power of imagination, the audacity of knowing, the courage of action, the sadness of failure, the indestructability of hope, the joy of success, the existence of non-satiation, the living essence of thinking and the force of reasoning, the rise of problems and the search for solutions.

b) The ***objective*** of the monograph is to develop a logically consistent unified explanatory and prescriptive theories to explain the rise and growth of areas of knowing through the partitioning of the space of problem-solution dualities into categories for intra-categorial and inter-categorical give-and-take-sharing modes to explain the categorial knowing of inter-dependencies that impose demands for cooperation as derived from the natural law of input-output relational structures with give-and-take sharing modes and demand-supply chain dynamics viewed in the space of input-output dualities under the general principle of opposites. The other objective is to show how the principle of opposites, consisting of duality and polarity with relational continuum and unity, acts as powerful analytical tools in understanding the current, emerging and possible areas of knowing defined over the space of problem-solution dualities that occupies the epicenter of human mind and cognition to establish the diversity and unity of areas of knowing and science under the conditions of the problem and solution parent-offspring process. Another basic objective is to reveal and emphasize the basic premise that all socio-natural varieties exist in dualistic-polar states, and transformations do not destroyed the basic proposition of variety dualistic-polar universal existence, where variety information is indestructible no much how one tries to overlook or hides it. It is also to show the important role that the fuzzy paradigm of thought plays in constructing areas of knowing as interconnected categories of problem-solution dualities, where all information structures within the space of exact-inexact dualities are connected to the space of absolute-relative dualities to make explicit the philosophical and analytical debate on interdisciplinary knowing and learning over space of problem-solution dualities in the dynamics of research universities over the space of quality-quantity space, where the framework provides the dynamics of research universities and institutions. The objectives are also amplified by the usefulness of dualistic-polar games as methods and techniques to connect the principle of opposites to the space of relativity, the construct of categories of areas of knowing to develop the theory of epistemic fields and how the theory of epistemic fields connects to information and relational fields in the understanding of diversity and unity of knowing and science as one examines the epistemic progress over the space of ignorance-knowledge dualities and the role of transversality conditions in separating zones of ignorance and knowledge and their applications in establishing the zones of influence of duals, poles and conflicts in the dualistic-polar games.

c) The *rationale* for writing this monograph is derived from multiple epistemic conditions of the relationships among the concepts of possibility, possible world, probability, imagination, reality, and paradigms of thought as they relate to the path of history of knowing, uncertainty, risk, and information-knowledge production composed of telescopic past, telescopic present, and the telescopic future where:

c1) There is a need to establish a system of cognitive spaces over the space of activities of cognitive agents and to show their interdependences, interconnectedness and give-and-and-take sharing modes in the space of variety mutual existence-destruction dualities, where these spaces may be seen in terms of organization of production-consumption processes in such a way that each space is a collection of categories to justify multidisciplinary programs in research, teaching and learning in our education systems toward restructuring the whole educational foundations for redefining research, teaching and learning universities and colleges in terms of problem-solution dualities.

c2) There is also a need to reexamine the concepts of information and knowledge in relation to imagination and reality and their connectivity to explanatory and prescriptive theories and to establish the spaces, where cognitive activities occur as well as allow the examination of the relationality between variety imaginations and variety realities and how they are connected to national personalities, exceptionalism and social philosophical consciencism in the space of neuro-decision-choice actions expressed over the space of problem-solution dualities. The success and usefulness of examination and reexamination will require an establishment of analytical distinctions and a separations between the phenomenon of information-knowledge transmission from the phenomenon of the meaning and production of information-knowledge systems in relation to the space of ignorance-knowledge dualities in the cognitive journeys of knowing from zone of ignorance to zone of knowledge.

c3) There is an important need to show the space in which nature operates relative to the space in which society operates in relation to their differences and similarities relative to natural knowing-transformation and social knowing-transformation processes over the space of direct-indirect dualities.

c4) There is a need to provide a framework of epistemic analytics for the idea that all knowing are undertaken in the same space of ignorance-knowledge dualities under the space of ignorance-knowledge dualities as generated by ontic-epistemic differences in variety characteristic dispositions which result from the multiplicities of cognitive capacity limitations as defined in different categories of knowing.

c5) There is also a need to show that categories of problem-solution dualities and ignorance-knowledge dualities are cognitive factories and plants of production-consumption dualities with interchangeability of input-output dynamics under the neuro-decision-choice actions over the space of fundamental-applied dualities that support social dynamics, where phenomena generate words, words generate concepts, concepts generate thoughts and thoughts generate actions over the space of neuro-decision-choice actions in infinite disequilibrium processes.

d) The ***monograph***: Given the aim, objective and rationale in relation to the interactive conditions among the space of neuro-decision-choice activities, the action space, the space of question-answer dualities and the space of problem-solution dualities as viewed in relative relations to categories of knowing and science through the space of ignorance-knowledge dualities, the monograph is set to provide answers to several questions through the development of the *theory of epistemic fields*, where any epistemic field is composed of categories of problem-solution dualities. The categories come to us as primary and derived categories with defined characteristic dispositions which are related to institutional and organizational conditions of knowing and information-knowledge production. The questions also come to us as primary and derived and are center around the role of how neuro-decision-choice actions over the different categories of problem-solution dualities

to bring about knowing and how the knowing processes generate information-knowledge accumulation in disequilibrium dynamics irrespective of the category of knowing in terms of science and non-science, social science and non-social science. The aggregates of the stock-flow disequilibrium conditions are related to categorial dynamic of the problem-solution and question-answer dualities.

The process of knowing, the knowledge accumulation process and information-knowledge stock-flow disequilibrium dynamics journey through spaces, starting from the space of acquaintance as its ignorance beginning to the space of description as its knowledge destination. The conditions of each space and their connectivity types are defined and analytically related to the space of give-and-take sharing modes, where special analytics of their connectivity to the space of possible-impossible world duality are provided and examined over the space of ignorance-knowledge dualities. The conceptual difficulties under socio-natural decision-choice actions over the all spaces and the space of problem-solution dualities creating increasing possibilities with variety necessities and anticipations and the space of probabilities with variety expectations and freedoms are analytically examined. The differences and similarities of the concepts of actual and reality are analytically structured to allow for rational discussions on intentionality of the concepts of misinformation, disinformation, news, fake news, and propaganda in communications over the space of source-destination dualities and their inputs into the space of neuro-decision-choice actions to affect the structure of cost-benefit distributions over the space of incentive-disincentive dualities. Similarly, the differences and similarities of the concepts of potential and imagination are examined which is then related to critical examinations of the concepts of belief, goals, objectives, vision, hope, creativity, courage, thinking and reasoning in the activities of neuro-decision-choice agents in the knowing and information-knowledge accumulation processes .

Every area of knowing in the space of imagination-reality dualities is a category of problem-solution dualities connected to the space of actual-potential dualities under neuro-decision-choice actions passing through a series of epistemological spaces. The constructed spaces start from the *ontological space* which is composed of the *space of actuals* and the *space of potentials* to present the *space of actual-potential dualities* that presents the *ontic space* as the primary category of identity which is connected to the *space of acquaintance* through the ontological characteristic-signal dispositions, and then to the epistemological space, where every variety is defined by the dualistic-polar existence in characteristic disposition over the space of negative-positive dualities. At the level of nature there are two important spaces that are of important in the knowing process. They are the spaces of actuals and potentials under the principle of opposites with relational continuum and unity. Analytically, the space of actuals contains *what there was* and *what there is*, while the space of potentials contains *what there would be.* In Sankofa-anoma representation, *what there was,* is the *telescopic past*, what there is, is *telescopic present* and *what there would be* is the *telescopic future.* The activities of nature are direct between the space of actuals and the space of potentials in the dualistic-polar negation and negation within the space of actual-potential dualities.

At the level of sociality, there are two important spaces and sub-spaces that are of important in the knowing processes as cognitive journeys. They are the spaces of *imaginations* and *realities* also under the principle of opposites with relational continuum and unity. Analytically, the *space of imaginations* is a derived space from the space of potentials. The space of *reality* is a derived and destination space from the space of actual-potential dualities for the epistemic journeys to know *what there was* and *what there is* in the space of actuals and to manufacture *what there would be* from the space of the potentials. The cognitive journeys by neuro-decision-choice agents are indirect to construct the telescopic past and telescopic present contained in the space of actuals and telescopic future contained in the space of potentials. The indirect journeys are undertaken in the space of acquaintance and the space of imaginations to the space of actuals to obtain the space of ontic-epistemic dualities to establish the spaces of certainty, uncertainty, and certainty-uncertainty dualities over the space of imagination-reality dualities contained in the space of actual-potential dualities.

The space of imaginations provides an indirect pathway to the space of actual through the space of imagination-reality dualities since direct epistemic journeys are not available to cognitive agents, where the space of imaginations plays a dominant pathway containing the spaces of *possibility, impossibility, possibility-impossibility dualities, probability, improbability, probability-improbability dualities, certainty, uncertainty, certainty-uncertainty dualities, necessity, anticipation, freedom, expectation, necessity-freedom dualities, anticipation-expectation dualities, possible world, impossible world*, and *possible-world-impossible-world dualities* relative to the space of ontic-epistemic dualities that establishes and connects the space of reality to the space of actuals. The journeys through these spaces are undertaken with the cognitive vehicles of thinking and reasoning over the space of success-failure dualities across the space of ignorance-knowledge dualities by crossing the transversality conditions that establish dualistic-polar zones of influences and conflicts of claims. The thinking-reasoning processes are then used in examining the roles played by creativity, courage, hope, despair, curiosity, innovation, and anomalies relative to the conditions of cognitive capacity limitations over the space of certainty-uncertainty dualities. The important question regarding neuro-decision-choice actions in Sankofa-anoma conceptual tradition is simply, what information can cognitive agents assess from the telescopic past, telescopic present, and telescopic future to assist in setting solutions against problems or answers against questions. This Sankofa-anoma conceptual tradition defines forecasting from the past information and discounting from the future information to construct the present information as an input into neuro-decision-choice actions as guided by the ruling social philosophical consciencism. In this epistemic process, the concepts of time and information trinities in their dualistic-polar structures are introduces as instruments for analytics.

e) The ***abstract*** provides a general summary of the contents of the book and a thematic structure, where the topical developments of the argument are given in the ***table of contents.***

2) THE GENERAL REFLECTIONS ABOUT THE MONOGRAPH

This monograph is a follow up of the previous monograph on the theory of problem-solution dualities and its connection to other previous works on information, knowledge, and decision in the space of production-consumption dualities. This monograph is devoted to the development of the theory that will support category formation and utilization of categorial converters and functors in the space of identification-transformation dualities relative to the general space of problem-solution dualities The development is based on the fundamental idea that neuro-decision-choice actions are the foundation of life which is anchored in the space of problem-solution dualities from which the path of social history is established by success-failure outcomes under cognitive capacity limitations. The space of problem-solution dualities is partitioned into the spaces of variety *identification problem-solution dualities* and *variety transformation problem-solution dualities* and human actions as they relate to changes in relations and properties. The development of institutions and organizations of institutions is related to categories of problem-solution dualities to simplify and create efficiencies for cognitive actions over the space of problem-solution dualities, where there are relational continua and unity. The theory of the epistemic fields with relational and information fields is developed to provide logical and rational foundations for development of institutions and organizations and the conceptual foundations are explicated to connect to the tactics and strategies of decision-choice actions over the space of problem-solution dualities. The degrees of efficiencies of the tactics and strategies will depend on intra-categorial and inter-categorial relations and interdependencies over the space of give-and-take sharing modes. The concepts of a problem and a solution are defined by cost-benefit conditions in the space of relativity rather than in the space of absoluteness, where the fuzzy paradigm of thought in semantics and neuro-decision-choice actions is used with conditions of principle of opposite in continuum and unity over categories and categorial dynamics.

All analytical spaces and concepts are specified and defined with characteristic dispositions in verbal and mathematical representations for easiness of analysis and synthesis in the space of relativity to provide an infinite framework and space of comparisons, differences, commonness and ranking under individual, collective, community and social preference orderings over the space of cost-benefit dualities for establishing sequences of visions, goals, objectives for intra-categorial and inter-categorial understanding of give-and-take-sharing modes under the socioeconomic principles of non-satiation. The concepts of educational, research and teaching diversities and the unity of general knowing through organization of institutions of social set-up are structured around the nature of variety problem-solution dualities over the space of static-dynamic dualities in infinite processes. The differences of neuro-decision-choice behaviors through the differences of social philosophical consciencism help to explain the differences in personalities, exceptionalism, preferences, and success-failure outcomes across nations and over generations within nations over epistemic fields with relational and information fields, where the space of the problem-solution dualities will vary over epistemic fields with differences in relational and information fields that induce differences and conflicts over the international neuro-decision-choice space under the asantrofi-anoma conditions of cost-benefit rationality.

The monograph is used to explore the dynamic framework for institutional development and organization of institutions as strategy and tactics of neuro-decision-choice actions over the space of problem-solution dualities contained in the space of static-dynamic dualities. The theory of epistemic fields with supporting relational and information fields provides descriptions and explanations of the cognitive journey over the space of ignorance-knowledge dualities, where collaborations and conflicts may arise. Each category in any epistemic field houses a fuzzy class of problem-solution dualities. Each space houses transitional stations of the cognitive intra-categorial and inter-categorial journeys in relational continuum and unity over the epistemic field. The transitional stations of the epistemic journey allow critical examinations of factors that affect the continuity of the neuro-decision-choice actions in tactics, strategies and possible restructuring in the knowing journeys to ensure effective feedback processes that generate self-correction system over the spaces. The transitional stations are temporary equilibrium states over the space of identification-transformation dualities.

The conceptual system of the theory is developed under the fuzzy paradigm of thought over the space of relativity that is characterized by the principle of opposites composed of duality and polarity where every variety is characterized by a series of dualistic-polar conditions over the space of cost-benefit dualities. The neuro-decision-choice actions over the space of problem-solution dualities are such that neuro-decision-choice actions in each space encounter the variety destruction-replacement processes guided by preference orderings under real cost-benefit conditions and rationality of cognitive agents with information-knowledge of quality control as provided by a system of entropic relational tests within the interactive spaces of categorial certainty-uncertainty dualities, doubt-surety dualities and decidability-undecidability dualities and success-failure dualities. The system of preference orderings varies over nations in line with experiential information as contained in the social philosophical consciencism from which visions are formed and goals and objectives are set.

3. THE SYNOPSES

A) For the general trust of the book see the attached abstract.

B) For the main topics see the table of contents

4. ORGANIZATION OF THE MONOGRAPH

The **monograph** is organized with a preamble, a prologue, a preface, an acknowledgement, seven chapters, an epilogue, and comprehensive cross-disciplinary references.

The *preamble* is composed of series of quotations from important works of different authors in philosophy, science, mathematics, economics, decision-choice theories, information theory and other relevant areas to the information-knowledge question-answer dualities about general sciences and art of knowing under cognitive capacity limitations that generate problem-solution dualities over the space of certainty-uncertainty dualities derived from the space of ignorance-knowledge dualities and projected to the spaces of doubt-surety dualities and decidability-undecidability dualities. The preamble is used to introduce the conceptual disagreements and agreements on the idea that all information-knowledge systems are generated by knowing actions irrespective of the area of human endeavor through neuro-information-decision-choice processes over the categories of the space of problem-solution dualities within the ontological-epistemological duality relative to the space of imagination-reality dualities contained in the space of actual-potential dualities and how one moves from the space of acquaintance to the space of description as one journeys over the space of variety ignorance-knowledge dualities.

The *prologue* of this monograph is used to indicate the entry point into the development of this monograph concerning the theory of epistemic fields as establishing the theoretical and applied frameworks for examining fuzzy and non-fuzzy semantic conditions over the space of success-failure dualities as well as its continuity with the epilogue of my monograph entitled *The Theory of Problem-Solution Dualities and Polarities*. The prologue points to the essential concepts, and phenomena and their fuzzy semantics structures relative to the classical semantic structures that will be central to the development of the theory and the requirements for its applications in the institutional and departmental zones of knowing and knowledge development over the space of problem-solution dualities for basic and applied conditions in the theory of the epistemic fields supported by the relational and information fields and how the epistemic, relational and information fields are related. In this respect, a basic reflection is undertaken followed by discussions on ontological, epistemological, imagination, ignorance, reality, and knowledge interact over the space of possibility-probability dualities, where the concept and nature of knowledge economy is introduced for future development.

The *preface* introduces the motivation for writing the monograph, what it is about and how the conceptual framework of the monograph is connected to other previous monographs by the same author on information, knowledge, data, evidence, economics, and decision-choice actions. The general conceptual frame makes explicit the structure of the theory of problem-solution duality and polarity and how it forms the foundations of knowing and information-knowledge accumulation as well as provides us with foundational links to information technology and relevant computer connections. The analytical connectedness of these monographs is found in the consistent use of the principle of opposites, fuzzy paradigm of thought with its mathematics, logic and corresponding toolbox of techniques and methods of analyses, where knowing is seen as a production of intellectual investment flows in the discovery of knowledge, and information-knowledge accumulation is seen as a production of intellectual capital stocks of discoveries providing us with a system of intellectual industries and factories of production with decision-choice actions over the space of problem-solution dualities. Together, these monographs present the meaning and essence of my works and research agenda, where information-knowledge stock-flow disequilibrium dynamics is seen as an economic production and related to continual neuro-decision-choice dynamics over the space of problem-solution dualities and, where stock-flow equilibrium and disequilibrium dynamics are related to matter-energy states and information-knowledge states in the universal existence. The information stock-flow disequilibrium takes place within the matter-energy stock-flow equilibrium. The preface provides an organization of paremiology of the essential spaces and sub-spaces contained in the development of the epistemic fields for neuro-decision choice actions under the general principle of opposites with relational continua and unity. Every neuro-decision-choice action is seen as problem-solving within the space relation-property dualities and every problem-solving action is seen as production-consumption process in the space of input-output.

Chapter 1 of this monograph is devoted to the development of the theory of *epistemic field* and corresponding *relational field* based on *information field* relevant in understanding the creations of specific and multidisciplinary areas of study, foundations of the creation of departmental systems of research, learning and teaching, and the creation of different institutions of research for information-knowledge and problem-solution transformational strategies to achieve desired social goals and objectives. It, further, provides the logical foundations for decomposition, analysis, multidisciplinary formations, institutional foundations, interdisciplinary approaches to neuro-decision-choice activities, over the space of problem-solution dualities, which are viewed as variety elements in the space of ontological-epistemological duality relative to the spaces of negative-positive and cost-benefit dualities. It is about the epistemic understanding of *diversity and unity, diversity and disunity, oneness and dividedness, and disorganization and organization* of knowing through neuro-decision-choice actions over the space of problem-solution dualities, where nothing is knowable without neuro-decision-choice action and the information field, and nothing is transformable without the interactive processes of the space of neuro-decision-choice actions and the space of problem-solution dualities. The spaces of neuro-decision-choice actions and problem-solution dualities form the foundation of the development of information-knowledge systems as stock-flow disequilibrium dynamics over the spaces of input-output dualities, parent-offspring dualities and simplicity-complexity dualities through knowing and information-knowledge accumulation seen in terms of production-consumption processes. The complete system of spaces and sub-spaces is presented in Fig IIIA in the preface.

Chapter 2 provides the understanding of the interactive relations among the phenomena of categorial dividedness, general analytical space, characteristic-signal fields, and their connectivity to the epistemic fields as supported by the corresponding relational and information fields. The objective is to show how thoughts and practices are integrated in the space of categorial dividedness to relate variety imaginations to variety realities with further understanding of how the possibility space and probability space are contained in the space of imaginations to present the space of necessity-freedom dualities and how are they related to knowing and information-knowledge processes as stock-flow disequilibrium dynamics. The chapter provides relevant definition and characteristic disposition of information-knowledge system from the space of problem-solution dualities and how the indicator functions are used to construct categories of areas of knowing with interdisciplinary dynamics as well as epistemic spaces. From the indicator functions three epistemic fields are presented as $((\sigma)\text{-field or }(\sigma)\text{-algebra})$ induced by the classical indicator function and $((\alpha\text{-field})\text{ or }(\alpha\text{-algebra}))$ induced by fuzzy indicator function and $((\gamma\text{-field})\text{-})$ $(\gamma\text{-algebra})$ induced from the characteristic-signal disposition with hybrids of $((\sigma\text{-}\gamma)\text{-field})$,and $((\alpha\text{-}\gamma)\text{-field})$. The chapter provides a framework for understanding the diversity-unity relationships in knowing and information-knowledge development, institutional constructs, and departmental creations in relation to all areas within the space of actual-potential dualities.

The chapter 3 deals with the conceptual and analytical foundations of the phenomena and conditions of diversity and unity in knowing and science through neuro-decision-choice actions over the space of problem-solution dualities and polarities, where the primary principle of unity of knowing and the primary principle of unity of science and non-science are established for identity and difference. This is followed by the development of supporting derived principles. Categorial connectivity analytics are undertaken to show how all the knowing and science are in relation to neuro-decision-choice actions over the space of problem-solution dualities under social philosophical consciencism at generation points, where the dynamics of the social philosophical consciencism shapes the dynamics of generational preferences over the space of problem-solution dualities and polarities. From the space of problem-solution dualities and the dynamics of social philosophical consciencism, methods and techniques of thinking and reasoning are introduced as concepts of meta-paradigms and distinguished from the concept of organic paradigm of thought. All neuro-decision-choice actions and the structure of variety problem-solution balances are defined over the space of cost-benefit dualities under the socio-psychological principle of non-satiation. Under these conditions, the knowledge by acquaintance as the primary category and the knowledge by description as derived category are introduced and their relationships with organic paradigm of thought and meta-paradigm of thought are established in the epistemological space. The chapter is ended with discussions on differences and similarities of information, knowledge, data, fact and evidence.

Chapter 4 of this monograph presents two essential elements for the construct of the principles of diversity and unity of knowing as well as principles of diversity and unity of science, where the latter is seen in relation to the former which is the parent in a parent-offspring process in the epistemic field and characteristic-signal field. Knowing is presented as a primary foundation in all human actions over the space of production-consumption dualities, where the space of consumption-production dualities is intimately connected to the space of neuro-decision-choice actions in all areas of human existence, where any category of knowledge as a production of intellectual stock is a derivative from category of knowing as the production of intellectual investment flow. The construct of the diversity and unity principles of knowing and science is based on the fundamental notion that all human activities are neuro-decision-choice actions to solve *problems* in *relation* to *things,* analytically called varieties with characteristic dispositions that establish variety identities, commonness and differences called *information-knowledge conditions*, where all neuro-decision-choice action are driven by existential needs and preference non-satiation for wants. This foundational framework allows for the development of an epistemic fields as a mathematical and philosophical entities to examine the thinking-reasoning paths to knowing, the diversity and unity of knowing and the partition of the space of knowing into categories of knowing in terms of disciplines and inter-disciplines, where the study within disciplines are intra-categorial studies and across disciplines are inter-categorial studies. This chapter examines extra factors in support of the primary and derived principles of diversities and unity in knowing and then questions the idea of disunity of intellectual investments and accumulation of information-knowledge stocks. It sets the stage for examination of organization of knowing as the production of *intellectual investment flows* and conditions of variety information knowledge claims as the production of *intellectual capital stocks* over the space of ignorance-knowledge dualities.

Chapter 5 is devoted to the development of the understanding of the nature of fuzzy paradigm of thought as an organic (macro-) paradigm for thinking and reasoning in the development of the set of fuzzy meta-paradigms for the analysis of the conditions in the spaces of ignorance-knowledge and problem-solution dualities. From the cognitive framework of fuzzy paradigm of thought, techniques and methods of fuzzy optimization are used to analyze the optimal conditions of neuro-decision-choice varieties in the space of ignorance-knowledge dualities, where the variety ignorance and the variety knowledge are mutually constraining for abstracting the fuzzy optimality in defining the transversality conditions in the journey from the zone of ignorance to the zone of knowledge for specifying the variety identity for knowing. The conditions of optimality are used to decompose the knowing space into the zones of knowledge and ignorance, and into zones of dualistic-polar influence and conflicts in specifying the optimal knowledge and the optimality of fuzzy decomposition. Discussions are made with examples in the use of fuzzy Numbers for optimizing ignorance-knowledge relation and the use of the conditions of fuzzy optimality for decomposition of the space of ignorance-knowledge dualities. The cumulative results of the chapter are the development of a theory of ignorance-knowledge dualities that allows the understanding of the phenomenon of specificationism for the conditions of zones of ignorance and knowledge in claims of confidence of variety identities as known for additions to the accumulated information-knowledge stocks.

In chapter 6, we explore certain phenomena of conceptual and epistemic anomalies such as paradox, phantomness, possible world, impossible world, imagination, fantasy, curiosity, and creativity in relation to reality, actuality, potentiality and other such analytical structures that emerge in knowing and information-knowledge accumulation as production of intellectual stock-flow activities under cognitive capacity limitations. These concepts are placed in a system of neuro-decision-choice actions in in relation to elements in the spaces of decidability-undecidability dualities and input-output dualities. The epistemic concern here is on how these anomalies relate to the space of possibility-probability dualities that houses the spaces of necessity-freedom dualities and risk-benefit dualities as one journeys between variety ignorance and variety knowledge within the space of ignorance-knowledge dualities in the general space of intellectual investment flows and information-capital accumulations over the space of individual-collective dualities. The discussions on the anomalies are then linked to the discussions of categorial conversions and functor actions in [R17.15] and to the discussions of social philosophical consciencism in [R17.16], where the contents in philosophical consciencism include the anomalies with conceptual weapons that generate structural conditions in the spaces of justice-injustice dualities, war-peace dualities, violence-nonviolence dualities, and other dualistic-polar structures. The space of imagination-reality dualities, within the epistemological space, is introduced as an analytical space and shown to be a sub-space of the space of actual-potential dualities within the ontological space. The space of imaginations is shown as a derived space from the space of the potentials and them shown to contain the possibility and probability spaces, while the space of realities is shown to be contained by the space of the actuals and then shown to contain disinformation, misinformation, lies, phantomness, illusions and other similar varieties. The concepts and phenomena of possibility and possibility are defined and explicated with comparative differences and similarities differently from the existing variety realities. The space of imagination-reality dualities is argued to generate all the intellectual anomalies, the structures of which are discussed.

Chapter 7 follows chapter 6 to reflect on the concepts of possibility, probability space, impossibility space and their relationships to possible and impossible worlds relative to the telescopic conditions over the space of identification-transformation dualities. The space of imaginations is shown as a derived space from the space of the potential and them shown to contain the possibility and probability spaces, which then give rise to the concepts and philosophy of possible and impossible worlds over the space of neuro-decision-choice actions. The concepts and phenomena of possibility and possibility are defined and explicated with comparative differences and similarities differently from the existing ones in literature. Their similarities and commonness of the relationships to the concepts and phenomena of possible and impossible worlds are discussed relative to the spaces of true-false and information-knowledge dualities. The chapter is concluded with examples of dualistic-polar membership functions that may be used under dualistic-polar analytics with the use of the fuzzy paradigm of thought.

The *epilogue* of the monograph provides a connectivity with the prologue and epistemic development of theories on knowing as social production of intellectual investment flows, information-knowledge accumulations as social production of intellectual capital stocks and projections of future actions on organizational analytics as central role of simultaneity of oneness, diversity, unity and disunity, where knowing and information-knowledge accumulation are seen in terms stock-flow disequilibrium dynamics under neuro-decision-choice actions. The fuzzy rationality, principle of opposites with relational continua and unity, the theory of info-statics, identification problem-solution dualities, the theory of info-dynamics, transformation problem-solution dualities, the theory of problem-solution dualities as production-consumption dynamics over the spaces of input-output dualities and destruction-construction dualities, all are connected to the space of varieties as contained in the space of imagination-reality dualities as a derived category by neuro-decision-choice actions from the space of actual-potential dualities, where the space of problem-solution dualities is represented as an epistemic fields containing categories with functors. The connectivity systems of actions for variety identifications and transformations are done through the development of organizations and sub-organizations containing institutions which are relevant for neuro-decision-choice actions and policies relative to dynamic categories of problem-solution dualities. The system of ideas and theories, as outlined so far, points to the development of the theory organizational analytics and institutional development within the processes of explanatory and prescriptive sciences to provide pathways of understanding of parent-offspring dynamics of departmental analytics within the space of problem-solution dualities. We proceed with reflections on conclusions, discussions on organization of institutions and categories of the space of problem-solution dualities through the economic concepts of production, consumption and input-output processes of information-knowledge system. In this respect, an introduction to a prelude to the theory of neuro-decision-choice system is advanced for future development, where the nature of the roles of imagination, reality, belief, hope, courage in the space of imagination-reality dualities is examined in the journeys over the space of ignorance-knowledge dualities and the developments of organization that may contain the relevant institutions.

PROFESSOR EMERITUS KOFI KISSI DOMPERE

Contents

CHAPTER 1

THE THEORY OF EPISTEMIC FIELDS:

Principles of Diversity, Unity and interdisciplinary Areas of Knowing

ABSTRACT

Chapter 1 of this monograph is devoted to the development of the theory of *epistemic fields* and corresponding *relational fields* based on *information fields* relevant in understanding the creations of specific and multidisciplinary areas of study, foundations of the creation of departmental systems of research, learning and teaching, and the creation of different institutions of research for information-knowledge and problem-solution transformation strategies to achieve desired social goals and objectives. It, further, provides the logical foundations for decomposition, analysis, multidisciplinary formations, institutional foundations, and interdisciplinary approaches to neuro-decision-choice activities over the space of problem-solution dualities which are viewed as variety elements in the space of ontological-epistemological duality under the spaces of negative-positive and cost-benefit dualities. It is about the epistemic understanding of *diversity and unity, diversity and disunity, oneness and dividedness and disorganization and organization* of knowing through neuro-decision-choice actions over the space of problem-solution dualities, where nothing is knowable without neuro-decision-choice action and the information field and nothing is transformable without the interactive processes of the space of neuro-decision-choice actions and the space of problem-solution dualities that form the foundation of the development of information-knowledge systems as stock-flow disequilibrium dynamics over the spaces of input-output dualities and simplicity-complexity dualities through knowing and information-knowledge accumulation. The complete system of spaces and sub-spaces, as applicable to natural dualities

K. K. Dompere, *The Theory of Epistemic Fields*, Studies in Systems, Decision and Control 498, https://doi.org/10.1007/978-3-031-42470-0_1

and polarities, and social dualities and polarities with similarities and differences, is presented in Fig IIIA in the preface.

1.1 RELATED PAST WORKS IN RETROSPECTIVE

In the monograph entitled *The Theory of Problem-Solution Duality and Polarity: Information-Decision-Choice Foundations of the Unity of Knowing and the Unity of Science,* discussions were made to establish the relational structures between problem-solution processes and information-knowledge processes and how these processes are relationally connected to epistemic journeys of cognitive agents with neuro-decision-choice actions as the vehicles to travel between duals in the space of opposites, especially between the variety ignorance and the variety knowledge within the space of ignorance-knowledge dualities connected to the space of production-consumption dualities as a sub-space of imagination-reality dualities[R8.16b]. In the discussions, several dualistic-polar concepts were introduced with corresponding distinctions for useful communications over the space of source-destination dualities. Among these concepts are the concepts and definitions of problem-solution duality and the space of problem-solution dualities, and problem-solution polarity and the space of problem-solution polarities on the basis of which the knowing and knowledge production processes are explained as stock-flow disequilibrium dynamics. From the definitions of the concepts involving the problem-solution duality and polarity and the corresponding spaces, the principles of diversity and the unity of knowing and science were advanced to establish the conceptual and applied relationalities of variety diversities within the universal unity of knowing and knowledge. In support of these principles are the propositions that the infinite universe with the multiplicity of galaxies exists in a universal unity composed of varieties of diversities under the same laws of transformation and change in relations and properties. These laws find expressions in information-knowledge disequilibrium dynamics under the matter-energy equilibrium dynamics with continual matter-energy variety transformations under the same laws in the spaces of construction-destruction dualities, input-output dualities and cost-benefit dualities, where the information-knowledge stock-flow disequilibrium processes are understood through the principles of diversity and unity of knowing as the general approach to the understanding of universal variety connectedness over the space of oneness of the information-knowledge system in conceptualizing the oneness and the dividedness of the universe, where one of the relations is the system of the food chain in addition to the system of parent-offspring chains over the space of genetic connectedness.

There are two inter-supportive principles for the unity of knowing and unity of science that make-up the intellectual capital stock-flow system over the space of input-output dualities which are based on socio-natural variety input-output processes. They are the *primary principle* and the *derived principle* based on the *general principle of opposites,* where the universe is seen as matter-energy-information integrated, containing varieties and categorial varieties that form the basis on which all languages are formed and developed as the variety knowing increases and the size of the information-knowledge accumulation expands under the matter-energy equilibrium and variety transformation disequilibrium. These varieties and categorial varieties are seen in terms of diversities in categories in a relational interdependence, continuum and unity that constitute the universal system of ontological existence under the principle of opposites, for which the epistemological existence is a simple derivative by collective neuro-decision-choice actions of cognitive agents over the space of imagination-reality dualities contained in the space of actual-potential dualities that must be related to the space of ontological-epistemological dualities with extensions represented by a geometry of analytical organization in paremiology as in Fig IIIA in the preface.

Let us keep in mind that all the spaces and varieties exist in dualistic-polar conditions with systems of characteristic-signal dispositions without which variety identifications and transformations are not possible. At the level of knowing, cognitive agents are preoccupied with telescopic past-present-future (Sankofa-anoma) varieties in knowing and the acquisition of information-knowledge stocks to manage their existence through problem-solution activities over the space of imagination-reality dualities. The space of variety imaginations contains the unrealized variety characteristic dispositions of realities, and the space of realities contains the variety characteristic dispositions of actuals and illusions in the space of imaginations. The demands of neuro-decision-choice actions over the space of problem-solution dualities are the flows of services from the information-knowledge stock systems. The complex system of spaces and sub-spaces must be familiar and understood if cognitive neuro-decision-choice actions are to be explained in theory and understood relative to the meaning and the practice of the essence of life for the maintenance of existence under the principle of non-satiation over the space of production-consumption dualities.

The space of actuals contains varieties of telescopic past and telescopic present, while the space of potentials contains varieties of telescopic future. The varieties of the telescopic past contain *what there was,* the varieties of the telescopic present contain *what there is* and the varieties of the telescopic future contain *what there would be.* The space of actual-potential dualities is the dualistic-polar game space of natural variety transformations from variety actuals to variety potentials and from the variety potentials to the variety actuals. The space of realities contains varieties of *what is claimed to be known* variety actuals from the telescopic past and telescopic present of the space of the actuals with cognitive conditionality to cover for cognitive capacity limitations, where the identities of the variety realities are established by cognitive characteristic dispositions. The space of imaginations contains varieties of *what would become known* or transformed from the space of potentials as telescopic future realities with hope, dreams, creativity, courage, and other actualizing tools. The varieties of the telescopic past contain *what there was,* the varieties of the telescopic present contain *what there is* and the varieties of the telescopic future contain *what would be,* while the space of imagination-reality dualities contains the game space of neuro-decision-choice activities of cognitive agents for variety identifications from the space of actuals and variety transformations from the space of imaginations to the space of realities, and from the space of variety realities to the space variety imaginations and then to the space of variety potentials. All the characteristic-signal dispositions are reduceable to cost and benefit characteristic dispositions for the neuro-decision-choice actions over the space of imagination-reality dualities.

1.2 BASIC REFLECTIONS ON THE EPISTEMIC FIELD

Essentially, the *theory of epistemic fields* involves the development of analytical partitioning of the ontological space based on the *information field* (info-field) into areas of knowing, where each area of knowing constitutes a fuzzy collection of problem-solution dualities under a *corresponding relational fields* [R8.16a] with common characteristic dispositions in variety identification and transformation processes, where variety problems are replaced by variety solutions which then generate variety problems, where the nature of the new problems will depend on the manner in which the solutions are obtained. The information field, as defined, provides inputs for variety identifications and transformations. The variety identifications and transformations reside in the space of imagination-reality dualities as a sub-space of the space of actual-potential dualities. The corresponding relational field provides a multidimensional relational connectivity of the areas of knowing in terms of inter-supportive relations that facilitate or hinder the successes or failures of variety substitutions in the space of problem-solution dualities as connected to the space of variety identification-transformation dualities. In pure analytical essence, the *theory of epistemic fields*, with *relational fields* under *information fields*, is developed to provide foundations that the general concept of the space of problem-solution dualities defines the *unity principles of knowing and information-knowledge accumulation* while different sub-spaces of problem-solution dualities define the *diversity principle of knowing* and variety information-knowledge accumulation under a multilateral relational connectivity of the classes of the problem-solution structures under the *principle of opposites* with relational continuum and unity and *cost-benefit rationality* as the guiding principle of all neuro-decision-choice actions with relevant conditions from the information field.

The socio-natural destruction-construction processes are the foundations of knowing and transformation in the substitution-replacement process in the space of relativity but not in the space of absoluteness. Let us remind ourselves that every variety exists in multiple dualities and polarities of negative-positive varieties, problem-solution varieties, and cost-benefit varieties in relative dispositions. Every variety exists with internal dualistic-polar conditions under internal relational conflicts and tensions for internal changes in relations and properties, and external dualistic-polar conditions under external relational conflicts and tensions for external changes in relations and properties. The changes at the level of nature are under the principle of environmental rationality, while the changes at the level of society are under the principles of cost-benefit rationality. At the level of society, all neuro-decision-choice actions are about variety identification and variety transformation in continual processes over the space of finite-infinite dualities.

1.2.1 The Core of the Theory of Epistemic Fields

The central core of the theory of the epistemic field is to create an *epistemic toolbox* for developing an analytical understanding of the factors that are necessary to establish the relational conditions of the diversities of the areas of knowing (diversity principle of knowing) , the unity of areas of knowing (unity principle of knowing), and to strengthen the argument in support of an interdisciplinary approach to areas of knowing (called interdisciplinary principle of problem-solution dualities) as well as to provide an epistemic *framework* to guide the development of departments of learning, research, teaching and social institutions of neuro-decision-choice actions for dealing with the dynamics of problem-solution dualities in all areas of human endeavor. The framework of the epistemic field will also help to provide justified pathways to curriculum developments of all areas of learning social institutional formations for directed actions over the space of problem-solution dualities and to encourage meaningful and continual collaborative neuro-decision-choice efforts within the search process of knowing in the enterprise of teaching, learning and research in the institutions of higher education in support of national and international stock-flows of information-knowledge accumulation (interdisciplinary principle of problem solving in the epistemic fields) providing analytical pathways for categorial transformative conditions in relation to the dynamics of research universities as the space of variety problem-solution dualities expands in response to the changing conditions of variety relations and properties in the space of categorial dynamics, where thought without action is empty in the space of variety realities and rests in the space of imaginations, and action without thought is blind and rests in the in illusions space of variety imaginations. The relational diversity is accomplished by analytical decomposition, while the relational unity is accomplished by analytical composition to ensure the organic principle of opposites in relational continuum and unity.

The theory of knowing, within this analytical structure, is composed of the epistemic field, the relational field, and the information field to relate varieties in the epistemological space to varieties in the ontological space under cognitive neuro-decision-choice actions. The information field is an indispensable connector for the existence of the ontological-epistemological duality that is connected to knowing through the neuro-decision-choice actions over the space of variety problem-solution dualities as reflected from the space of acquaintance. The theory of knowledge is the accumulation of results of neuro-decision-choice activities over the space of problem-solution dualities through the results of knowing from the space of success-failure dualities. The accumulation of global categorial results of individual and collective neuro-decision-choice actions over the space of general problem-solution dualities is the global intellectual heritage of humanity. The theory of knowing and the theory knowledge constitute the input-output processes which are explainable with the logical framework of neuro-decision-choice actions in the space of variety production-consumption dualities. From the economic logic of neuro-decision-choice activities over the space of production-consumption dualities, the theory of knowing is the explanation of the production of intellectual investment flows as outputs and the flow of services from the intellectual investments as inputs into the production of information-knowledge accumulation to produce intellectual capital stocks the outputs of which become inputs into knowing.

The theory of knowledge is the explanation of the production of intellectual knowledge accumulations as outputs and with inputs from the intellectual investment flows as inputs. In this way, the theory of knowing and the theory of knowledge jointly constitute a theory of variety input-output dynamics in explanation and predictions on one hand over the space of actuals through the space of variety realities, and prescriptions and instructions on the other hand over the space of potentials through the space of imaginations. The neuro-decision-choice actions are connected to the space of potentials through the space of imaginations, while the same actions are connected to the space of actuals through the space of realities. Generally, the neuro-decision-choice actions, therefore, connect the space of imagination-reality dualities and polarities to the space actual-potential dualities and polarities in all areas of human endeavors. It is through this analytic process that a clear understanding, clarification, and distinction may be exercised between explanatory science with explanatory theories and prescriptive science and prescriptive theories as they relate to thought and action, where thought resides in the space of imaginations and action resides in the space of realities.

1.2.2 The Usefulness of the Core of the Theory of Epistemic Fields

The usefulness of the central core of *the theory of the epistemic fields* relates to the neuro-decision-choice actions over the space of problem-solution dualities relative to the spaces of unity-disunity dualities and oneness-dividedness dualities in the search for information-knowledge varieties, the understanding of diversity in general knowing and interdependencies of relationality in intra-categorial and inter-categorial information-knowledge systems, and science in particular, to provide justifications for epistemic collaboration in the knowing processes and information-knowledge development processes within the space of input-output dynamics in inter-categories and inter-population geography. The reasons for the interdisciplinary approach in the search for knowledge are abstracted from a general variety encounter over the space of acquaintance regarding the continual call for repeated experimentations in sciences and other possible areas of knowing, requiring the need for an increasing number of witnesses in neuro-decision-choice actions, judicial decisions, the development of information-data systems and many others to find justifications in the fuzzy paradigm of thought under the principle of opposites with relational continua and unity. There are varieties in the space of varieties where experimentations are not possible, and there are some neuro-decision-choice actions which are irreversible in the space of reversibility-irreversibility and convertibility-inconvertibility dualities.

The variety *relational continua* find expressions through give-and-take-sharing processes in dualistic-polar structures to establish a dynamic multidimensionality of interdependencies in the information-knowledge search over the space of problem-solution dualities to deal with cognitive capacity limitations over the space of variety imagination-reality dualities that creates multiplicity of vagueness and penumbral conditions over the space of acquaintance and linguistic structures. The variety *relational unity* finds expressions through the mutual existence and destruction over the space of identification-transformation dualities. An objective, here, is to increase the understanding of the relational diversities to maintain conditions of dividedness in oneness as strategies for neuro-decision-choice actions as well as to understand the relational unity that maintains the wholeness in variety knowing over the ontological space within the space of ontological-epistemological dualities (see[R4.13] for discussions on the interactions and dependencies between exact and inexact sciences in dualistic-polar structures). The increase in collaborative human efforts over the space of production-consumption dualities in applied human problem solving such as the practice of industrial production are here projected to intellectual efforts in the space of the epistemic field with sub-fields to provide an analytical framework for redesigning research universities, institutions of information-knowledge development and the inputs into neuro-decision-choice activities in a system of socio-economic production-consumption dualities which includes information-knowledge production and the space of problem-solution dualities over the space of input-output dualities.

As the global information-knowledge accumulation expands and increasing categories of problems are resolved, they will give rise to a space of new categories of problem-solution dualities that will require increasing inter-disciplinary approaches in their understanding and treatment in such a way that the concentration on the specific-area approach of neuro-decision-choice activities over the space of problem-solution dualities will give way to a multi-field approach in dealing with the elements of problem-solution dualities and their continual transformations under the general principle of opposites. In fact, the development of technologies, industrial production and socio-economic development are driven by multi-disciplinary collaborative approaches in dealing with their corresponding problem-solution dualities over the penumbral and non-penumbral areas of neuro-decision-choice actions. The known history of human existence, with its multiplicity of civilization, achievements, survivability and continual transformation, reveals the strength of the collaborative power over all areas of production-consumption activities through systems of inter-supportive institutions, departments and categorial give-and-take-sharing relations, where the principle of *tikro-nko-agina* (one head does not constitute a counsel) rules over the space of problem-solution dualities by appealing to sharing, collaboration and coordination. This collaborative power, enshrined in human relations, must be examined and judiciously extended to neuro-decision-choice activities over the design of institutions of areas of knowing and the organization of these institutions, to improve the efficiency of the collective abilities in the enterprise of knowing and information-knowledge accumulation in support of individual and collective neuro-decision-choice actions over the space of problem-solution dualities at all levels of human actions.

The individual cognitive limitations are such that the progress of knowing is retarded without a collaborative system of knowing which enhances the social ability to know by reducing general uncertainty in the collective experiential information and risk from acquaintance in the use of knowledge by description as inputs into the space of neuro-decision-choice actions. In other words, the increasing successes in social actions find strength in collective actions under specialization and specialized skills and talents in the space of problem-solution dualities under problem-solution disequilibrium dynamics which are also information-knowledge disequilibrium dynamics. The same relaxation of individual cognitive constraints also applies to geographical and generational constraints in knowing where the conditions of give-and-take sharing may modify the conditions of the cost-benefit space under social philosophical consciencism [R17.16]. There are some side effects to the system of collaborations and the give-and-take sharing processes. The side effects are the rise of misinformation, disinformation, news fakeness and propaganda over the give-and-takes sharing processes over the source-destination duality to abstract either personal, collective, or national rent. The effectiveness of these side effects will depend on collaborative structures and the generational social philosophical consciencism.

The increasing successes in cognitive problem solving in the space of individual-collective dualities will give rise to an increasing number of new problems, and thus expand the zone of difficulties in solving new problems. Here, we must keep in mind that problems and solutions present themselves as varieties under complex dualistic-polar conditions, where a variety solution generates a variety problem which may be a variety solution as viewed in the space of goals, objectives, and visions. The expansion of the zone of problem-solution difficulties will force the neuro-decision-choice agents to abandon the traditional disciplinary exactness and acquire disciplinary inexactness into accepting the multidisciplinary approach in the creation of departments in the institutions of higher learning, teaching and research to increase the efficiencies in all classes of tactics and strategies over the space of problem-solution dualities, when we understand that knowing, learning and teaching are about neuro decision-choice actions over the space of problem-solution dualities under cognitive capacity limitations of individuals and collective, social and generations. Every problem-solution interchange brings in a new identification-transformation duality over the space of problem-solution dualities in the knowing and information-knowledge processes and generates a new neuro-decision-choice action and response on its path of occurrence for variety identification and transformation in the knowing process.

The new identification-transformation decision process in knowing will define a golden age of research, teaching, department formations and institutional developments not in separate existence but in mutually inter-supportive approaches to the information-knowledge process and decision-choice actions over the socio-natural space, where an individual must be equipped with tools of categorial multiplicity to function and communicate to deal with inter-categorial generations of elements in the space of problem-solution dualities, in the sense that a problem-solution duality in one category may find its replacement in another category. In this respect, there is the collaboration and interdisciplinary processes over the space of static-dynamic dualities. The static collaborations are cross-sectional in skills, individuals, and nations, while the dynamic collaborations are intergenerational through all kinds of preservations, storage, and transmissions of generational and intergenerational information-knowledge stock-flows such as books and libraries.

The multidisciplinary approach to the institutional creation for knowing and information-knowledge accumulation through neuro-decision-choice activities over the space of problem-solution dualities will affect the distribution of social resources in the information-knowledge accumulation over the space of production-consumption as the general space of socio-economic activities in support of existence. The rise of interdisciplinary or multidisciplinary regimes to design institutions of information-knowledge processes will provide an improved collective and individual thinking in information-knowledge development, reduce qualitative uncertainty and individual neuro-decision-choice inputs where the unity principle of knowing and science is seen in terms of neuro-decision-choice activities over the space of problem-solution dualities under an information regime, a paradigm of thought and a cost-benefit rationality.

The thesis that is advanced here is that the principle of unity of knowing is general and finds expression in the cognitive activities over the spaces of problem-solution dualities and ignorance-knowledge dualities with connectivity from conditions of relationality that establish a set of give-and-take sharing processes under the principles of diversity and unity with categorial problem-solution dynamics. This unity principle of knowing is the same as the unity principles of science and non-science. Every area of knowing is about a reduction of ignorance; and increasing knowledge is about either a variety or categorial varieties in terms of identity, behavior, and transformation. The reduction in ignorance and the increased knowledge finds different kinds of expressions over the space of problem-solution dualities irrespective of whether the category of knowing is designated as science or non-science, as explanatory science or prescriptive science and other categories for knowing. It is better to see these groupings as an approach to simplify the complexity of social decision-choice actions over the space of problem-solution dualities for variety identity and transformation under cost-benefit rationality.

The knowing is also about identification problem-solution dualities that relate to variety identities, and transformations which relate to variety behavior and change, where either success or failure identification of a variety problem-solution duality generates new information and the success or failure transformation of the variety problem-solution duality into a new variety problem-solution duality generates new information to increase the info-stocks of information-knowledge accumulation in a continual disequilibrium to show the structural process of the laws of info-dynamics [R17.18]. The dynamics of the problem-solution dualities, the principle of opposites, the cost-benefit conditions (*Asantrofi-anoma retionality,* cost-benefit duality), the individual-community duality (*Anoma-kokone-kone problem*: conflicts in collective decision-choice actions) and the relational continuum and unity, (the *Funtummereku-Denkyemmeriku problem:* A relational conflict problem of a two-headed crocodile connected to one stomach in processes of eating), ensure the infinite decision-choice actions under the decision-choice principle of non-satiation towards unreachable perfection in nature and society guaranteeing the infinite permanence of the universe with a continual variety creation, where gradualism, limiting processes, approximations and error-corrections processes are foundational neuro-decision-choice practices moving elements in the space of relativity to the space of absoluteness. These are also the foundational techniques in differential marginalism in the space of minimum-maximum dualities [R3.26] [R4.9] [R17.15] [R17.16].

1.2.3 Diversity and Unity in the Development of the Theory of Epistemic Fields

Generally, one cannot speak of unity without the acknowledgment of diversity. Additionally, we cannot speak of diversity without acknowledgement of information, and we cannot speak of diversity and unity without the acknowledgement of either a simple or a complex relationality. Our universe exists in various forms of unity of matter-energy varieties with multiplicity of give-and-take-sharing variety relations that establish matter-energy identities and transformations from which complexities over the space of variety of existence-nonexistence dualities emerge. In fact, the universe exists as a relational oneness in a multiplicity of galaxies and dividedness in oneness in the space of oneness-dividedness duality under the principle of opposites with a multiplicity of symbiotic relationships for mutual existence in creation and destruction. Our understanding of the relational unity does not provide us with the understanding of whether matter-energy unity is primary, or matter-energy dividedness is derived and vice versa. The universe as infinite may also exist as an organizational-dis-organizational duality. The opposites and relationality of these opposites help to explain life-death dualities, new-old dualities, wellness-sickness dualities, desert-forest dualities, stability-instability dualities and such duals and poles in the space of dualistic-polar games to define the space of continual variety dynamics in relations and properties through the matter-energy processes.

How do diversities arise from unity in the knowing process in the understanding of the hidden elements of the universe and how does unity emerge out of diversities? These questions relate to the conditions and elemental behaviors within the variety unity-disunity dualities under the laws of opposites with contradictions, tensions, and conflicts. These questions also relate to the concept that the universe is one and a whole which encompasses variety dynamic dualities and polarities without the loss of the unity and the diversities over the space of static-dynamic dualities where all planets and galaxies are governed by the laws of opposites. Similarly, all energy generations, all power emergence, all works, all transformations in the universe are under dualistic-polar games where progress and retrogress, construction and destruction, transformation, and non-transformation over the space of relation-property dualities and many other such behaviors are impossible without the laws of opposites. The theory of the epistemic field draws on the concept of universal information-knowledge unity that exists and may be obtained through processes of diversity, commonness, oneness, and dividedness to relate it to the space of problem-solution dualities where each category constitutes fuzzy elements in the epistemic fields.

1.2.4 Diversity, Unity, Paradigm of Thought, and the Theory of Epistemic Fields

The knowing process is through the neuro-decision-choice activities of cognitive agents with information obtained as knowledge by acquaintance or experiential information that serves as a primary input from the acquaintance space. The nature of the information input passes through the space of doubt-surety dualities. This input is then processed by a cognitive machine of neuro-decision-choice agents to create an output called knowledge by description. The structure of the cognitive machine will depend on the nature of the accepted information input from the space of acquaintance There are two information input structures. The two information input structures from the space of acquaintance are perfect and imperfect information structures. The perfect information structure relates to natural conditions of variety characteristic dispositions within the space of identification-transformation dualities as established over the space of absoluteness. The imperfect information structure relates to social conditions of knowing from cognitively abstracted variety characteristic dispositions from variety natural signal dispositions that present two analytically alternative spaces of absoluteness and relativity which are claimed by neuro-decision-choice agents.

The imperfect information structure has sub-structures of exactness and inexactness over the space of absoluteness-relativity dualities. When the conditions of exactness are supported by the principle of opposites with relational separation and an excluded middle, then the information substructure is moved to the space of absoluteness. The conditions of inexactness with vagueness, relational continuum, unity, and give-and-take-sharing mode of the information sub-structure find expressions over both the spaces of absolutes and relativity within the space of absoluteness-relativity dualities. The cognitive machines are the paradigms of thought, thinking, reasoning and neuro-decision-choice action for processing the information inputs. The paradigms of thought are themselves the creations of the neuro-decision-choice actions that are collectively accepted as rules of thinking and reasoning to arrive at a conclusion within the space of decidability-undecidability dualities.

In the knowing process, there are two paradigmatic situations currently available for critical examination for use in ascertaining knowledge items under the conditions of the principle of opposites. One paradigmatic situation, corresponding to the *classical paradigm of thought*, is where there is an epistemic unity-disunity duality and polarity in all opposites with relational separation, complete excluded middle among the areas of knowing, non-relational connectivity of diversities and no system of give-and-take-sharing modes in the general field of information-knowledge production. The other paradigmatic situation, corresponding to the *fuzzy paradigm of thought*, is an epistemic unity-disunity duality and polarity in all opposites with relational continuum and unity with dynamic multi-dimensional relations of a system of give-and-take-sharing modes among the areas of knowing of diversities. Within the classical paradigm of thought with no acceptance of contradictions, one can speak of categorial disunity and separate existence of areas of problem-solution dualities without relational connectivity, where diversities exist in relational non-connectedness. Within the fuzzy paradigm of thought, one can speak of unity, diversity, and disunity with conditions of a relational field, where the areas of problem-solution dualities have dynamic multiplicity of relational connectivity from the relational field connecting variety diversities to one another and to the whole.

The search for the answers to questions concerning the areas of diversity knowing, unity and the conditions of similarities and differences to the understanding of disunity, continuum and unity of knowing is the subject matter of this chapter by establishing the conditions for *epistemic fields* that will provide us a platform to discuss the notions of possibility, impossibility, possibility-impossibility duality, possible world, impossible world, possible-world-impossible world duality, epistemic space, probability, improbability, probability-improbability duality, all in relation to neuro-decision-choice actions over the space of imagination-reality dualities, the space of imaginations and the space of realities and how they relate to the spaces of actuals, potentials and actual-potential dualities in the process of knowing and development of information-knowledge stocks.

There are many misconceptions about fuzzy logic in the sense of thinking and reasoning relative to decision-choice actions over the space of problem-solution dualities. *Fuzzy logic* and *fuzzy mathematics* composed of the *fuzzy paradigm of thought* for processing fuzzy information, where vagueness, subjectivity, volume limitations, value judgements, approximations and dualistic-polar gradualism with asymptotic processes of error corrections and limiting processes in cognitive journeys are essential elements in neuro-decision-choice actions in the processes of knowing and the development of the information-knowledge system in support of dynamic self-exiting and self-correcting fuzzy neuro-decision-choice activities over the spaces of actions and relativity [R4.7] This is the *fuzzy process* and the *fuzzy cognitive frame* of thought in all areas of fundamental and applied systems of human cognition and social progress which is gradual and error corrective in nature.

In comparison, the classical approach to neuro-decision-choice actions. over the space of problem-solution dualities, is such that the classical logic and mathematics are contained in the classical paradigm of thought for processing classical information, where vagueness, subjectivity and value judgement are not essential elements in the neuro-decision-choice actions in the processes of knowing and the development of the information-knowledge system in support of decision-choice activities over the action space, where the conditions of exactness, non-contradictions and the principle of an excluded middle operate over the space of absoluteness without dualistic-polar gradualism for error correction. This is the *classical process* and the *classical cognitive frame* of thought in all areas of fundamental and applied system of human cognition and social progress, where opposites cannot exist in the same space and time and give-and-take-sharing conditions between opposites are not dealt with. Here, the fuzzy paradigm is an attempt to fill in the conditions of the neuro-decision-choice elements of the classical excluded middle that contains simultaneity of duals and contradictions, and epistemic journeys over the space of ignorance-knowledge dualities are gradual with error correction. The discussions of the structures of the classical paradigm of thought and fuzzy paradigm of thought may be found in Russell [R3.86], [R18.58], Brouwer [R3.13], [R3.15], Black,[R3.6], [R18.4], [R18.9], Lakoff [R18.39] and Dompere, [R4.7], [R4.10] ,[R8.16].

1.3 CATEGORIAL VARIETIES AS DIVERSITIES IN THE AREAS OF KNOWING

Let us now turn our attention to the development of on analytical platform for examining the diversity and unity of knowing and information-knowledge irrespective of whether science or non-science is under neuro-decision-choice action. The knowing and information-knowledge processes are viewed as neuro-decision-choice activities over the space of problem-solution dualities where these activities are embedded over the space of social production-consumption dualities The universe exists as a collection of matter-energy varieties with fuzzy similarities and differences for all different dimensions of acquaintance to establish diversities and a multiplicity of complex relations for mutual existence and nonexistence with unbreakable multiplicities of give-and-take relational modes such as the food chain. Each variety exists with *characteristic disposition* $(\mathbb{X}_\nu)$ that defines the intrinsic nature of the variety. The nature of the variety is made available to other varieties in the universe through its corresponding *variety signal disposition* $(\mathbb{S}_\nu)$ such that the information on a variety $(\nu \in \mathbb{V})$ is given as $(\mathbb{Z}_\nu = (\mathbb{X}_\nu \otimes \mathbb{S}_\nu))$. Our acquaintance with $(\nu \in \mathbb{V}_\Omega)$ in the epistemological space $(\mathfrak{E})$ is through the signal disposition $(\mathbb{S}_\nu)$ that allows one to record the observational data called the *experiential information*, *knowledge by acquaintance* or *epistemological information* which may appear as soft data, hard data or both in qualitative, quantitative dispositions or both. It is this experiential information that provides input for continual refinements with a paradigm of thought to develop *knowledge by description* about any variety in terms of identification and transformation irrespective of whether science or non-science is in the space of interest.

The nature of the varieties is such that some may present themselves in multiple acquaintances by different cognitive agents, while some may present themselves as one event to an individual in the acquaintance space without a repeated occurrence. In this respect, the identification of stages of economic development is no different from the identification of stages of a capsule on the flight to another planet with stages. They all reside in the space of identification problem-solution dualities in knowing that begins from the zone of ignorance and a journey to the zone of knowledge. Similarly, the movement of stage-to-stage of economic development which, as a *qualitative motion,* is no different from a movement of a capsule from stage-to-stage which is a *quantitative motion.* They. all reside in the space of transformation problem-solution dualities in knowing that begins with ignorance and a journey to the zone of variety transformation. The study of these two different types of motions may require different technologies and methods due to the differences of the characteristic dispositions. Qualitative motion is in relation to changes in property, while quantitative motion is in relation to changes in relation.

The discussions on the *variety identification problems,* the corresponding solutions and formation of categories in the universal system of existence are carried out in *the theory of info-statics* [R17.17]. The discussions on the variety-transformation problems and the generation of new information, data, evidence, and knowledge are carried out in *the theory of info-dynamics* [R17.18]. Every process of variety knowing whether classified as science or non-science must encounter the *principle of acquaintance* at the first level of knowledge acquisition. The space of acquaintance has a multiplicity of pathways to enter. The cognitive interaction of the space of acquaintance is with *ontological variety signal disposition* to generate *knowledge by acquaintance* as a solution to the *primary category* of variety information-knowledge over the space of problem-solution dualities from which variety *knowledge by description* is derived with an acceptable paradigm of thought over the same space of problem-solution dualities. In this respect, knowledge by description is the derived category of information-knowledge through the space of problem-solution dualities to check the validity of knowledge by acquaintance. It is a position through all these discussions that knowledge by acquaintance, epistemological information or experiential information pass through the spaces of subjective-objective and quality-quantity dualities with vagueness, soft-hard, exact-inexact characteristics which are produced by cognitive deficiencies and observational limitations involving all areas of knowing, to give rise to qualitative and quantitative uncertainties in information-knowledge structures and risks in outcomes of cognitions and neuro-decision-choice activities over the space of problem-solution dualities embedded in the space of imagination-reality dualities relative to the cognitive journey over the space of ignorance-knowledge dualities.

As it has been pointed out, there are no problems or solutions as seen by natural forces, only transformations of varieties and forms of continual restructuring over the ontological space that contains the space of actual-potential dualities. How do we classify earthquakes, volcanos, and other natural events in the ontological space? Are they problems, solutions or natural transformations over the space of ontological property-relation dualities? These natural transformations are seen as problem-solution dualities when they emerge at the presence of cognitive agents under differential preference schemes when they alter cognitive cost-benefit relations with greater net costs over benefits within the space of imagination-reality dualities contained in the space of actual-potential dualities which resides in the ontological space as the primary category of existence relative to the epistemological space as a derived category of existence.

1.3.1 Identification, transformation, and Variety Experiential Information

All epistemological information, knowledge by acquaintance or experiential information from the space of acquaintance are naturally fuzzy-stochastic which are here classified as *fuzzy-stochastic information* involving possibilistic dispositions over the developmental process of knowledge from acquaintance, and probabilistic dispositions over the constructional process of knowledge by description and combined possibilistic-probabilistic disposition over the space of acceptance-rejection dualities by neuro-decision-choice action for the information-knowledge stock-flow system. The interdependent stages of information-knowledge development are part of the principle of unity of all categories of knowing under the principle of diversity over the epistemic field, where the multi-categorial relational connectivity through give-and-take sharing modes draws its foundations from the dynamics of the relational field given the information field. In this respect, the study, and the development of the arguments in support of epistemic unity, disunity, oneness, dividedness, and diversity, and in the space of unity-disunity dualities, must take place over the *epistemic field* and the corresponding *relational field* and *information field* (ERI).

The unity and disunity, like true-false dualities, are dualistic-polar extremes and between them are dividedness and diversities in terms of varieties with give-and-take-sharing relations from the epistemic field. Let us keep in mind that the epistemic field shows *what there is, what there was and what would be,* all in diversities and categories; the relational field shows the actual and potential *categorial relationality* that reveals the give-and-take-sharing principle, while the information field establishes variety identities to allow the identifications of varieties, categories, and relational multiplicities to be revealed in static and dynamic states. The epistemic field with the relational field and the information field is a complex organic unity composed of categories welded together by give-and-take-sharing modes just like the totality of the human body with parts which are welded together by complex conditions of give-and-take sharing modes. In fact, the whole of information-knowledge stock-flow system is like a giant tree with common roots producing the intellectual investment flows and the stem accumulating information-knowledge and the many branches producing intellectual capital stocks.

The existence of differences in areas of knowing does not suggest disunity in areas of knowing and by extension, areas of science. It shows categories in the space of problem-solution dualities. Within the unity-disunity dualities, we find diversities, and not a disunity in the space of problem-solution dualities. Diversity does not imply disunity and disunity does not imply diversity. The observed differences in categorial diversities, therefore, cannot be used to justify disunity at the fundamental level of variety and categorial knowing. Categorial diversities imply differences in parts with a multiplicity of relational continuum and unity with give-and-take-sharing structures, just like the human body in unity with different parts in continual give-and-take-sharing modes that provide it with livelihood over the space of life-death dualities. Categorial disunity implies a separation of parts without a relational continuum and possibility of a give-and-take-sharing structure in such a way as to deprive an entity of its existence over the space of life-death dualities. The areas of knowing are established by cognitive agents for dealing with the best approach to find solutions to information-knowledge accumulation problems about the universal existence, and the use of the accumulated information-knowledge structure in transformation decision-choice actions relative to problems in relation to an element in the space of goals and objectives. The search for information-knowledge items and the use of such information-knowledge items are neuro-decision-choice responsibilities defined over the space of problem-solution dualities towards elements in the space of goals, objectives, and visions. In this respect the system of knowing and information-knowledge accumulation is a living organism welded together with diverse varieties in complex relations of categories and sub-categories.

The number of areas of learning, just like the information-knowledge accumulation, is always in a temporary equilibrium under general disequilibrium dynamics. The areas of knowing correspond to some categorial varieties, where each category corresponds to an area of learning or a possible future area of learning. Each category of varieties is a sub-space of problem-solution dualities and hence, we have infinite categories of problem-solution dualities, each of which belongs to the *general epistemic field* to be defined and explicated, where the ontological size of the general epistemic field is infinite while the epistemological size is finite and always limited by cognitive capacity limitations of neuro-decision-choice agents. From the viewpoint of knowing, the epistemic categories are formed, and may be seen in terms of a problem-solving strategy defined as a sequential or compartmental problem solving involving fuzzy categories as areas of knowing in the epistemic field with an α-algebra (*alpha-algebra*). The alpha-algebra means that the epistemic field is defined by fuzzy information for processing and is equipped with all the fuzzy mathematical and logical operations, the construction of which we now turn our attention.

1.3.2 The Universe, Partitions, Categorial Varieties, Areas of Knowing and the Development Of Epistemic Fields

The structure and discussions on the universe and categories have been presented in [R17.17] with further discussions in [R17.18] related to the space of acceptance-rejection duality in [R21.15] and then further related to the space of problem-solution dualities in [R8.16]. The family of categories has been linked to the universal existence and possible areas of knowing or the *epistemic field.* The problem is to transform the epistemological space into a form that justifies the existence of the space of problem-solution dualities under the general principle of opposites which may allow the creation of an *analytical space* for all epistemic studies, problem-solution dualities and question-answer dualities under relational continua and unity for neuro-decision-choice actions over the social space. The *epistemological space* for cognitive activities in developing knowing and information-knowledge systems through the decision-choice activities over the space of problem-solution dualities is what is being called the *epistemic field* with a corresponding *relational field* and *information field*. Let us always keep in mind that knowledge on varieties and their relations in a multiplicity of connectedness and non-connectedness is not available without information as input in the space of neuro-decision-choice activities.

The concept and phenomenon of information are defined in terms of characteristic disposition to satisfy all the areas of variety existence and knowing. the information is defined in terms of characteristic dispositions This definition ensures the variety identities and their existence in the space of identification-transformation dualities. Without this definition every variety is indistinguishable and variety identification and transformation are pure illusions in the space of neuro-decision-choice actions by cognitive agents. The neuro-decision-choice activities are about variety identifications and transformations as *derived behaviors* in the epistemological space derived from the natural transformations as the *primary behaviors* from the ontological space. In its bare essential, the neuro-decision-choice behaviors in all areas of activities are actions over the space of identification-transformation dualities. Let us turn our attention to essential definitions for the development of the theory of epistemic fields that will help the understanding of the relational connectivity between the ontological and epistemological spaces and between the nature and social spaces, where all natural actions reside in the ontological space and all social actions reside in the epistemological space.

DEFINITION 1.3.2.1: The Epistemic Field$(\mathcal{C})$

Given the infinite ontological space$(\Omega)$with an information structure $[\mathbb{Z}_\Omega=(\mathbb{X}_\Omega \otimes \mathbb{S}_\Omega)]$, an epistemological space$(\mathfrak{C})$ with an information structure $[\mathbb{Z}_{\mathfrak{C}}=(\mathbb{X}_{\mathfrak{C}} \otimes \mathbb{S}_{\mathfrak{C}})]$, the space of actual and possible ontological varieties$(\mathbb{V}_\Omega)$ as infinite space also with information structure$\left[\mathbb{Z}_{\mathbb{V}_\Omega}=\left(\mathbb{X}_{\mathbb{V}_\Omega} \otimes \mathbb{S}_{\mathbb{V}_\Omega}\right)\right]$and the space of epistemological varieties$(\mathbb{V}_{\mathfrak{C}})$ as infinite space also with characteristic-signal dispositions$\left[\mathbb{Z}_{\mathbb{V}_{\mathfrak{C}}}=\left(\mathbb{X}_{\mathbb{V}_{\mathfrak{C}}} \otimes \mathbb{S}_{\mathbb{V}_{\mathfrak{C}}}\right)\right]$, it is useful to distinguish the family$(\mathcal{C})$ of categorial varieties of the form$(\mathbb{C})$ from$(\Omega,\ \mathfrak{C}$ and$\mathbb{V})$with their corresponding information structures and information field$\left(\mathcal{J}=\left[\mathbb{Z}_\Omega \cup = \mathbb{Z}_{\mathfrak{C}} \cup \mathbb{Z}_{\mathbb{V}_\Omega} \cup \mathbb{Z}_{\mathbb{V}_{\mathfrak{C}}}\right]\right)$. The family of categories, $(\mathcal{C})$ is called the *epistemic field* if:

1. $(\varnothing,\Omega)\in\mathcal{C}.\mathfrak{C}\subset\Omega$

2. $\mathcal{C}=\bigcup_{i\in\mathbb{I}^\infty}\mathbb{C}_i,\ \forall\mathbb{C}_i\neq\varnothing$

3. If $\mathbb{C}_i\in\mathcal{C}\Rightarrow\mathbb{C}_i'\in\mathcal{C}\ \forall i\in\mathbb{I}^\infty$, where $\mathbb{C}_i'$ is the complement of $\mathbb{C}_i$

4. If$\{\mathbb{C}_i, i\in\mathbb{I}\}$is finitely countable family of categories with an expanding finite index set $\mathbb{I}$, then

$$\left(\bigcup_{i\in\mathbb{I}}\mathbb{C}_i\right)\in\mathcal{C},\ \bigcap_{i\in\mathbb{I}}\mathbb{C}_i\neq\varnothing\ ,\ \{\mathbb{C}_i\nabla\mathbb{C}_j,\ \forall i,j\in\mathbb{I}\}\subset\mathcal{C}\text{ and }\left(\bigcap_{i\in\mathbb{I}}\mathbb{C}_i\neq\varnothing\right)\subset\mathcal{C}$$

5. If $(\mathcal{C})$is an *epistemic field,* then there exist a *relational field* $\mathcal{R}$ of relational functions $\left(\mathcal{R}=\{r_i \mid i\in\mathbb{I},\ \mathcal{R}(\mathbb{C}_i.\mathbb{C}_j,\forall i,j\in\mathbb{I})\}\right)$that establishes a continuum and unity in the epistemic field with give-and-take-sharing structures among the categories$\{\mathbb{C}_i, i\in\mathbb{I}\}$, and there exist an

information field$(\mathscr{J})$ that allows the identities of all the elements in $(\mathscr{C})$ and $(\mathscr{R})$ to be established, distinguished and identified.

6. If $(\mathscr{C})$ is an *epistemic field,* then there exists a fuzzy-stochastic indicator function

$$\mathcal{I}_{\tilde{\mathcal{C}}}(\omega_i)=\begin{Bmatrix} 1, \text{ if } \omega_i\in\tilde{\mathcal{C}}=\{(\omega_i,\mu_{\tilde{\mathcal{C}}}(\omega_i))\,|\,\mu_{\tilde{\mathcal{C}}}(\omega_i)\geq\alpha^*\in[0,1])\},\ \tilde{\mathcal{C}}\subset\Omega \\ 0, \text{ if } \omega_i\in\tilde{\mathcal{C}}^C=\{(\omega_i,\mu_{\tilde{\mathcal{C}}}(\omega_i))\,|\,\mu_{\tilde{\mathcal{C}}}(\omega_i)<\alpha^*\in[0,1])\},\ \tilde{\mathcal{C}}^C\subset\Omega \end{Bmatrix},\ i\in\mathbb{I}_{\tilde{\mathcal{C}}},\ \Omega=\tilde{\mathcal{C}}\cup\tilde{\mathcal{C}}^C$$ for the

generation of the fuzzy categorial series $\{\tilde{\mathcal{C}}_i, i\in\mathbb{I}^\infty\}$, where the α^* is obtained as an optimal fuzzy value for a solution to the dualistic-polar conflict problem, where the epistemic field is called a *fuzzy epistemic field* or simply an (α-algebra) with relational continua among the categorial diversities with give-and-take-sharing relations. Alternatively, the fuzzy-stochastic indicator function may be replaced by the classical indicator function of the form:

$$\mathcal{I}_{\mathcal{C}}(\omega_i)=\begin{Bmatrix} 1, \text{ if } \omega_i\in\mathcal{C}=\{(\omega,\mu_{\mathcal{C}}(\omega_i)\,|\,\mu_{\mathcal{C}}(\omega_i)=1),\forall\omega\in\mathcal{C}\subset\Omega\} \\ 0, \text{ if } \omega_i\in\mathcal{C}^C=\{(\omega,\mu_{\mathcal{C}}(\omega)\,|\,\mu_{\mathcal{C}}(\omega)=0),\forall\omega\notin\mathcal{C}\subset\Omega\} \end{Bmatrix},\ i\in\mathbb{I}_{\mathcal{C}},\Omega=\mathcal{C}\cup\mathcal{C}^C$$ to generate

the disunity categorial series $\{\mathcal{C}_i, i\in\mathbb{I}^\infty\}$,where $\left(\bigcap_{i\in\mathbb{I}^\infty}\mathcal{C}_i=\varnothing\right)\subset\mathscr{C}$ and the corresponding field is called the σ-algebra or the *classical epistemic field* with relational separation among categorial diversities and no-give-and-take sharing relations. Every category in the epistemic field is equipped with a functor or categorial converter in the space of identification-transformation dualities.

1.3.3 Reflections on the Epistemic Field $(\mathscr{C})$, Paradigms of Thought and Knowing

It may be noted that the difference between the fuzzy-stochastic indicator function and the classical indicator function lies in the way neuro-decision-choice actions are introduced into their formulations before their applications to the selection process to generate categories that form the epistemic field with the corresponding relational field, that contains the variety and categorial give-and-take-sharing structures, and the information field which also contains the conditions of variety and category of identities, distinctions and unity. It also depends on the general conception of the information field which provides conditions for all neuro-decision-choice actions over the space of imagination-reality dualities. The fuzzy information field is inexact, vague, and messy from which each member of the family of ordinary languages (FOL) is constructed, dictionaries are developed, and normal and real definitions are structured with synonyms and antonyms for source-destination communications. The use of either the fuzzy-stochastic indicator function or the classical indicator function generates an algebra on the universal object set $\{\Omega\}$ such that while the classical indicator function generates a σ-algebra (sigma-algebra) under the *exact classical information field* with volume limitation, the fuzzy-stochastic indicator function generates an α-algebra under the *fuzzy information field* with inexactness, vagueness, and volume-limitation in terms of α-level category formations, where the (alphas) α's are obtained by fuzzy optimization decision of the dualistic-polar-game interactions between the duals of a dualities which may be different for each category formation, and where every element of the epistemic field is a collection of problem-solution dualities with common characteristics which are obtained by normalizing the alpha-cuts from the fuzzy-truth table.

1.3.3.1 The Classical Indicator Function, Category Formation, and the Epistemic Field

In the case of the *classical indicator function*, the choice from the binary elements $\{0,1\}$ for assignments into any $\{\mathcal{C}_i \subset \mathscr{C} \mid i \in \mathbb{I}^{\infty}\}$ is done arbitrarily without indications of what determines

$\{0,1\}$ which follows the classical truth table under the classical paradigm of thought with the excluded middle and acceptance of contradictions. Here, the principle of opposites operates with relational separation and disunity without channels of a give-and-take sharing process to generate either crisp category, sub-sets and groups under an exact information field. The excluded middle in the principle of opposites, simply deprives the system of the simultaneous existence of opposites in the same space and time and in the same variety as well as the existence of contradiction. This creates a lack of an explicit decision-choice action in the construct of the classical indicator function that always leads to assumptions of the statement "given an arbitrary set, an arbitrary non-empty collection of sets and many others." In this situation, no epistemic relational distinctions are placed on the categories and the collection of the categories. Every variety exists as either negative or positive but not both, good or bad but not both and cost or benefit but not both, all of which lead to contradictions in nature and classical logic.

Under the classical paradigmatic thinking and reasoning the sub-sets have no intersectionality and hence for any classical categories $\{\mathcal{C}_i, i \in \mathbb{I}^\infty\}$ from Ω, the intersections must lead to $\left(\bigcap_{i\in\mathbb{I}^\infty} \mathcal{C}_i = \varnothing\right) \subset \mathscr{C}$. The condition $\{(\mathcal{C}_i \cap \mathcal{C}_j) \neq \varnothing,\ \text{for some } i, j \in \mathbb{I}^\infty \text{ or } \forall i, j \in \mathbb{I}^\infty\}$ will lead to an acceptance of contradictions in the classical paradigm which will violate the logical excluded middle while the condition $\left(\bigcup_{i\in\mathbb{I}^\infty} \mathcal{C}_i = \Omega\right) \subset \mathscr{C}$ leads to unity with no relational continuum under the principle of opposites. In the classical epistemic field, every variety $(\nu \in \mathbb{V})$ cannot exist as a fuzzy variety where every variety is a problem-solution duality and polarity with the problem duality residing in the problem polarity and the solution duality residing in the solution polarity for continual variety ignorance-knowledge transformations. In the classical epistemic field, every variety exists as a problem or a solution in the space of absoluteness, and hence, has no internal conflict relative to a goal. In this respect, the cost-benefit analytics is an ad-hoc process that has no logical foundation because cost and benefit cannot exist in the same variety where the neuro-decision-choice is a cost-benefit balance relative to a goal or an objective.

The use of the construct of the classical indicator function for the *Borel field* leads to a σ-algebra (sigma-algebra) on Ω, where $\left(\bigcap_{i\in\mathbb{I}^\infty} \mathcal{C}_i = \varnothing\right) \subset \mathscr{C}$. By accepting the condition $\left(\bigcup_{i\in\mathbb{I}^\infty} \mathcal{C}_i = \Omega\right) \subset \mathscr{C}$, the classical paradigm of thought with the classical indicator function, while defining crisp analytical categories in the abstract space of absoluteness, accepts some form of unity in the *epistemic field.* The form of the connectivity or relational connectivity is difficult to abstract from the analytical structure or from the classical paradigm of thought while increasing difficulties are encountered in some areas of applications. This difficulty in establishing a global connectivity is an artificial foundation in the analytical argument in support of disunity in knowing and science. This logical difficulty may also help to explain the rise of *paradoxes* in thinking and *phantom problems* and others in the space of problem-solution dualities as the classical paradigm is extended into conditions of increasing relativity. The practice of thinking and reasoning in everyday life over the space of problem-solution dualities accepts the simultaneous existence of opposites in the same variety. For example, economic analysis is based on dualistic-polar conditions, such as supply-demand duality in the space of relativity of supply disposition and demand disposition in relative dominance, to give rise to supply surplus or supply shortage and the inventory phenomenon under inexact fuzzy market information but not exact classical information. Similarly, the development of synonyms and antonyms draws its foundation from the principle of opposites with relational continuum and unity over the linguistic space of nominalism.

1.3.3.2 The Fuzzy Indicator Function, Category Formation, and the Epistemic Field

In the case of the *fuzzy indicator function*, the choice within $[0,1]$ for assignments into any of the set $\{\mathcal{C}_i \subset \mathscr{C} \mid i \in \mathbb{I}^\infty\}$ is infinite and is done with an optimal fuzzy decision-choice action of what determines the infinite values of $\alpha \in [0,1]$ in the conflicts between the duals of duality, where the acceptance of any conflict relationality of the duals follows the fuzzy truth table under the fuzzy paradigm of thought with relational continuum and unity without the conditions of the excluded middle under the fuzzy information field. In the case of the fuzzy indicator function, the processing of the categorial formations into the epistemic field admits of subjective judgement corrections irrespective of the user, where the principle of opposites operates without relational separation and disunity but with the establishment of mutual channels of give-and-take sharing processes. It is these optimal fuzzy decision-choice actions that always lead to the statement "relational continuum and unity, set interdependency, non-empty intersections, diversity in unity, possibility-probability, interdisciplinary, actual-potential interdependencies, imagination-reality interdependencies, the past-present-future relational inter-connectedness cost-benefit conditions" and many other such dualistic-polar conditions over the universal existence of varieties over the space of composition-decomposition dualities.

In this way, the epistemic distinctions with relational connectedness, placed on the categories, sub-sets, groups and the collections of the categories, sub-sets, and groups in time and over time, are neuro-decision-choice determined which is independent of the user with optimal epistemic conditionality. The optimal epistemic conditionality takes the form of optimal α-level cuts that show the optimal fuzzy decision-choice partitions, fuzzy separations, and connectivity over the space of acceptance-rejection dualities for the constructed categories. Under the fuzzy paradigmatic thinking and reasoning, the categories have intersectionality with relational interconnectedness and hence for any set of fuzzy categories $\{\mathcal{C}_i, i \in \mathbb{I}^{\infty}\}$ from Ω, the intersections must lead to $\left(\bigcap_{i \in \mathbb{I}^{\infty}} \mathcal{C}_i \neq \varnothing\right) \subset \mathscr{C}$. The condition $\{(\mathcal{C}_i \cap \mathcal{C}_j) \neq \varnothing,$ for some $i, j \in \mathbb{I}^{\infty}$ or $\forall i, j \in \mathbb{I}^{\infty}\}$ leads to the acceptance of contradictions, where opposites exist in the same space and time in the fuzzy paradigm of thought such that $\left(\bigcup_{i \in \mathbb{I}^{\infty}} \mathcal{C}_i = \Omega\right) \subset \mathscr{C}$ leads to a relational continuum and unity under the principle of opposites, where every variety $(v \in \mathbb{V})$ exists simultaneously as a problem-solution duality and polarity, and where the problem duality resides in the problem polarity and the solution duality resides in the solution polarity in the sense that every variety resides as a negative-positive duality which is translated into problem-solution duality. Every variety negative is a cost-benefit duality and polarity, and every variety positive is also a cost-benefit duality and polarity over the decision-choice space under the fuzzy information field relative to the space of imagination-reality dualities.

The use of the fuzzy indicator function for the construct of the *epistemic field* leads to an α-algebra (alpha-algebra) that generates the epistemic categories and areas of knowing that are subject to change under the changing fuzzy information field which is generated by the relational dynamics of neuro-decision-choice actions and the space of problem-solution dualities. By accepting the condition $\left(\bigcup_{i \in \mathbb{I}^{\infty}} \mathcal{C}_i = \Omega\right) \subset \mathscr{C}$, the fuzzy paradigm of thought with the fuzzy indicator function, where $\left(\bigcap_{i\in\mathbb{I}^{\infty}} \mathcal{C}_i \neq \varnothing\right) \subset \mathscr{C}$, while defining crisp analytical categories under the fuzzy-stochastic decision-choice conditionality, accepts relational continua and unity among categories in the epistemic field. The relational connectivity is established through the α-algebra (alpha algebra) as a foundation in the analytical argument to support the relational diversity and unity in knowing and science. The extensive development of the *theory of epistemic fields* and the corresponding α-algebra (alpha algebra) with the relevant properties will be taken up in a separate monograph.

What is being presented here is necessary to link together, my analytical views on the information-knowledge stock-flow process, neuro-decision-choice actions, problem-solution dynamics, diversity, and the unity of knowing in the variety space under differential paradigms of thought. The combined structures of the fuzzy paradigm of thought, principle of opposites with relational continuum and unity help to explain the existence of the infinite space of problem-solution dualities, unintended consequences and collateral damages and to present a critical understanding of the dynamics of variety problems to variety solutions, variety solutions to variety problems, variety costs to variety benefits, variety benefits to variety costs, variety destruction to variety construction and many other such dualistic-polar structures in the universe. The theory of the epistemic field presents the general notion that the economic systems are and have always been the *knowledge economic system*, the continual development of which depends on an increasing knowledge through neuro-decision-choice actions over the space of problem-solution dualities where the results of the neuro-decision-choice actions may be enhanced by curiosity, creativity, innovation, hard work and perseverance as abstracted from the social philosophical consciencism that may constrain individual, collective and national personalities.

The importantly implicit observation in this analytical process is that all symbolic representations are representations of explicit and implicit characteristic dispositions of varieties that provide the contents of real and abstract elements in the universal object set as they relate to the spaces of problem-solution dualities, real cost-benefit dualities, identification-transformation dualities, exact-inexact dualities, and many other dualistic-polar structures under the principle of opposites with relational continuum and unity in the enterprise of knowing and information-knowledge accumulation. Similarly, an important implicit and explicit condition in the epistemic structure is that every cognitive neuro-decision-choice activity is about an action over the space of problem-solution dualities to substitute one variety with greater real benefit disposition for another variety with greater real cost disposition towards a preferred state of being. The characteristic dispositions give value to linguistic meanings and identifications of varieties and categorial varieties to establish differences and similarities required in category formation, set-group formations, and rank order in decision-choice actions of all forms over the spaces of source-destination dualities and polarities and languages.

In this respect, the study of any language may be viewed as the study of variety representations for recognition of variety identifications and the understanding of variety transformation through information may be defined as variety characteristic-signal disposition that gives support to the epistemic field with the relational field, and where verbal connectivity and non-verbal connectivity are viewed as varieties. As it has been argued, a major change in the internal distribution of a variety's characteristic disposition leads to a new variety and hence a movement from the initial category of belonging to a new category consistent with the new identity [R17.18], and hence a movement from a variety problem to a variety solution, or a variety ignorance to a variety knowledge constitutes a qualitative or quantitative transformation. The variety transformation is a fuzzy process, the dynamics of which must deal with *categorial transversality conditions* and entropic actions on the quality control from the information field.

1.3.3.3 Information Support of the Epistemic Fields

Information is an indispensable element in all areas of the space of human endeavors which demand continual neuro-decision-choice actions in understanding of the conditions of general existence in organization and management in the space of life-death dualities supported by multiplicities of variety properties and relations. Information is the third dimension of our universal existence. It resides in every variety and is indispensable in the space of identification-transformation dualities and polarities under the principle of opposites with relational continua under a multiplicity of give-and-sharing processes. Information provides the conditions for the spaces of relativity and absoluteness for triangular interdependencies of varieties, relations and information regarding oneness, dividedness, organization, disorganization, diversity and unity over the universal existence and the space of destruction-construction dualities and polarities. The information defined as characteristics in the general universe is divided into a multiplicity of characteristic dispositions that are assigned to varieties in their identity support over the space of identification-transformation dualities such that every variety has its information support over the space of source-destination dualities of an infinite system of communication interchanges in give-and-take-sharing modes through the corresponding multiplicity of signal dispositions, where the characteristic dispositions allow us to speak of the *ontic space* which is attached to ontology with perfect information and the *epistemic space* which is attached to epistemology with imperfect information [R8.16][R17.17][R17.18][R21.5].

LEMMA 1.3.3.1: The Epistemic Information Support.

Any epistemic field of the form $(\mathscr{E})$ with the corresponding relational field $(\mathscr{R})$ has an epistemic information support structure of the information field $(\mathscr{I})$ in the form $\{\mathbb{Z}_{\mathscr{E}}=(\mathbb{X}_{\mathscr{E}}\otimes\mathbb{S}_{\mathscr{E}})\}$ such that

$$\mathbb{Z}_{\mathscr{E}}=(\mathbb{X}_{\mathscr{E}}\otimes\mathbb{S}_{\mathscr{E}})=\left\{\mathbb{Z}_{\mathcal{C}_i}=\left[\begin{array}{c}(\mathbb{X}_{\mathcal{C}}\otimes\mathbb{S}_{\mathcal{C}_i})|,(\mathbb{X}_{\mathcal{C}}\subset\mathbb{X}_{\mathscr{E}}),(\mathbb{S}_{\mathcal{C}_i}\subset\mathbb{S}_{\mathscr{E}}),\left(\mathbb{X}_{\mathscr{E}}=\bigcup_{i\in\mathbb{I}}\mathbb{X}_{\mathcal{C}_i}\right),\\ \left(\bigcap_{i\in\mathbb{I}}\mathbb{X}_{\mathcal{C}_i}\neq\varnothing\right),\left(\mathbb{S}_{\mathscr{E}}=\bigcup_{i\in\mathbb{I}}S_{\mathcal{C}_i}\right),\left(\bigcap_{i\in\mathbb{I}}\mathbb{S}_{\mathcal{C}_i}\neq\varnothing\right),(i\in\mathbb{I}^{\infty})\end{array}\right]\right\}$$

Proof of the Lemma 1.3.3.1

If $(\mathscr{E})$ is an epistemic field with the corresponding relational field $(\mathscr{R})$ then there exist a family of categories $\{\mathcal{C}_i, i\in\mathbb{I}\}$, with a family of information structures called the *information field* $(\mathscr{I})$ by definition 1.3.2.1 given by

$$\mathbb{Z}_{E}=(\mathbb{X}_{\mathfrak{E}}\otimes\mathbb{S}_{\mathfrak{E}})=\left\{\mathbb{Z}_{\mathcal{C}_i}=(\mathbb{X}_{\mathcal{C}_i}\otimes\mathbb{S}_{\mathcal{C}_i})\mid \mathbb{X}_{\mathcal{C}_i}\subset\mathbb{X}_{\mathfrak{E}},\mathbb{S}_{\mathcal{C}_i}\subset\mathbb{S}_{\mathfrak{E}},\forall i\in\mathbb{I},\right\}$$ with auxiliary conditions

$$\mathscr{E}=\left(\bigcup_{i\in\mathbb{I}}\mathcal{C}_i,\ \varnothing,\Omega\right)\subset\mathscr{E}\Rightarrow\ \left(\mathbb{X}_{\mathscr{E}}=\bigcup_{i\in\mathbb{I}}\mathbb{X}_{\mathcal{C}_i}\right),\left(\bigcap_{i\in\mathbb{I}}\mathbb{X}_{\mathcal{C}_i}\neq\varnothing\right),\left(\mathbb{S}_{\mathscr{E}}=\bigcup_{i\in\mathbb{I}}\mathbb{S}_{\mathcal{C}_i}\right),\left(\bigcap_{i\in\mathbb{I}}\mathbb{S}_{\mathcal{C}_i}\neq\varnothing\right),(i\in\mathbb{I})$$

by characteristic-signal definition of variety identity and category formation with a fuzzy indicator function constrained by a fuzzy-stochastic conditionality, such that each $\{\mathbb{C}_i, i \in \mathbb{I}\}$ is an epistemic category representing an actual or potential area of knowing with a class of problem-solution dualities $\{\mathfrak{R}_{\mathbb{C}_i}, i \in \mathbb{I}\}$. The family of characteristic-signal dispositions $\{\mathbb{Z}_{\mathbb{C}_i} = (\mathbb{X}_{\mathbb{C}_i} \otimes \mathbb{S}_{\mathbb{C}_i}), i \in \mathbb{I}\}$ as its information support constitutes the information field $(\mathscr{I})$ □.

NOTE 1.3.3.3.1: Reflections on the LEMMA 1.3.3.1:

The definition of the epistemic field $(\mathscr{C})$, with its relational field $(\mathscr{R})$, where the categories of the epistemic field are formed as areas of knowing, and the lemma suggest to us that the categorial relationality of give-and-take-sharing structure in the epistemic processes is such that the classical indicator function in category formation cannot justify disunity in the epistemic field when the concept of information is well defined in the areas of knowing, and when category formations are neuro-decision-choice determined in the information environment of vagueness and volume limitations which result in production of inexactness and uncertainties from *cognitive capacity limitations* of decision-choice agents. We must keep in mind that every categorial area of knowing has intersectionality with at least two other categories constrained by a fuzzy-stochastic conditionality such that with the corresponding relational field $(\mathscr{R})$, the information structure of the form, $(\mathbb{Z}_{\mathscr{C}} \subset \mathscr{I})$ establishes relationally epistemic diversities and intersectionality with relational continua and give-and-take-sharing modes rather than relationally epistemic individualities and disunities with excluded middles and without relational give-and-take-sharing modes.

The relationship between the epistemic field and the space of problem-solution dualities is that the alpha algebra (α-level cuts) on the universal object set Ω is also the same alpha algebra on the spaces of problem-solution dualities, information $(\mathbb{Z}_{\Omega})$ and varieties $(\mathbb{V}_{\Omega})$. The implication of the existence of epistemic intersectionality is that groups of problem-solution dualities in areas of knowing interact in give-and-take-sharing structures under the dynamic relational field $(\mathscr{R})$ such that, for example, successful solutions of some problems in the social sciences may facilitate the search for solutions for some problems in the natural sciences and vice versa. The groups of problem-solution dualities are fuzzy-categorial dividedness or rational decomposition, where each category represents an area of research on the space of problem-solution dualities. Here, the relational field presents not only the relational interdependencies of categories in the process of knowing but presents relational problem-solution interdependencies in the same category such that a variety solution generates variety problems and vice versa. In fact, the efficiency of the space of epistemic behaviors of problem-solution dualities over the category of natural science finds meaning in the category of social sciences, while the gains in efficiency of epistemic activities in the category of natural science facilitate the efficiency in organization, operations, and politico-economic management of knowing and information-knowledge stock-flow dynamics through the utilizations of gains from both natural and social sciences.

In this respect, the problem-solving strategies are such that the diversities in matter-energy varieties within the universal existence are cognitively imposed by the logic of fuzzy decomposition, differentiation, and category formation based on *fuzzy cluster analysis* to establish variety differences and similarities in the collection of varieties that correspond to the distribution of the universal characteristic-signal dispositions. The similarities of characteristic-signal dispositions of varieties constitute the formation of categories, while the differences of characteristic-signal dispositions form the essential basis of inter-categorial differences, identities, and formation. The differentiations and similarities within the formation of categories and different categories are established by human neuro-decision-choice actions with *fuzzy logic* that admits of vagueness, volume-incompleteness, inexactness, error-corrections with continual limiting processes and linguistic impurities due to cognitive capacity limitations of neuro-decision-choice agents over the space of acquaintances linked to nominalism and every kind of language development. Within the framework of neuro-decision-choice actions, for knowing and information-knowledge accumulation, the processes of thinking and reasoning are defined to balance dualistic-polar forces from the space of relativity to the space of absoluteness under the principle of gradualism, uncertainty and risk with intertemporal give-and-take-sharing processes as established by needs, goals and objectives of duals and poles of the general system of dualistic-polar structures of the universal existence.

The objective of the fuzzy differentiation, fuzzy clustering and fuzzy partitioning is to decompose the general space of the knowledge acquisition problem-solution dualities of the universe into a collection of manageable categories of sub-problem-solution dualities called areas of knowing of many specialized attentions in the research over the space of problem-solution dualities at the level of epistemology to reduce and ease the knowing difficulties confronted by cognitive agents with cognitive capacity limitations. The space of problem-solution dualities is then partitioned into sub-spaces of problem-solution dualities with fuzzy-stochastic indicator functions producing relational continua and unity in flexibility of thought and practice. Each category, then, becomes an area or a sub-set of problem-solution dualities under the *general principle of opposites* with a multiplicity of relational continua and unity in knowing where categories may represent institutions and sub-categories may represent departments over the space of problem-solution dualities. We may now speak of *categorial problem-solution dualities* equipped with a functor space or space of categorial converters , *intra-categorial problem-solution dualities* and *inter-categorial problem-solution dualities* in the understanding of the factors that are continually shaping the conditions in the space of unity-disunity dualities of knowing, and by logical extension the factors that are shaping the conditions of the space of unity-disunity dualities of science and the rise of different methods and techniques for problem-solving in the sub-spaces of problem-solution dualities with categorial diversities as well as shaping the paths of socio-natural transformations of varieties and categories.

1.3.3.4 Categories, Functors, Categorial Converters, Transformations and Epistemic Fields

In the process of developing an analytical framework for the understanding of variety relations, diversity and the unity of knowing in all areas or cognitive activities as economic production-consumption dualities of intellectual investment flows and information-knowledge accumulations as production-consumption dualities of intellectual capital stocks, we introduced the concepts of categories, functors, categorial conversions and variety characteristic changes over the space of identification-transformation dualities relative to the space of ignorance-knowledge dualities. The concepts of categorial conversion, functor space, conformal mapping, variety characteristic changes and transformations imply motions over the space of quality-quantity dualities, where every variety characteristic disposition is in a temporary equilibrium and in the state of change, and similarly, every variety signal disposition is in a temporary equilibrium and in the state of change over the space of property-relation dualities relative to the space of information-relation dualities. These motions relate to changes of properties and relations of identities of categories and varieties in intra-categorial and inter-categorial conditions. Under the general and specific principles of opposites, the motions may be seen as journeys among categories in terms of mapping from one category to another over the relational and information fields.

The mappings are special categorial functions called functors or categorial converters that take varieties from categories of their belongings, alter their characteristic dispositions or their relational dispositions or both and place them in other categories of new belongings. The categories are formed with either a classical indicator function to create a σ-algebra with a *classical functor* and *functor space* or a fuzzy indicator function to create α-algebra with a *fuzzy functor* and *fuzzy functor space*. Let us specify such categorial mappings keeping in mind the structure of the set of categories in the epistemic field with the supporting relational and information fields and that every transformation is a creation of new information without destructing the old information. Every transformation is either intra-categorial or inter-categorial categorial-convertor of functor mapping which are conformal to maintain the path of genetic identities.

DEFINITION 1.3.2.2A: Functor $(\mathfrak{f})$ and the Space of Functors $(\mathbf{F})$ (Verbal)

In categorial conversions and variety transformations over the epistemic field as a family of categories, a *functor* $(\mathfrak{f})$ is a mapping from a category to a category or categories in an epistemic field altering old identities by changing the characteristic disposition and relational disposition of varieties in categories of belonging and acquiring new characteristic dispositions and relational variety dispositions for the new identities in order to enter into categories of new belongings. The collection of all functors over the epistemic field is the space of functors or *functor space* $(\mathbf{F})$

. The degree of the efficiency of *functor action* in variety transformation among categories is the *functor moment* $(\mathbf{m} \in [0,1])$ which is the strength of the functor, and the collection of all functor moments is the *space of functor moments* $\{\mathbf{F_m} \mid \mathfrak{f}_\mathbf{m} \in \mathbf{F_m}\}$.

DEFINITION 1.3.2.2B: Functor $(\mathfrak{f})$ and the Space of Functors $(\mathbf{F})$ (Algebraic)

If $\{\mathcal{C}_i . \mathcal{C}_j \in (\mathscr{C}), i \neq j \in \mathbb{I}\}$ is the set of $(i-j)$ categories belonging to the epistemic field $(\mathscr{C})$, then a functor $(\mathfrak{f})$ from $\{\mathcal{C}_i, i \in \mathbb{I}\}$ to $\{\mathcal{C}_j, j \in \mathbb{I}\}$ is a characteristic-relational mapping that transforms the identity $(\mathbf{I}_{v_{\mathcal{C}_i}})$ of a variety $\{v \in \mathcal{C}_i, i \in \mathbb{I}\}$ to a variety identity $(\mathbf{I}_{v_{\mathcal{C}_j}})$ of $\{\mathfrak{f}(v) \in \mathcal{C}_j, j \in \mathbb{I}\}$ in the sense that $(\mathfrak{f} : \mathbb{X}_{v_{\mathcal{C}_i}} \to \mathbb{X}_{v_{\mathcal{C}_j}})$, where $\mathfrak{f}(\mathbf{I}_{v_{\mathcal{C}_i}}) = \mathbf{I}_{\mathfrak{f}(\mathbf{I}_{v_{\mathcal{C}_i}})}$ and the symbol $(\mathbf{I})$ represents a set of variety identities which are established by the corresponding characteristic dispositions $(\mathbb{X}_{v_{\mathcal{C}}})$, where the categories in the epistemic field are established by either classical or fuzzy indicator function. The collection of all functors over the epistemic field is the *space of functors* or *functor space* $\mathbf{F} = \{\mathfrak{f}_i(\cdot) \mid \forall i \in \mathbb{I}^\infty\}$ where $(\mathbf{m}_i \in [0,1], i \in \mathbb{I}^\infty)$ is the degree of efficiency associated with the *functor action* of $(\mathfrak{f}_i(v) \mid v \in \mathbb{V})$ in variety transformation among categories in $(\mathscr{C})$ as the *functor moment* $(\mathbf{m}_i \in [0,1])$ and the collection of all associated functor moments is the space of *functor moments* $\mathbf{F_m} = \{\mathbf{m}_i \mid \forall i \in \mathbb{I}^\infty\}$. The general functor space with the functor moments may be written as $\mathbf{F}_{\mathbf{m}_i} = \{(\mathfrak{f}_i(\cdot), \mathbf{m}_i) \mid \mathbf{m}_i [0,1], \forall i \in \mathbb{I}^\infty\}$ where every category is equipped with a functor operator and specified as $\{(\mathcal{C}_i, \mathfrak{f}_{\mathbf{m}_i}), i \in \mathbb{I}\}$.

DEFINITION 1.3.2.3A: Fuzzy Functor $\left(\mathfrak{F}, \mu_{\mathbf{F}}(\bullet)\right)$ and the Space of Fuzzy Functors $(\mathbf{F}, \mathbf{M})$ (Verbal)

In categorial conversions and variety transformations in relational continuum and unity, a fuzzy functor $\left(\mathfrak{F}, \mu_{\mathbf{F}}(\bullet) \in [0,1]\right)$ is a mapping equipped with a distribution of degrees of belonging $\mu_{\mathbf{F}}(\bullet) \in [0.1]$ to the space of categorial conversions among fuzzy categories in the epistemic field constructed with a fuzzy indicator function altering the existing characteristic dispositions and relational dispositions of varieties, thus changing their identities in the fuzzy categories of their belonging and acquiring new characteristic dispositions and relational variety dispositions with new identities in order to enter into fuzzy categories of new belongings consistent with their new identities. The collection of all fuzzy functors over the epistemic field $(\mathscr{C})$ with an $\alpha-$algebra is the space of fuzzy functors or fuzzy functor space $(\mathbf{F}, \mathbf{M})$ where $(\mathbf{M})$ is the space of fuzzy membership functions and $\left(\mu_{\mathbf{F}}(\bullet) \in [0,1]\right)$ is the fuzzy membership characteristic function of belonging to the *fuzzy functor space* indicating the degree of which a functor belongs to the space of identification-transformation and problem-solution dualities.

DEFINITION 1.3.2.3B: The Fuzzy Functor $\left(\mathfrak{F}, \mu_{\mathfrak{F}}(\bullet)\right)$ and the Space of Fuzzy Functors $(\mathbf{F}, \mathbf{M})$ (Algebraic)

If $\left\{\mathcal{C}_i . \mathcal{C}_j \in (\mathscr{C}), i \neq j \in \mathbb{I}\right\}$ is the set of $(i-j)$ categories belonging to the epistemic field $(\mathscr{C})$ then a fuzzy functor $\left(\mathfrak{F}, \mu_{\mathfrak{F}}(x) \in [0,1]\right)$ from $\left\{\mathcal{C}_i, i \in \mathbb{I}\right\}$ to $\left\{\mathcal{C}_j, j \in \mathbb{I}\right\}$ formed with a fuzzy indicator function, is a characteristic-relational mapping that transforms the identities $\left(\mathbf{I}_{v_{e_i}}\right)$ of a varieties in a fuzzy category $\{v \in \mathcal{C}_i, i \in \mathbb{I}\}$ to a variety identity $\left(\mathbf{I}_{v_{e_j}}\right)$ of $\{\mathfrak{f}(v) \in \mathcal{C}_j, j \in \mathbb{I}\}$ into another fuzzy category in the sense that $\left(\mathfrak{f}: \mathbb{X}_{v_{e_i}} \to \mathbb{X}_{v_{e_j}}\right)$, where $\mathfrak{f}\left(\mathbf{I}_{v_{e_i}}\right) = \mathbf{I}_{\mathfrak{f}\left(\mathbf{I}_{v_{e_i}}\right)}$ and the symbol $(\mathbf{I})$ represents the variety identities as established by the corresponding characteristic dispositions with a fuzzy indicator function[R8.16], where $\left\{\mathfrak{f}\left(\mathbf{I}_{v_{e_i}}\right) = \left(\mathbf{I}_{v_{e_i}}\right) \Rightarrow \mathfrak{f}\left(v_{e_i}\right) = \left(v_{e_i}\right)\right\}$ and $\mu_{\mathfrak{f}}(x)$ is the degree to which $(\mathfrak{f})$ the ordinary functor space $(\mathbf{F})$ and $(\mathbf{M})$ is the space of fuzzy membership functions and $(\mathbf{F}, \mathbf{M})$ is fuzzy functor space where $\{\mathfrak{f} \in \mathbf{F} \text{ and } \mu_{\mathfrak{f}}(x) \in \mathbf{M}\}$. The general fuzzy functor space with the functor moments may be written as $\mathbf{F}_{\mathbf{m}_i} = \left\{\left(\mathfrak{f}_i(\cdot), \mu_{\mathfrak{f}_{\mathbf{m}_i}}(x)\right) \mid \mathbf{m}_i \in \mathbf{M}, \mathfrak{f} \in \mathbf{F}, \ \forall i \in \mathbb{I}^{\infty}\right\}$, where every fuzzy category is equipped with a fuzzy functor operator and specified as $\left\{\left(\mathcal{C}_i, \mathfrak{f}_{\mu_{\mathfrak{f}_{\mathbf{m}_i}}}\right), i \in \mathbb{I}\right\}$.

NOTE1.3.3.4.1: On Functors, Categorial Converters and Moments

In categorial conversions over the space of problem-solution dualities and polarities, we are concerned with categorial convertors as functors with effective moments that have the capability and efficiency to convert a variety in one category to another category in the epistemic field supported by relational and information fields. It is important to note that every variety has a corresponding characteristic disposition that reveals its categorial identity and this categorial identity rests in the conditions of intra-categorial and inter-categorial relationality, continuum, and unity in the relational field, where the variety identity and relationality in continuum and unity are established by characteristic dispositions in the information field. The categorial convertors are to alter the variety identities by either altering the characteristic dispositions or key elements of the characteristic dispositions changing their identities, thus transforming their relational continua and unity within categories contained in the epistemic field. The category from which an element is transformed from is called the *primary category* or *parent category* and the category that receives the transformed category is called the *derived category* or *offspring category,* where every category resides in the space of *primary-derived duality* or space of *parent-offspring dualities* with relational continua and unity. In other words, the categorial convertors and the functors are instruments of transformation actions in the space of primary-derived categories containing varieties.

The transformations are processes over the space of input-output dualities where the categorial convertors and functors take the inputs from the primary category, change their characteristic dispositions, and create new characteristic dispositions to change variety identities to enter into derived categories. The principle of opposites with diversities in relational continua and unity is such that every category belongs to the space of primary-derived dualities or parent-offspring dualities, where every category serves simultaneously as primary and derived categories in the space of input-output transformation processes over the space of parent-offspring dualities. In the space of problem-solution dualities, the variety problem serves initially as the primary while the variety solution becomes as derived, and then serves as primary from which a problem emerges as a derived. The understanding of functor actions, functor space, functor moment, space of functors and the space of functor moments in categorial conversion as elements over the production-consumption dualities may be obtained from the corresponding analytical elements in the space of categorial conversions such as categorial convertor actions, moments and the corresponding spaces over the space of problem-solution dualities [R8.16] [R17.17][R17.18] [R21.5] supported by the fuzzy paradigm of thought [R4.7][R4.10].

It is important to note that the theories of knowing as the production of intellectual investment flows and information-knowledge accumulation as the production of intellectual capital stocks see the logic of categorial conversion over the epistemic field in the same way as the logic of functor within the mathematical theory of categories with conformal and non-conformal mapping over the system of categories within the spaces of input-output dualities, destruction-construction dualities relative to the spaces of problem-solution dualities and ignorance-knowledge dualities, where categories encompass real and nominal varieties in their formations into the epistemic field. The categories may also be taken as epistemic spaces and departments with inter-categorial connectivity under restrictions of fuzzy conditionalities, where intra-categorial and inter-categorial journeys over the space of qualitative-quantitative dualities must satisfy specific sets of transversality conditions moving between dualistic-polar zones. The functor and categorial convertor are varieties to categorial mapping. They may also be viewed as categorial variety identity mapping. Additionally, the functor may also be a said to be a categorial mapping.

1.4: SOME ANALYTICS ON THE EPISTEMIC FIELDS AND CATEGORIES OF KNOWING

There are some examples of the epistemic field that are useful to reflect on. The *simplest epistemic field* is $\left(\mathscr{C}=\{\Omega,\varnothing\}\right)$ with a *relational field* $\left(\mathscr{R}\right)$ that connects the empty category of problem-solution dualities and the category of all actual and potential problem-solution dualities (Ω) and the *information field* $\left(\mathscr{I}\right)$ which provides the conditions for variety and categorial identities and connectivity. This is the nothingness-somethingness duality or the beginning-end duality of universal existence. In this respect, we have intra-categorial relationship with $(\mathscr{C} = \{\Omega\})$, intra-categorial relationship also with $(\mathscr{C} = \{\varnothing\})$ and inter-categorial relationship on $(\mathscr{C} = \{\Omega, \varnothing\})$ as established by the conditions of the relational field and distributed by the conditions of the information field. The structure $(\mathscr{C} = \{\Omega, \varnothing\})$ is the *smallest epistemic field* which is a trivial one, and which simply says that the universe is the area of study which constitutes the collection of all problem-solution duality in oneness, and $(\mathscr{C} = \{\Omega, \varnothing\})$ constitutes the nothingness-somethingness polarity [R17.19], where the beginning is in the end and the end is in the beginning, nothingness is in somethingness and somethingness is in nothingness which is a variety solution to variety somethingness. Similarly, the variety somethingness is a variety solution to the variety problem of nothingness. In this way, the dualistic-polar conditions of the beginning-end structure is the same as the dualistic-polar conditions of the nothingness-somethingness structure that must be related to existence-nonexistence dualities and zero-infinity $(0, +\infty)$ and $(-\infty, 0)$ dualistic-polar conditions in our mathematical constructs which in categorial dynamics represent a postulate that from nothingness emerges somethingness and from somethingness, everything will vanish into nothingness in the space of nothingness-somethingness dualities.

The challenges of cognitive agents in the knowing process over the space of acquaintance are that the fullness and exactness of the characteristic disposition of somethingness $(\mathscr{C} = \{\Omega, \mathbb{X}_{\Omega}\} = \{\infty, \mathbb{X}_{\infty}\})$ through its signal disposition $(\mathscr{C} = \{\Omega, \mathbb{S}_{\Omega}\} = \{\infty, \mathbb{X}_{\infty}\})$ are unavailable to cognitive agents except through speculations from the space of imagination-reality duality. It is this lack of fullness and exactness of the that creates disparities between characteristic dispositions of the ontic varieties and the epistemic varieties to generate the space of uncertainty-certainty dualities that is contained in the epistemological space in which cognitive agents operate. Notice that the trivial epistemic field of nothingness-somethingness duality as the beginning-end duality presents the conditions for debates over the space of theism-antitheism duality. The signal disposition $\left(\mathscr{C}=\{\Omega,\mathbb{S}_{\Omega}\}\right)$ is infinitely large and overwhelms the cognitive capacity of cognitive agents in a way that the corresponding characteristic disposition $\left(\mathscr{C}=\{\Omega,\mathbb{X}_{\Omega}\}\right)$ is always in a fuzzy-stochastic space with a small revelation of it. Similarly, the signal disposition $\left(\mathscr{C}=\{\varnothing,\mathbb{S}_{\varnothing}\}\right)$ is infinitely amorphous and overwhelms the cognitive capacity of cognitive agents in a way that the corresponding characteristic disposition $\left(\mathscr{C}=\{\varnothing,\mathbb{X}_{\varnothing}\}\right)$ is always in the fuzzy-stochastic space with a small revelation of it such that $\left(\mathscr{C}=\{\varnothing\subset\Omega\subset\varnothing\}\right)$, where nothingness resides in somethingness and somethingness resides in nothingness with the end defined in the beginning which is also defined in the end in the space of beginning-end dualities.

Mathematics is a language belonging to the family of abstract languages (FAL) as such to be helpful in storage, communication, and all analytics over the space of problem-solution dualities, it must relate to nominal and real varieties as well as the characteristics that present some meaning in the space understanding. In relation to the universal existence, it is useful to note that $(-\infty,0)\bigcup(+\infty,0)=(-\infty,0,+\infty)$ represents the somethingness process and $(-\infty,0)\bigcap(+\infty,0)=(0)$ represents the nothingness process which, together represent the beginning-end process or the somethingness-nothingness process of universal existence for the process of knowing and information-knowledge accumulation as intellectual activities over the spaces of varieties and relations within the space of identification-transformation dualities. All attempts to the increasing telescopic efficiencies are to help the identifications and understanding of existential conditions of varieties in $(-\infty,0)$ of the telescopic past where, (0) represents initial conditions and necessity. Similarly, all attempts to the increasing telescopic future efficiencies are to help the understanding of varieties in $(0,+\infty)$ of telescopic future, where (0) represents the state of initial conditions and freedom.

Given the minimal epistemic field, there is also the *maximal epistemic field* which contains all the specific varieties, where each category has one element in the form $\left(\mathscr{C}=\{\varnothing,\mathbb{C}_i=\{v_i\}\,\forall i\in\mathbb{I}^{\infty}\}\right)$ with a relational field $(\mathscr{R})$ and with pairwise and multilateral variety relational connectivity. If there are only two categories of areas of knowing of form, science and non-science, then the epistemic field is $\left(\mathscr{C}=\{\varnothing,\mathbb{C}_1=\{v_1\},\mathbb{C}_2=\{v_2\}\}\right)$ and $(\mathscr{R})$ is a relational field establishing a relational connectivity with an optimal fuzzy-stochastic conditionality in categorial formation between the categorial-variety structure of *science* $\mathbb{C}_1=\{v_1\}$ and the categorial-variety structure of *non-science* $\mathbb{C}_2=\{v_2\}$ under an information field $(\mathscr{I})$. In the simple epistemic field, all problem-solution dualities are considered for neuro-decision-choice actions such that there is only one category of problem-solution dualities over which neuro-decision-choice actions may be exercised. In the maximal epistemic field, the variety problem-solution dualities are considered separately for neuro-decision-choice actions, such that each category is a singleton set of a variety problem-solution duality and the variety problem is its own variety solution and the variety solution is its own variety problem with relational separation to the space of goals, objective, and vision, where the space of acquaintance is defined and established by all human senses composed of the varieties in the space of basic-supporting dualities, These human senses may be enhanced by external factors of technologies to reduce cognitive capacity limitations.

The nature of relationship and connectivity of $\{\mathcal{C}_1 = \{\nu_1\}, \mathcal{C}_2 = \{\nu_2\}\}$ has been explained through organizational give-and-take-sharing process [R 4.10]. The nature of relational unity for $\{\varnothing, \mathcal{C}_1 = \{\nu_1\}, \mathcal{C}_2 = \{\nu_2\}\}$ has also been explained by the existence of the space of acquaintance, the space of problem-solution dualities and the methodological constructionism-reductionism duality. It is analytically useful to notice the relational interdependency between the area of knowing $(\mathcal{C}_1 = \{\nu_1\})$ and the area of knowing $(\mathcal{C}_2 = \{\nu_2\})$ that can mutually propel their behavior over the space of progress-retrogress dualities. The efficient neuro-decision-choice operations on the problem-solutions dualities in the scientific area of knowing depend essentially on the efficient neuro-decision-choice operations over the space of problem-solution dualities in the non-scientific area of knowing while the gains in scientific area of knowing enhances the development of the non-scientific area of knowing and activities over the space of problem-solution dualities. The usefulness of the results of scientific area of knowing to the space of individual-community dualities depends on the organizational efficiency within the space of the individual-community dualities. In other words, the real benefits and costs of the neuro-decision-choice results of the inter-categorial process depend on the understanding of science-non-science relationality in thinking and organization, where the general information-knowledge system is an input into neuro-decision-choice actions, the outcomes of which are information-knowledge items as outputs to update the information-knowledge stocks, the flows of their services of which become inputs into neuro-decision-choice actions in the knowing process over the space of identification-transformation dualities.

The relational field $(\mathscr{R})$ of the epistemic field $\left(\mathscr{C} = \{\varnothing, \mathcal{C}_i = \{\nu_i\}\ \forall i \in \mathbb{I}^\infty\}\right)$ with an alpha-algebra $(\alpha\text{-algebra on } \Omega)$ leads to a situation of multilateral categorial interdependencies in continua and unity under the fuzzy paradigm of thought with a fuzzy-stochastic conditionality, where the categories are formed with fuzzy-stochastic indicator function [R17.17]. The relational field under the information field generates a family of fuzzy categorial relations under transformational conditions over the space of reversibility-irreversibility dualities. The elements of the relational field depend on the nature of categorial inter-dependencies relative to general efficiencies. The space of reversibility-irreversibility dualities helps to understand reversible and irreversible socio-natural processes, especially social decision-choice processes on cognitive activities between opposites of duals and poles and between dualities and opposites over the spaces such as climatological-geomorphological dualities and destruction-construction dualities and life-death dualities, where both reversibility and irreversibility depend on knowhows of functors or categorial convertors as technologies for characteristic transformations.

PROPOSITION 1.4.1: Inter-Categorial Efficiency in Knowing

If $\mathcal{C}_i$, $\mathcal{C}_j, \mathcal{C}_\ell \in \mathscr{C}, \forall i, j, \ell \in \mathbb{I}^\infty$, then $\mathcal{C}_i$ and $\mathcal{C}_j$ are in an efficient relation in a give-and-take-sharing process if the efficient operations on problem-solution dualities of$\left(\mathcal{C}_i\right)$enhance the efficient operations on the problem-solution dualities of $\left(\mathcal{C}_j\right)$and if the efficient operation in the problem-solution dualities of$\left(\mathcal{C}_j\right)$ enhance the efficient operations in the problem-solution dualities of $\left(\mathcal{C}_\ell\right)$, then the efficient operations on problem-solution dualities of$\left(\mathcal{C}_i\right)$enhance the efficient operations in the problem-solution dualities of $\left(\mathcal{C}_\ell\right)$ and vice versa.

NOTE 1.4.1: Reflections on Proposition 1.3.1, Inter-Categorial Efficiencies

It must be understood that operations in intra-categories and inter-categories are identifications of intra-categorial problems and search for their solutions as seen in terms of intra-categorial and inter-categorial actions on varieties. In terms of the theory of knowing, there are the mega-categories such as social sciences, natural sciences, biological sciences, medical sciences, and further mega-categories such as fundamental sciences and applied sciences and other such areas of knowing that may be constructed. There are also minor categories, where categories are areas or departments of knowing through research, teaching and problem-solving in the mega-categories, such as economics, physics, chemistry, civil engineering, electrical engineering, and other specific areas of knowing. Intra-categorial problem-solution dualities are knowing area-specific problem-solution dualities within categories such as areas of problem-solution dualities with departments. Over the space of problem-solution dualities, there may be intra-categorial vertical and horizontal problem-solution structures for the variety problem-solution replacement process. It is also possible to have inter-categorial vertical and horizontal problem-solution replacement process in terms of efficiency.

Between the two extremes of the trivial epistemic field $\left(\mathscr{C}=\{\Omega,\varnothing\}\right)$ and the infinite categorial single varieties $\left[\mathscr{C}=\{\varnothing,\mathcal{C}_i=\{v_i\}\forall i\in\mathbb{I}^{\infty}\}\right]$, there are intermediate epistemic fields that may be constructed alongside the relational field $(\mathscr{R})$ and the information field $(\mathscr{I})$. In the practice of areas of knowing, therefore, new categorial areas of knowing (departments of learning or problem-solution dualities) may be constructed through cross-categorial formations (interdisciplinary formation) with a fuzzy-stochastic indicator function such as economic engineering, economic history, physical chemistry, biological engineering, political economy, economics of law, political geography and many more since there is $(\alpha\text{-algebra on }\Omega)$. The alpha-level constructions will always depend on knowledge needs and the evolving space of problem-solution dualities. The use of the fuzzy-stochastic indicator function in category formations and the construction of the epistemic field, indicate imprecise categorial boundaries due to cognitive capacity limitations of neuro-decision-choice agents, socio-natural relational connectivity through give-and-take-sharing modes and the rise of vagueness and imprecision in all elements in the family of ordinary languages (FOL) while the whole idea of explication and the construct of elements in the family of abstract languages are to minimize vagueness and imprecision in information-knowledge representations of variety characteristic dispositions. The cross-categorial formations with the fuzzy-stochastic indicator function allow for the following definition and theorem for the inclusion of small areas of knowing, extension of basic knowledge to applied knowledge and explanatory science to prescriptive science over the *analytical space.*

DEFINITION 1.4.1: The Basic Analytical Space $(\mathscr{A})$ in the Epistemic Field $(\mathscr{E})$

The basic analytical space $(\mathscr{A})$ is composed of the epistemic field $(\mathscr{E})$, relational field $(\mathscr{R})$ and information field $(\mathscr{I})$ equipped with a paradigm of thought $(\amalg)$. In other words, $(\mathscr{A} = (\mathscr{E}, \mathscr{R}, \mathscr{I}, \amalg))$, where the analytical space is said to be a *fuzzy analytical space* $\left[\mathscr{A}_{\mathfrak{F}} = (\mathscr{E}, \mathscr{R}, \mathscr{I}_{\mathfrak{F}}, \amalg_{\mathfrak{F}})\right]$ if it is equipped with the fuzzy information $(\mathscr{I}_{\mathfrak{F}})$ and fuzzy paradigm of thought $(\amalg_{\mathfrak{F}})$ with relational continuum and unity and acceptance of contradictions. It is said to be a *classical analytical space* $\left[\mathscr{A}_{\mathfrak{C}} = (\mathscr{E}, \mathscr{R}, \mathscr{I}_{\mathfrak{C}}, \amalg_{\mathfrak{C}})\right]$ if it is equipped with the classical information $(\mathscr{I}_{\mathfrak{C}})$ and classical paradigm of thought $(\amalg_{\mathfrak{C}})$ with relational separations, excluded middles and non-acceptance of contradictions.

THEOREM 1.4.1: Epistemic Field-Embedding

Given any epistemic field $(\mathscr{D})$ of categories with a relational field $(\mathscr{G})$ and information field $(\mathscr{H})$ of (Ω) also with a relational field $(\mathscr{R})$ and information field $(\mathscr{I})$, there is always another epistemic field $(\mathscr{C}(\mathscr{D}))$ that contains all the categories in $(\mathscr{D})$ with a relational field $(\mathscr{R}(\mathscr{G}))$ and information field $(\mathscr{I}(\mathscr{H}))$ such that for any epistemic field $(\mathscr{C}_1)$ that contains $(\mathscr{D})$, it is the case that $(\mathscr{C}(\mathscr{D}) \subset \mathscr{C}_1)$, $(\mathscr{R}(\mathscr{G}) \subset \mathscr{R}_1)$ and $(\mathscr{I}(\mathscr{H}) \subset \mathscr{I}_1)$ where $(\mathscr{R}_1)$ is a relational field for $(\mathscr{C}_1)$ and $(\mathscr{I}_1)$ is the information field for $(\mathscr{C}_1)$.

PROOF

It is observed that there is an epistemic field such that we take all epistemic fields containing all categories of $(\mathscr{D})$ as well as the epistemic field of all categories of Ω. The next step is to form the intersection of the classes of all epistemic fields. Take all the epistemic fields containing $(\mathscr{D})$ and retain only the categories common to all the epistemic fields and call it the family of epistemic fields $\mathscr{C}(\mathscr{D})$. Clearly $\mathscr{D} \subset \mathscr{C}(\mathscr{D})$, $\mathscr{G} \subset (\mathscr{R}(\mathscr{G}))$ and $\mathscr{I} \subset (\mathscr{I}(\mathscr{H}))$ as well as $\{\mathscr{C}(\mathscr{D}) \subset \mathscr{C}_1\}$, $\{\mathscr{R}(\mathscr{G}) \subset \mathscr{R}_1\}$ and $\{\mathscr{I}(\mathscr{H}) \subset \mathscr{I}_1\}$, where $(\mathscr{C}_1)$ is any epistemic field containing $(\mathscr{D})$, $(\mathscr{R}_1)$ any relational field containing $(\mathscr{R})$ and $(\mathscr{I}_1)$ any information field containing $(\mathscr{I})$. The proof that $\mathscr{C}(\mathscr{D})$ is an epistemic field is carried on by direct verification against the definition given by (1.3.2.1). Note that if $\{\mathbb{C}_i, i \in \mathbb{I}\} \subset \mathscr{C}(\mathscr{D})$ then $\{\mathbb{C}_i, i \in \mathbb{I}\} \subset \mathscr{C}_1$ for any given epistemic field $(\mathscr{C}_1)$ that contains $(\mathscr{D})$. But since $(\mathscr{C}_1)$ is an epistemic field,

$\left(\bigcup_{i\in\mathbb{I}^\infty}\mathcal{C}_i\subset\mathscr{C}_1\right)$ which is also contained in any epistemic field of $(\mathscr{C}_1)$, hence, it belongs to $\left[\mathscr{C}(\mathscr{D})\right]$, and hence the corresponding relational field $(\mathscr{R}(\mathscr{G})\subset\mathscr{R}_1)$ and the information relation $(\mathscr{I}(\mathscr{H})\subset\mathscr{I}_1)$ □

NOTE 1.4.2: Reflections on the Theorem 1.4.1 of Epistemic Field-embedding.

It is useful to give a meaning to the theorem which simply states that there always exists a smallest epistemic field $\mathscr{C}(\mathscr{D})$ that contains a given family of categories $(\mathscr{D})$ which are categories of (Ω) as an epistemic field embedding. The epistemic field $(\mathscr{D})$ may be seen as composed of categories of prescriptive science which apply information-knowledge accumulation for cognitive creations such as engineering of all forms including social planning, and other applications of knowledge which collectively constitute the *applied information-knowledge system* as an input into decision-choice actions over the space of problem-solution dualities abstracted from the space of imagination-reality dualities. The epistemic field $(\mathscr{C})$ may be the categories of explanatory science which applies to the information-knowledge accumulation of identification and behaviors of varieties which collectively constitute the *fundamental information-knowledge system* relative to the basic research. The explanatory science includes basic knowledge across all the spectra of knowing and the prescriptive science includes all the applied sciences, planning and engineering sciences of multiplicity of forms.

The universal object set of (Ω) endowed with the epistemic field $(\mathscr{C})$ of categories, the relational field $(\mathscr{R})$ of categorial connectivity and the information field $(\mathscr{I})$ for identities, is the socio-natural frame for assigning dynamic areas of knowing as well as creating categories of areas of social, technological and engineering problem-solution dualities that may accommodate inter-categorial collaborations in research, teaching, learning and neuro-decision-choice actions over the space of problem-solution dualities, where variety solutions and variety problems are interchangeable relative to goals, objectives and visions. The epistemic embedding allows us to extend the analytical space of fundamental problem-solution dualities to areas of problem-solution dualities for applications, technology, engineering, management of society and other areas of application such as social games in dualistic-polar conflicts.

PROPOSITION 1.4.2: Characteristic-Signal Fields

Given the set of paradigms of thought $(\amalg)$, we shall represent the fundamental information-knowledge space as the universe (Ω), epistemic field, $(\mathscr{E})$, relational field $(\mathscr{R})$, and information field $(\mathscr{I})$ in the form $(\Omega, \amalg, \mathscr{E}, \mathscr{R}, \mathscr{I})$. Every information-knowledge space is equipped with a variety field $(\mathscr{V})$ which is a collection of categorial varieties of the form $(\mathscr{V} = \{\varnothing, \{\mathbb{V}_i\}, \forall i \in \mathbb{I}^\infty\})$, a *characteristic field* which is a collection of categories of characteristic dispositions in the form $(\mathfrak{X} = \{\varnothing, \{\mathbb{X}_i\}, \forall i \in \mathbb{I}^\infty\})$, a *signal field* which is a collection of categories of signal dispositions $(\mathbb{S} = \{\varnothing, \{\mathbb{S}_i\}, \forall i \in \mathbb{I}^\infty\})$ and a *characteristic-signal field* which is a collection of characteristic-signal dispositions of the form $(\mathfrak{Z} = \{\varnothing, \{\mathbb{Z}_i\} \forall i \in \mathbb{I}^\infty\})$. Given the space of the problem-solution dualities, the representation of the information-knowledge space for knowing as the production of intellectual investment flows and information-knowledge accumulations as the production of intellectual capital stocks may be transformed into a *characteristic-signal field* which is also called the (gamma field)$\left(\gamma\text{-}\text{field or }\gamma\text{-algebra}\right)$ in the form:

$$\mathscr{G}=\left(\Omega,\amalg,\mathscr{C},\mathscr{R},\mathscr{I}\right)=\left(\mathfrak{Z},\mathscr{V},\mathfrak{K},\mathbb{S},\amalg\mathscr{R}\right)=\left(\left\{\left\{\mathbb{Z}_i\right\}\forall i\in\mathbb{I}^{\infty}\right\},\left\{\left\{\mathbb{V}_i\right\}\forall i\in\mathbb{I}^{\infty}\right\},\left\{\left\{\mathbb{C}_i\right\}\forall i\in\mathbb{I}^{\infty}\right\}\left\{\left\{\mathbb{S}_i\right\}\forall i\in\mathbb{I}^{\infty}\right\},\amalg,\mathscr{R}\right)$$

NOTE I.4.3: ON PROPOSITION 1.4.2 of the Characteristic-Signal Field

The characteristic-signal field represents the varieties in the space of actual-potential dualities in variety categories with the corresponding system of characteristic-signal dispositions for neuro-decision-choice actions over the space of problem-solution dualities, communication actions over the space of source-destination dualities and knowing action over the space of identification-transformation dualities as well as over the action space. The characteristic-signal field differs from the epistemic field in that it includes signal dispositions that correspond to the characteristic dispositions for variety identification and transformation in the spaces of production-consumption dualities and input-output dualities. The characteristic-signal field provides the connectivity between the ontological space and the epistemological space revealing the differences between ontic variety characteristic dispositions and epistemic characteristic dispositions. It provides the pathways of neuro-decision-choice communication over the space of source-destination dualities to present all forms of communication games within the space of imagination-reality dualities, to affect the neuro-decision-choice directions and outcomes, and their effects over the spaces of distributions of advantage-disadvantage and cost-benefit dualities.

In this respect, the $\left(\gamma\text{-field or }\gamma\text{-algebra}\right)$ may be analytically combined with the the $\left(\alpha\text{-}\text{field or }\alpha\text{-algebra}\right)$ to create an $\left(\left(\alpha\text{-}\gamma\right)\text{-}\text{field or }\left(\alpha\text{-}\gamma\right)\text{-algebra}\right)$. Alternatively, the $\left(\gamma\text{-}\text{field or }\gamma\text{-algebra}\right)$ may be combined with the $\left(\sigma\text{-}\text{field or }\sigma\text{-algebra}\right)$ to create the

$((\sigma\text{-}\gamma)$- field or $(\sigma\text{-}\gamma)$-algebra$)$ for the study of knowing as a production of intellectual investment flows and the production of information-knowledge accumulation as intellectual stocks. The $((\alpha\text{-}\gamma)$- field or $(\alpha\text{-}\gamma)$-algebra$)$ is in relation to the fuzzy information and the fuzzy paradigm of thought. The $((\sigma\text{-}\gamma)$- field or $(\sigma\text{-}\gamma)$-algebra$)$ is in relation to the classical information and the classical paradigm of thought. In both cases, the corresponding spaces involve varieties, and characteristic-signal dispositions in the space of imagination-reality dualities.

The characteristic-signal field contains the epistemic field with the information and relational fields in addition to the paradigm of thought as the information processor. It also contains a family of sub-spaces of individual and collective imagination-reality dualities which accommodates the action spaces. The characteristic disposition is always the contents of the variety communication while the signal disposition is always the message of the variety which must be interpreted to reveal the contents relevant to identify the identity of the variety. The characteristic-signal field allows for the examination of conditions of illusion, misinformation, disinformation, actual-reality differences, paradoxes, deceptions, propaganda, truth, false and many such phenomena through the interpretive relations between variety signal dispositions and variety characteristic dispositions. Further discussions will be carried on in the chapters that follow.

CHAPTER 2

CATEGORIAL DIVIDEDNESS, GENERAL ANALYTICAL SPACE, THE CHARACTERISTIC-SIGNAL FIELDS, THE EPISTEMIC FIELDS, THE RELATIONAL FIELDS, AND THE INFORMATION FIELDS

ABSTRACT

Chapter 2 provides the understanding of the interactive relations among the phenomena of categorial dividedness, the general analytical space, the characteristic-signal field, and its connectivity to the epistemic field as supported by the corresponding relational and information fields. The objective is to show how thought and practice are integrated to relate variety imaginations to variety realities with further understanding of how the possibility space and probability space are contained in the space of imagination to present the space of necessity-freedom dualities and how they are related to knowing and the information-knowledge processes as stock-flow disequilibrium dynamics. The chapter provides relevant definition and characteristic dispositions of the information-knowledge system from the space of problem-solution dualities and how the indicator functions are used to construct categories of areas of knowing with interdisciplinary dynamics as well as epistemic spaces. From the indicator functions three epistemic fields are presented as ((σ)-field or (σ)-algebra) induced by the classical indicator function ((α-field) or (α-algebra)) induced by the fuzzy indicator function and((γ-field)-)or (γ-algebra). or (gamma-algebra))induced from the characteristic-signal disposition with hybrids of ((σ-γ)-field), and ((α-γ)-field) . The

K. K. Dompere, *The Theory of Epistemic Fields*, Studies in Systems, Decision and Control 498, https://doi.org/10.1007/978-3-031-42470-0_2

chapter provides a framework for understanding diversity and unity in knowing and information-knowledge development, institutional constructs, and departmental creations in relation to all areas within the space problem-solution dualities contained in the space of actual-potential dualities.

2.1 THE CHARACTERISTIC-SIGNAL FIELD AND GENERAL ANALYTICS

In the concluding discussions in chapter one, discussions were made regarding the factors that may arise from the characteristic-signal field. It was pointed out that the knowing processes are complex relations among horizontal and vertical categories of problem-solution dualities. The characteristic-signal field contains the epistemic field. The epistemic field contains categories of variety problem-solution dualities which are related to the space of imagination-reality dualities which is a cognitive abstraction from the space actual-potential dualities. The space of the actual-potential dualities is a natural construct which is contained in the ontological space. The space of imagination-reality dualities is a social construct which is contained in the epistemological space. In this respect actual and potential reside in natural existence in the ontological space, while reality and imagination reside in social existence in the epistemological space. The space of imagination-reality dualities is linked to the space of the actual-potential dualities by neuro-decision-choice activities from the action space to fulfil the notion that *practice without thought is blind; thought without practice is empty* [R17.44, p.78]. Thought finds expressions in the space of imaginations and practice finds expressions in the space of realities while the space of thought-practice dualities resides in the space of imagination-reality dualities such that social progress is the result of the interactions of neuro-decision-choice actions and the space of thought-practice dualities within the space of imagination-reality dualities as a derived category of existence which is contained in the space of actual-potential dualities as the primary category of existence, where human actions are derived from the space of theory-application dualities contained in the space of ignorance-knowledge dualities such that theories reside in the space of emptiness when there are no supporting applications, and applications reside in the space of blindness when there are no supporting theories. The emptiness-blindness statement is in support of the idea that every practice is anchored in the space of theories, and every theory is anchored in the space of applications over the space of identification-transformation dualities.

In the space of neuro-decision-choice activities, the emptiness relates to the loss of freedom in the space of realities which also contains the results of variety probabilities in $\left((\sigma\text{-}\gamma)\text{-field or }(\sigma\text{-}\gamma)\text{-algebra}\right)$ from practice, and blindness relates to the loss of necessity in the space of imaginations which also contains the variety possibilities in $\left((\alpha\text{-}\gamma)-\text{field or}(\alpha\text{-}\gamma)-\text{algebra}\right)$ for thought. Thought finds expressions in both the epistemic field, and practice find expressions in the action space, where the neuro-decision-choice activities on thought and practice find efficiencies in the general analytical space. The important thing to keep in mind is that not only does the space of thought-practice dualities find expressions in the space of neuro-decision-choice actions but that, the spaces of neuro-decision-choice actions and the thought-practice dualities depend on the social philosophical consciencism at all points of cognition that must pass through the *general analytical space* [R8.16][R17.16] [R17.18].

DEFINITION 2.1.1: The General Analytical Space, $(\boldsymbol{\mathcal{A}})$

The general analytical space $(\boldsymbol{\mathcal{A}})$ is composed of the epistemic field $\mathscr{E}(\mathscr{D})$, relational field $\mathscr{R}(\mathscr{G})$ and information field $\mathscr{I}(\mathscr{H})$ equipped with a paradigm of thought $(\amalg)$. In other words, $(\boldsymbol{\mathcal{A}}=(\mathscr{E}(\mathscr{D}), \mathscr{R}(\mathscr{G}), \mathscr{I}(\mathscr{H}), \amalg))$ where the general analytical space is said to be a *fuzzy general analytical space* $\left[\boldsymbol{\mathcal{A}}_{\mathfrak{F}}=(\mathscr{E}(\mathscr{D}), \mathscr{R}(\mathscr{G}), \mathscr{I}_{\mathfrak{F}}(\mathscr{H}_{\mathfrak{F}}), \amalg_{\mathfrak{F}})\right]$ if it is equipped with the fuzzy information $\mathscr{I}_{\mathfrak{F}}(\mathscr{H}_{\mathfrak{F}})$ and fuzzy paradigm of thought $(\amalg_{\mathfrak{F}})$ with relational continuum unity, give-and-take sharing mode and contradiction in the space of relativity. It is, however, said to be a *classical general analytical space* $\left[\boldsymbol{\mathcal{A}}_{\mathfrak{C}}=(\mathscr{E}(\mathscr{D}), \mathscr{R}(\mathscr{G}), \mathscr{I}_{\mathfrak{C}}(\mathscr{H}_{\mathfrak{C}}), \amalg_{\mathfrak{C}})\right]$ if it is equipped with the classical information $\mathscr{I}_{\mathfrak{C}}(\mathscr{H}_{\mathfrak{C}})$ and classical paradigm of thought $(\amalg_{\mathfrak{C}})$ with relational separations, excluded middles and non-acceptance of contradictions in the space of absoluteness.

PROPOSITION 2.1.1: Thought and Practice

Given the set of paradigms of thought $(\amalg)$, the applied information-knowledge system $(\boldsymbol{\Lambda})$ may be represented as the universe (Ω), epistemic field, $(\mathscr{D})$, relational field $(\mathscr{G})$, and information field $(\mathscr{H})$ in the form $(\boldsymbol{\Lambda})=(\Omega, \amalg, \mathscr{E}(\mathscr{D}), \mathscr{R}(\mathscr{G}), \mathscr{I}(\mathscr{H}))$.

Every applied information-knowledge system is equipped with a applied variety field $\mathscr{D}(\mathscr{V})$ which is a collection of categorial varieties of the form $\left(\mathscr{D}(\mathscr{V})=\left\{\varnothing,\left\{\mathscr{D}(\mathbb{V}_i)\right\},\forall i\in\mathbb{I}^\infty\right\}\right)$, a *characteristic field* $(\mathfrak{X}_{\mathscr{D}})$ which is a collection of categories of characteristic dispositions in the form $\left(\mathfrak{X}_{\mathscr{D}}=\left\{\varnothing,\left\{\mathscr{D}(\mathbb{X}_i)\right\},\forall i\in\mathbb{I}^\infty\right\}\right)$, a *signal field* $(\mathbb{S}_{\mathscr{D}})$ which is a collection of categories of signal dispositions in the form $\left(\mathbb{S}_{\mathscr{D}}=\left\{\varnothing,\left\{\mathscr{D}(\mathbb{S}_i)\right\},\forall i\in\mathbb{I}^\infty\right\}\right)$ and a *characteristic-signal field* $(\mathfrak{Z}_{\mathscr{D}})$ which is a collection of characteristic-signal dispositions of the form $\left(\mathfrak{Z}_{\mathscr{D}}=\left\{\varnothing,\left\{\mathscr{D}(\mathbb{Z}_i)\right\}\forall i\in\mathbb{I}^\infty\right\}\right)$, where the characteristic dispositions are the contents of the varieties, the signal dispositions are the messages of the varieties and the characteristic dispositions are the information of the varieties. The system is to combine thought and practice in the space of imagination-reality dualities to provide analytical connectivity of varieties in the space of fundamental-applied dualities as constituting the totality of the system of production-consumption and input-output processes.

DEFINITION 2.1.2: The Information-Knowledge System

Given the space of the problem-solution dualities $(\mathfrak{P}\otimes\mathfrak{S})$, the general information-knowledge system $(\mathbf{G})$ is the union of the fundamental information-knowledge system $(\mathbf{F})$ and the applied information-knowledge system $(\mathbf{A})$ such that the representation of the general information-knowledge system $\mathbf{G}=\left\{(\mathbf{F}\cup\mathbf{A})\mid\mathbf{F}\cap\mathbf{A}\neq\varnothing\right\}$ may be transformed into a *characteristic-signal field* containing the identification conditions in the form:

$$\mathbb{G}=\begin{cases}\left(\Omega,\amalg,\mathscr{C}(\mathscr{D}),\mathscr{R}(\mathscr{G}),\mathscr{I}(\mathscr{H})\right)=\left(\mathfrak{Z}_{\mathscr{D}},\mathscr{D}(\mathscr{V}),\mathfrak{K}_{\mathscr{D}},\mathfrak{S}_{\mathscr{D}},\amalg\mathscr{R}(\mathscr{G})\right)\\ =\left(\left\{\left\{\mathscr{D}(\mathbb{Z}_i)\right\}\forall i\in\mathbb{I}^{\infty}\right\},\left\{\left\{\mathscr{D}(\mathbb{V}_i)\right\}\forall i\in\mathbb{I}^{\infty}\right\},\left\{\left\{\mathscr{D}(\mathbb{C}_i)\right\}\forall i\in\mathbb{I}^{\infty}\right\}\left\{\left\{\mathscr{D}(\mathbb{S}_i)\right\}\forall i\in\mathbb{I}^{\infty}\right\},\amalg,\mathscr{R}(\mathscr{G})\right)\end{cases}$$

2.2: SEMANTIC REFLECTIONS ON THE INFORMATION-KNOWLEDGE SYSTEM

The morphology of the epistemic field is introduced in chapter one, to assist in the understanding of the abstract representation of a model of information-knowledge production as stock-flow disequilibrium dynamics through neuro-decision-choice actions over the space of problem-solution dualities as contained in the space of imagination-reality dualities. The epistemic production process is such that the space of imagination-reality dualities is contained in the space of actual-potential dualities in analytical complexities with relational connectedness to the past-present-future telescopic processes in relation to the space of goals, objectives, visions, and destinies. Here, every problem resides in the space of realities with practice processes as a sub-space of the space of the actuals, and every solution resides in the space of imaginations with thought processes as contained in the space of the potentials. The existence of the spaces of actual-potential dualities, actuals and potentials is naturally ontological and independent of cognitive agents. The existence of the spaces of imagination-reality dualities, imaginations, and realities through knowing and knowledge production is socially epistemological and is dependent on the existence of cognitive agents. The epistemic field, the characteristic-signal field and all the field embeddings are the creations in the epistemological space to understand the varieties and their behaviors in the space of identification-transformation dualities to develop information-knowledge stocks that will provide service flows into neuro-decision-choice activities to support variety dualistic-polar existence and transformations.

It is useful to note that $(\mathscr{C})$ is represented by $(\mathscr{O})$, the elements of which are identified by $(\mathcal{X})$ which sends the signals $(\mathbb{S})$ to generate information $(\mathfrak{I})$ which then becomes inputs into the processing modules with a paradigm of thought from $(\pi \in \amalg)$ as well as establishes relational continua, separation, diversity, and unity with $(\mathscr{R})$, where every epistemic field embedding such as $\mathscr{C}(\mathscr{D})$ has a corresponding relational field embedding such as $\mathscr{R}(\mathscr{G})$ and an information field embedding such as $\mathscr{I}(\mathscr{H})$. The paradigm space has a corresponding *space of toolboxes* $(\mathbb{T})$, where there is the fuzzy paradigmatic toolbox $(\mathfrak{F} \subset \mathbb{T})$ and a classical paradigmatic toolbox $(\mathfrak{C} \subset \mathbb{T})$ such that $(\mathfrak{F} \cup \mathfrak{C}) \subset \mathbb{T}$, $(\mathfrak{F} \cap \mathfrak{C}) \neq \varnothing \subset \mathbb{T}$ and $(\mathfrak{C} \subset \mathfrak{F})$. The knowing is said to be a fuzzy process if $(\pi = \mathfrak{F} \in \amalg)$ and is said to be a non-fuzzy process if $(\pi = \mathfrak{C} \in \amalg)$, where $(\pi \in \amalg)$ is a paradigm of thought. The analytics suggest that the classical paradigmatic toolbox is contained in the fuzzy paradigmatic toolbox, the classical information is contained in the fuzzy information such that $\left\{\mathscr{I}_{\mathfrak{C}}(\mathscr{H}_{\mathfrak{C}}) \subset \mathscr{I}_{\mathfrak{F}}(\mathscr{H}_{\mathfrak{F}})\right\}$ and the classical results are contained in the fuzzy results in the analytical space.

The system is such that every variety resides as an internal duality with a negative dual and a positive dual and as an external negative or positive pole corresponding to another external positive or negative pole of polarity in creating continual conflicts and resolutions to establish dualistic-polar games between internal duals and external poles, where the negative pole has a residing negative duality, and the positive pole has a residing positive duality. In the neuro-decision-choice space in relation to the space of problem-solution dualities, each dual of a duality resides in the space of cost-benefit dualities and each pole of a polarity resides in the space of cost-benefit duality relative to the elements in the space of goals, objectives and visions, where the pace of cost-benefit dualities contains all the elements in the space of disincentive-incentive dualities which provides motivations in the action space such that curiosity is the father of knowing and necessity is the mother of inventions and curiosity-necessity duality is the parent of hope-freedom duality.

2.2.1 The Information-Knowledge System and the Categorial Interdisciplinary of Knowledge Production

Let us remember that the fuzzy pairwise categories are such that given any two categories $\left\{ \mathcal{C}_i, \mathcal{C}_j \in \mathscr{C}\, \forall i \neq j \in \mathbb{I}^{\infty} \right\}$ under an $(\alpha\text{-field, or } \alpha\text{-algebra})$, it is the case that $\left\{ \mathbb{Z}_{\mathcal{C}_i} \cap \mathbb{Z}_{\mathcal{C}_j} \neq \varnothing,\ \forall i \neq j \in \mathbb{I} \right\}$ $\left\{ \mathbb{Z}_{\mathcal{C}_i} \cup \mathbb{Z}_{\mathcal{C}_j} \subset \mathbb{Z},\ \forall i \neq j \in \mathbb{I}^{\infty} \right\}$ and $\left\{ \mathbb{Z}_{\mathcal{C}_i} \nabla \mathbb{Z}_{\mathcal{C}_j} \subset \mathbb{Z},\ \forall i \neq j \in \mathbb{I}^{\infty} \right\}$, where the symbol (∇) represents a symmetric difference with fuzzy-stochastic conditionality (for the nature of fuzzy-stochastic conditionality see [R8.16][R17.18][R21.5]. The implication here is that the relational field $(\mathscr{R})$ may or may not exclude disunity relations over the epistemic field even though $(\varnothing \in \mathscr{C})$ and in this respect, the classical paradigm with an excluded middle deals with the endpoint conditions of the information-knowledge processes of dualities and polarities with excluded middles.

The analytical point that is being made in these discussions is that the universal system exists as unity under variety dualistic-polar conditions with dynamic relational continua of the varieties at the level of ontological existence, and hence the derived categories must maintain the relational continua and unity at the level of the epistemological existence for cognitive activities over the space of problem-solution dualities. Our logic and truth-false structures must acknowledge the relationality of continuum and unity, where the universal existence is a collection of categories of matter-energy varieties with relational continua and unity in a multiplicity of dualistic-polar directions from the infinite system of multiplicity of opposites presenting the conditions of internal transformations over the infinite universe.

The acquisition of knowledge about the matter-energy configuration of the ontological universe and the distinctions among the elements of various ontological varieties $(\nu \in \mathbb{V})$ take place at the level of epistemology by neuro-decision-choice agents working with substantial cognitive capacity limitations under distributional zones of variety ignorance that have relational continuum and unity with what is claimed to be known and what is unknown over the epistemological space relative to the infinite universe of socio-natural existence, where varieties continually pass through the space of destruction-construction dualities. Each category $\{\mathcal{C}_i \in \mathscr{C}, \forall i \in \mathbb{I}^\infty\}$ defines a subject area of knowing as well as an area of problem-solution dualities, while each intersection such as $\{\mathcal{C}_i \cap \mathcal{C}_j \in \mathscr{C}, \forall i \neq j \in \mathbb{I}^\infty\}$ defines an interdisciplinary subject area of knowing and areas of interactive problem-solution dualities for the problem-solution substitution-replacement process under neuro-decision-choice systems. The interdisciplinary subject areas of knowing may involve more areas than two such as

$\left\{ \left(\mathcal{C}_i \cap \mathcal{C}_j\right), \left(\mathcal{C}_\ell \cap \mathcal{C}_h\right), \left(\mathcal{C}_i \cap \mathcal{C}_j \cap \mathcal{C}_\ell \cap \mathcal{C}_h\right) \in \mathscr{C}, \forall i \neq j \neq \ell \neq h \in \mathbb{I}^\infty \right\}$ with a multiplicity of specialized persons and skill sets which give a current trend in the university departments, research systems and complex individual organization and socio-economic production over the social space.

Every area of knowing is a fuzzy category contained in the epistemic field with a supporting relational field, information field and a paradigm of thought to establish the characteristic-signal field which contains $\left((\sigma\text{-}\gamma)\text{-field or }(\sigma\text{-}\gamma)\text{-algebra}\right)$ in the space of absoluteness and $\left((\alpha\text{-}\gamma)\text{-field or }(\alpha\text{-}\gamma)\text{-algebra}\right)$ in the space of relativity. The creation of various interdisciplinary areas in institutions of teaching, research and learning production and consumption must be seen in terms of neuro-decision-choice strategies that motivate and drive the collective effort over the space of problem-solution dualities over the space of input-output processes. An example of this creativity of departmental formation for collaborative interdisciplinary learning over the space of problem solution dualities is the University of Chicago's Department of Comparative Human Development as an interdisciplinary micro-category within the macro-category of social sciences. There is also the engineering economic system as an interdisciplinary micro-category with the macro-category of prescriptive sciences.

Similarly, the theories and applications of medical sciences are based on complete interdisciplinary micro-macro categories of socio-natural sciences. As new variety problem-solution dualities emerge from the destruction-replacement process of old variety problem-solution dualities, the need for increasing efficiencies of neuro decision-choice actions over the space of problem-solution dualities will demand increasing inter-categorial approaches to research, teaching, learning and communications. The morphological dynamics of the inter-categorial approach will depend on the nature of problem-solution disequilibrium dynamics relative to the elements of the space of goals, objectives and visions contained in the space of imagination-reality dualities.

Every area of knowing in the epistemic field, whether one calls it science or non-science, devoted to solving a class of problems of variety identification-transformation ignorance in the relevant category of problem-solution dualities, takes place over the space of imagination-reality dualities within the epistemological space with neuro-decision-choice activities of cognitive agents using epistemological information under ontological-information constraints and multi-dimensional cognitive capacity limitations as constraints on efficiency in the space of success-failure dualities. Let us keep in mind that the infinite structure of the epistemic field with an infinite relational field and an infinite information field transformed to the general characteristic-signal field is such that the individual-community cognitive capacity limitations in static state are complemented by intergenerational cognitive capacity limitations in dynamic states such that one may see generations in terms of the parent-offspring process, where each generation resides in the space of primary-derived dualities in the sense that each generation is an offspring of the previous generation and a parent to the next generation within the time trinity of the *Sankofa-anoma conditions* of the past-present-future structure with information-knowledge stock-flow disequilibrium dynamics. The claim of the identification of any variety of what there was, what there is or what there would be is based on information-knowledge items abstracted from the characteristic-signal field as input into a neuro-decision-choice action over the epistemological space with a fuzzy-stochastic conditionality, where the fuzzy-stochastic conditionality presents degrees of trust over the space of acceptance-rejection dualities.

The sets of ontological knowing problems are infinite and present an infinitely difficult problem set to cognitive agents working over the epistemological space for the solution set. The infinite problem set appears as the space of problem-solution dualities where every variety problem has a residing variety solution over the space of identification-transformation dualities. The information-knowledge process involves comparative activities between the epistemic identities and the ontic identities with the use of methodological constructionism-reductionism duality, where the epistemic identities are reconciled with ontic identities within the space of methodological constructionism dualities. The methodological constructionism creates the epistemic results from the decoding of the ontic characteristic dispositions while the methodological reductionism verifies the epistemic results with the ontic identities. This forward-backward process for variety knowledge-ignorance conditions constitutes an element of unity of knowing in science and non-science, within science and within non-science, where neuro-decision-choice actions in categories of science and non-science are about knowing activities over the space of problem-solution dualities to develop the information-knowledge system which then become inputs into the space of neuro-decision-choice actions. Every space of the problem-solution dualities is defined in the space of imagination-reality dualities which is intimately connected to the space of variety identification-transformation dualities. At the level of statics, the knowing involves the problem-solution dualities of variety identification to provide inputs to the development of information-knowledge accumulation on varieties as well as defines the initial conditions and necessity to change.

At the level of dynamics, the knowing involves the problem-solution dualities of transformation to provide inputs into information-knowledge accumulation of variety transformations defining conditions of freedom and sufficiency to change. All knowing processes involve two things of variety identification and variety transformation, where any conceivable knowing is reducible either to identification, transformation, or both.

A claim of either knowledge or ignorance about a variety, just like a claim of either truth or falsity of statements and events, takes place over the epistemological space through the space of imagination-reality dualities with a complex system of uncertainties involving quantitative and qualitative dispositions. The quantitative uncertainty involves volume limitations, and the qualitative uncertainty involves vagueness, inexactness, imprecision, and subjectivity. The acceptances of all epistemic claims are based on either justified or unjustified belief based on some information defined in the space of socio-natural characteristic dispositions. These acceptances are decision-choice determined and always under some form of epistemic conditionality in terms of degrees of doubt in the zonal space of certainty-uncertainty dualities as unified forms of all areas of knowing. The claim of unity and arguments in support of disunity rest on how the epistemic categories are constructed and with what common characteristic dispositions in addition to what paradigm of thought is used in the thinking-reasoning system over the space of neuro-decision-choice activities within the space of variety quantity-quality dualities.

The construction of epistemic categories $\left\{ \mathcal{C}_i, \forall i \in \mathbb{I}^\infty \right\}$ relates to the clustering of neuro-decision-choice actions over the organic characteristic disposition $(\mathbb{X})$ with corresponding organic information $(\mathbb{Z})$ into a set of sub-characteristic dispositions

$\{\mathbb{X}_i, \forall i \in \mathbb{I}^\infty\}$ with corresponding information $\{\mathbb{Z}_i, \forall i \in \mathbb{I}^\infty\}$. The corresponding information is then used to establish identities of varieties that become inputs into the decomposition of the space of varieties $(\nu \in \mathbb{V})$ to establish categories $\{\mathcal{C}_i, \forall i \in \mathbb{I}^\infty\}$ for the construction of the epistemic field $(\mathscr{C})$ with the corresponding relational field $(\mathscr{R})$ and the information field $\mathscr{I} = \{\mathbb{Z}_{\mathcal{C}_i} \subset \mathbb{Z}, |\mathcal{C}_i \in \mathscr{C}, \forall i \in \mathbb{I}^\infty\}$, where the characteristic-signal field presents its foundation.

It is useful to understand the interactive and connectivity conditions of the fields of $((\alpha)$- field or (α)-algebra$)$, $((\sigma)$- field or (σ)-algebra$)$ and $((\gamma)$- field or (γ)-algebra$)$, $((\alpha\text{-}\gamma)$- field or $(\alpha\text{-}\gamma)$-algebra$)$ and $((\sigma\text{-}\gamma)$- field or $(\sigma\text{-}\gamma)$-algebra$)$ in the spaces of relativity, absoluteness, their union and intersections. The $((\alpha)$ and (σ)- fields$)$ relate directly to variety identities. The $((\gamma)$- field $)$ relates to variety characteristic-signal dispositions for decoding. The $((\alpha\text{-}\gamma)$ and $(\sigma\text{-}\gamma)$ fields $)$ are about variety identities and characteristic dispositions. The intersections of the categories present the conditions of the excluded middle of the duals of dualities and the poles of polarities with dualistic-polar relational separation, no-give-and-take sharing mode and disunity in the space of varieties. The union presents the conditions of dualistic-polar relational continuum, diversity with give-and-take sharing modes and unity of the duals of dualities and the poles of polarities in the space of varieties. The development and construction of all these fields depend on the system of decomposition decisions for categorial dividedness to create categories of similarities and differences within the space of problem-solution dualities.

2.2.2: Decomposition Decision, Categorial Dividedness and Areas of Knowing

It is now useful to establish relationships among the constructs of category formation, categorial dividedness and the epistemic field with the corresponding relational field, the information field, and the decomposition decision of the space of problem-solution dualities from the characteristic-signal field. The category formation, the epistemic field with the corresponding relational field, the information field and decomposition of the space of problem-solution dualities may be done with a thinking process under either the fuzzy paradigm of thought with relational continuum and unity or with the classical paradigm of thought with the excluded middle and relational separation and disunity with theoretical advantages and difficulties. In general, current socio-intellectual practice, the dominant analytical space is the classical analytical space and the corresponding paradigmatic space, which is that of the classical paradigm that constitutes the dominant analytical space in terms of usage over the space of absoluteness, where categories and fields are created with the classical indicator function. The classical non-fuzzy analytical space has been substantially developed and generally accepted. It occupies the dominant space at all institutions of thinking, learning, teaching, research, and common discussions from the beginning to the highest with the classical truth-value table where the varieties in the excluded middle and the space of relativity are examined and analyzed with some kind of classical approximations as an element in the intergenerational social philosophical consciencism.

Over the space of information-knowledge conditions, neuro-decision-choice agents are confronted with variety information over the complex space of exact-inexact dualities, enhanced by quality-quantity dualities and subjective-objective dualities. The assumption of complete exactness produces the extreme case of the information-knowledge structure while the assumption of complete inexactness produces another extreme case of the information-knowledge structure with the fuzzy paradigm of thought. The classical and fuzzy paradigms of thought working with appropriate information-knowledge structures will result in differential justifications for the unity and disunity of knowing and science producing differential claims and conclusions over the space the space of true-false dualities and acceptance-rejection dualities. The fuzzy paradigm of thought with its logic and mathematics, from which certain techniques and methods of reasoning, such as the fuzzy truth-value table and the fuzzy info-algorithms for optimal decision-choice actions, are designed to assist reasoning under the conditions of information-knowledge defectiveness with simultaneous existence of exactness, inexactness, vagueness, and clarity. Corresponding to the fuzzy paradigm of thought is the *fuzzy-stochastic thinking-reasoning* process.

The classical paradigm of thought with its logic and mathematics from which certain techniques and methods, such as the exact truth table, the exact info-algorithms for optimal decision-choice actions and other elements in the classical toolbox, is designed as general *thinking strategies* to assist an individual and collective reasoning under the conditions of information-knowledge exactness and non-contradiction including statements of theorems and proofs of theorems. This is the *classical thinking-reason system*. The simplicity of the classical paradigm of thought and the difficulties of developing an alternative to it are such that the assumption of exactness with completeness and incompleteness are imposed on different categorial information to fit the information structure to the demands of the classical paradigm of thought with its logic and mathematics.

One important thing to be noted is that the thinking system of the classical paradigm of thought is contained in the thinking system of the fuzzy paradigm of thought at the extremes, where the fuzzy-paradigmatic method of approximate thinking may be developed as a process to obtain the exact states of being. Another qualitative important feature of the fuzzy paradigm of thought is its emphasis on the integration of the user's subjective judgments in the neuro-decision-choice activities of cognitive agents as the dominant processes in innovation and technology over the epistemic field in order to generate self-correcting and self-improving conditions of knowing and transformations of social states with a continual establishment of dynamic fuzzy-stochastic conditionality at each neuro-decision-choice time-point through the use of categorial convertors or functors, where the categorial convertors or the functors are seen as technology with corresponding processors. In other words, cognitive agents, as neuro-decision-choice agents, are internalized in the structure of the paradigm in all the processes over the epistemic field and the corresponding relational field which gives the power for establishing the argument of unity of knowing and science through the relational field and information field with continuum and unity. The classical thinking system induced by the classical paradigm of thought *externalizes* the neuro-decision-choice agent as a user, while the *fuzzy thinking system* induced by the fuzzy paradigm of thought *internalizes* the neuro-decision-choice agent as a user over the space of problem-solution dualities.

All the essential elements of the fuzzy paradigm of thought are useful in the logic of fuzzy decomposition of any whole into category-group-set formations, from universal to particular, and composition of particular to universal, whether the universal and the particular are arbitrary entities or not. Working with the fuzzy paradigm, one maintains the relational continua and unity of all opposites in socio-natural states and processes, where conditions of contradictions are essential in generating forces of dualistic-polar conflicts for transformations and changes as well as allowing logical flexibilities in the thinking process of cognitive agents for creativity, mistakes, approximations, failure and success under the space of certainty-uncertainty dualities as generated by qualitative-quantitative dispositions and linguistic limitations of cognitive agents in the neuro-decision-choice actions over the space for problem-solution dualities. The neuro-decision-choice correction processes over the space of problem-solution dualities, where solutions to problems bring new problems, generate disequilibrium dynamics towards the creation of disequilibrium stock-flow dynamics of information-knowledge neuro-systems in knowing through critical balances of deciding factors in the space of subjective-objective dualities.

The logical flexibilities of the fuzzy paradigm of thought are found in the simplicity of the structure of the fuzzy paradigm of thought in dealing with the problems of aggregation and disaggregation through the fuzzy composition and decomposition of information into variety-specific and category-specific information as has been discussed in [R17.17] with further extensions in [R8.16]. The methods and techniques of fuzzy decomposition and composition with relational continua and unity as established by fuzzy-stochastic conditionality may be extended to deal with analytical conditions of the unity of knowing and science. These logical flexibilities are enhanced by the conditions that the subjective judgements of cognitive agents may be included in the thinking-reasoning process incorporating approximations using the available characteristic membership functions. Here, it may be suggested that certain paradoxes and unresolved problems in the classical paradigm may be due to the conditions of the excluded middle and hence may be resolved in the fuzzy paradigm of thought with its logic and mathematics as its development advances with increasing acceptance and applications.

Additionally, the classical paradigm of thought and all its results of thinking and reasoning are contained in the fuzzy paradigm of thought with fuzzy thinking and reasoning at the extremes as processes to produce exact states from inexact states of being at the epistemological space. In other words, the thinking-reasoning processes of variety information-knowledge claims proceed under the conditions of the principle of doubt with fuzzy approximation in limiting and gradualism of continual self-corrections towards conditions of the principle of surety in the space of either true-false dualities, the space of question-answer dualities or the space of problem-solution dualities. By the very nature of the laws of reasoning, the cognitive agents using the classical paradigm of thought and its algorithms of thinking have not integrated subjective judgement into the decision-choice activities in the process of category formation, the clustering activities, decomposition, composition, and the development of the epistemic field to generate the self-correcting, self-progressive conditions of knowing and transformations of social states of theoretical structures. Let us keep in mind that all types of connectivity within the space of imagination-reality dualities are generated by neuro-decision-choice activities. Similarly, all types of connectivity between the spaces of imagination and realities are through neuro-decision-choice actions generating information and using information for variety over the space of identification-transformation dualities.

In the space of certain-uncertain dualities, the classical restriction on acceptance is the stochastic conditionality due to quantitative volume limitation. In other words, cognitive agents are not internalized in all the neuro-decision-choice processes over the epistemic field and the corresponding relational and information fields which give the power for establishing the argument of exact separation of areas of knowing and science through the relational field of disunity. Working with the classical paradigm as established over the space of absoluteness, one maintains the relational separation, individuality, and disunity of all opposites at the extremes in socio-natural states and processes, where conditions of intersectionality and contradictions are not viewed as essential in generating forces of internal transformation and change. The fuzzy paradigm of thought is defined in the general space of relativity which includes the elements in the space of absoluteness for neuro-decision-choice actions over the problem-solution space to guide the process of knowing and generate the information-knowledge system.

In the classical paradigm of thought, there is not enough logical flexibility over the penumbral regions in the thinking process of cognitive agents for creativity, mistakes, failure, and success under the space of certainty-uncertainty dualities as generated by qualitative-quantitative limitations of cognitive agents using its analytics. Allowances in the classical paradigm are not made for judicious judgments in the space of subjective-objective dualities, where every action over the space of imagination-reality dualities is neuro-decision-choice determined, which is constrained by cognitive capacity limitations and decoding conditions of variety characteristic dispositions over the space of imaginations. The logical inflexibility in thinking is found in the simplicity of the classical paradigm of thought in the process of establishing exact separations in the space of problem-solution dualities, without relational connectivity, classical composition, and decomposition of information into variety-specific and category-specific information in establishing the conditions of belonging which have been discussed in [R17.17] [R17.18]. The methods and techniques of classical decomposition and composition with the excluded middle to produce separation and disunity in categorial formation may be extended to argue against unity and in favor of the disunity of knowing and science in such a way that there is either diversities in unity or disunity of diversities. The classical paradigm of thought is thus defined in the space of absoluteness for neuro-decision-choice actions over the problem-solution space to guide the process of knowing and true-false acceptance to generate the information-knowledge systems of classical types.

In the space of absoluteness, every representation must be logically and axiomatically complete to avoid conflicts that may arise in the space of consistency-inconsistency dualities. With reference to any phenomenon of knowing, where decidability, computability and predictability are logically sought without fuzzy-stochastic conditionality, the axiomatic model representation of any variety must meet the completeness conditions. These conditions are such that the system's propositions in the model must reflect the exactness of the variety characteristic dispositions, which are the exactness of the variety information; the rules of inference from the laws of the classical paradigm of thought must be made exactly as the objective laws governing the variety's behavior. Additionally, the defined initial conditions in the characteristic dispositions must correspond exactly to the variety identities and behaviors. These completeness conditions are equivalent to Gödel-completeness , and hence, if the initial system is solvable, decidable, and computable then all subsequent points of solvability, decidability, computability, and predictability must be derivable from the initial truth of the characteristic dispositions of solvability, decidability, computability, and predictability. To achieve this completeness, the knowledge by acquaintance must be exact, true, and empirically perfect to ontic exactness. The laws of thought must be the same as the objective laws governing the variety's behavior. These completeness conditions are the demands of the classical paradigm of thought in its logic and mathematics over the space of absoluteness. They require a perfect paradigm of thought, perfect language, and perfect information with no deficiencies of vagueness, subjectivity, and volume-limitations. The completeness conditions are unattainable due to the cognitive capacity limitations of cognitive agents because the varieties in the excluded middle constitute the bulk of the elements in the system of knowing and information-knowledge production that must pass through the spaces of possibility-impossibility dualities and probability-improbability dualities contained in the space of imaginations as a sub-space of potentials.

2.3 INDICATOR-FUNCTIONAL ANALYTICS, CATEGORIAL DIVIDEDNESS AND THE EPISTEMIC FIELDS

The ontological universe (Ω) which is a union of the actual and potential spaces $\Omega = (\mathfrak{A} \cup \mathfrak{U})$ or as actual-potential duality is analytically equipped with a grand membership characteristic function $\{\mu_{\Omega}(\omega) \in [0,1] \mid \omega \in \Omega\}$ and a family of membership characteristic functions of the form $\mathfrak{M} = \{\mu_{A_i}(\omega) \in [0,1] \mid \omega \in A_i \subset \Omega, i \in \mathbb{I}^{\infty}\}$ that may be used to decompose (Ω) into sub-sets, where the sub-sets may be viewed as categories transmitted to create the epistemological space, containing epistemological varieties and categories. The sub-sets are categories in that they are equipped with categorial convertors (converters) or functors for intra-categorial and inter-categorial mappings of the variety characteristic dispositions contained in them. One may also use

$\left\{\Omega_i = \mathrm{A}_i \ni \bigcup_{i\in\mathbb{I}^\infty}\Omega_i = \Omega = \bigcup_{i\in\mathbb{I}^\infty}\mathrm{A}_i\right\}$, where Ω_i's and A_i's are the categorial decompositions.

The ontological universe is viewed as a collection of actual and potential varieties which constitute the elements in the space of identification and transformation problem-solution dualities. In the case of the classical paradigm, the decomposition of (Ω) into both categories and the creation of categorial dividedness of areas of knowing and science may be undertaken with the *classical indicator function* with its categorial or set membership function of unit-uniform belonging of the form:

$$\mathcal{I}_{\mathrm{A}}(\omega_i)=\begin{cases}1,\ \text{if}\ \omega_i\in\mathrm{A}=\{(\omega,\mu_{\mathrm{A}}(\omega_i)\mid\mu_{\mathrm{A}}(\omega_i)=1),\forall\omega\in\mathrm{A}\subset\Omega\}\\0,\ \text{if}\ x_i\in\mathrm{A}^{\mathrm{C}}=\{(\omega,\mu_{\Omega}(x)\mid\mu_{\Omega}(\omega)=0),\forall\omega\notin\mathrm{A}\subset\Omega\}\end{cases},\ i\in\mathbb{I}_{\mathrm{A}},\Omega=\mathrm{A}\bigcup\mathrm{A}^{\mathrm{C}} \quad (2.3.1)$$

Similarly, in the case of the fuzzy paradigm of thought, the decomposition of (Ω) into both categories and the creation of categorial dividedness of areas of knowing and science may be undertaken with the use of the *fuzzy indicator function* $\mathcal{I}_{\tilde{\mathrm{A}}}(\omega_i)$ with a non-uniform membership function of belonging at a *fixed level upper-cut* of the form:

$$\mathcal{I}_{\tilde{\mathrm{A}}}(\omega_i)=\begin{cases}1,\ \text{if}\ \omega_i\in\tilde{\mathrm{A}}=\{(\omega_i,\mu_{\tilde{\mathrm{A}}}(\omega_i)\mid\mu_{\tilde{\mathrm{A}}}(\omega_i)\geq\alpha\in[0,1])\},\ \tilde{\mathrm{A}}\subset\Omega\\0,\ \text{if}\ \omega_i\in\tilde{\mathrm{A}}^{\mathrm{C}}=\{(\omega_i,\mu_{\tilde{\mathrm{A}}}(\omega_i)\mid\mu_{\tilde{\mathrm{A}}}(\omega_i)<\alpha\in[0,1])\},\ \tilde{\mathrm{A}}\subset\Omega\end{cases},\ i\in\mathbb{I}_{\tilde{\mathrm{A}}},\mu_{\tilde{\mathrm{A}}}(\omega_i)\in\mathfrak{M},\Omega=\tilde{\mathrm{A}}\bigcup\tilde{\mathrm{A}}^{\mathrm{C}} \quad (2.3.2)$$

For the dualistic-polar analysis as needed in conditions for defining the *general principle of opposites*, the fixed level upper-cut has a corresponding *fixed level lower-cut* of the form:

$$\mathcal{I}_{\tilde{A}}(\omega_i)=\begin{Bmatrix} 1, \text{if } \omega_i \in \tilde{A} = \{(\omega_i, \mu_{\tilde{A}}(\omega_i) \mid \mu_{\tilde{A}}(\omega_i) \leq \alpha \in [0,1])\}, \tilde{A} \subset \Omega \\ 0, \text{if } \omega_i \in \tilde{A}^C = \{(\omega_i, \mu_{\tilde{A}}(\omega_i) \mid \mu_{\tilde{A}}(\omega_i) > \alpha \in [0,1])\}, \tilde{A} \subset \Omega \end{Bmatrix}, i \in \mathbb{I}_{\tilde{A}}, \mu_{\tilde{A}}(\omega_i) \in \mathfrak{M}, \Omega = \tilde{A} \cup \tilde{A}^C \quad (2.3.3)$$

The fixed-level sets, the fixed-level categories and the fixed-level groups may be seen in terms of *categorial dividedness* with a set of constraints of fuzzy-stochastic conditionality that relates to conditions over the space of acceptance-rejection dualities, where all the α's are optimally constructed in fuzzy-stochastic information space. All formations of commonness, similarities, differences, disaggregation, aggregation, clustering and others in the space of oneness-dividedness dualities over the epistemological space are induced by cognitive actions in the space of neuro-decision-choice processes on the basis of which such concepts exist and present themselves as the real definitions, the nominal definitions or both in linguistic systems with established characteristic-signal dispositions for variety identities as well as the categorial identities with similarities and differences, The identity of each variety has its information support and similarly, the identity of each category has its information support. The information support is made up of characteristic dispositions as the set of contents and the signal disposition as the set of messages from the characteristic dispositions.

2.3.1: Lemma and Theorem on Categorial Dividedness and Indicator-functions

The category formation that establishes areas of knowing in the epistemic field with the corresponding relational field carries information contents as characteristic dispositions which reveal themselves as signal dispositions for solutions to either the identification, transformation problems or both of varieties as well as in the abstract sense of either static, dynamic or both. Let us keep in mind that the cognitive activities in the epistemic field with accompanying relational and information fields for the identification problems, transformation problems and combination of them relate to static problems and dynamic problems and the combination of both, where all neuro-decision-choice actions are about variety identification or transformation of both. The identification conditions relate to *initial conditions* and *necessity* in the system of categorial conversions in the epistemic field. The transformation conditions relate to *sufficient conditions* and *freedom* under the principle of *social philosophical consciencism* at each decision-choice time point relative to the elements in the space of oneness-dividedness dualities. We shall now state a lemma of Ω – dividedeness under fuzzy-stochastic information structure $\mathbb{Z}_{\Omega}$ that may be used for the category formation and the epistemic field. At the epistemological space, the epistemic image of Ω is a fuzzy set through $\mathbb{Z}_{\Omega} = \left(\mathbb{X}_{\Omega} \otimes \mathbb{S}_{\Omega}\right)$, where the fuzzy variety space $\mathbb{V}_{\Omega}$ is to be constructed from the acquaintance space in terms of decomposition of the varieties.

The dividedness decision assumes of the existence of universal oneness of the space of matter-energy varieties, where the variety dividedness is carried on through the space of general characteristic dispositions which present the general signal dispositions. The knowing process is seen in terms of a communication system of varieties. The socio-natural varieties constitute the *ontic varieties* and the source which send messages over the space of source-destination dualities, where the neuro-decision-choice agents are the destination entities receiving the messages, the *ontic* characteristic dispositions are the contents and the signal dispositions are the messages that must be decoded to identify the characteristic dispositions of the varieties to become the *epistemic* characteristic dispositions which then establish the *epistemic varieties*. The knowledge accumulation process is comparative analytics to verify and falsify the variety ontic-epistemic equality over the space of ignorance knowledge dualities by the method of reductionism. The accepted and rejected known varieties are then placed into the knowledge stock. The following things must be kept in mind within the space of knowing and information-knowledge production: there are the ontic characteristic dispositions, ontic signal dispositions and ontic varieties contained in the ontological space; there are also the epistemic characteristic dispositions decoded from the ontic characteristic dispositions, and epistemic varieties contained in the epistemological space. The theory of knowing is to produce epistemic variety characteristic dispositions as continual intellectual investment flows. The theory of knowledge is to verify the degrees of equality between the epistemic varieties and ontic varieties in order to produce intellectual capital stocks.

Lemma 2.3.1.1: Dividedness of the Universal Space (Ω)

The universal space (Ω) as a fuzzy set of elements with generic element $(\omega \in \Omega)$ may be written as $\Omega = \{(\omega, \mu_\Omega(\omega)) \mid \omega \in \Omega,\ \mu_\Omega(\omega) \in [0,1]\}$ which may also be constructed into a variety space $(\mathbb{V}_\Omega)$ and divided into a series of ordinary or classical sets at differential fixed-level membership values of belonging in fixed-level upper-cuts of the form $\Omega_{\alpha_i} = \left\{(\omega_i, \mu_{\Omega_{\alpha_i}}(\omega_i)) \mid \omega \in \Omega_{\alpha_i},\ \mu_\Omega(\omega_i) \geq \alpha_i \in [0,1] \forall i \in \mathbb{I}^\infty\right\}$ or fixed-level lower-cuts of the form $\Omega_{\alpha_i} = \left\{(\omega_i, \mu_{\Omega_{\alpha_i}}(x_i)) \mid \omega \in \Omega_{\alpha_i},\ \mu_\Omega(\omega_i) \leq \alpha_i \in [0,1] \forall i \in \mathbb{I}^\infty\right\}$,where $\mathbb{I}^\infty$ is a dynamic infinite index set for the categorial elements in (Ω).

DEMONSTRATION: Lemma 2.3.1.1

The fixed-level upper-cuts and the fixed-level lower-cuts are dualistic in in their nature states which allow for extended analysis under the principle of opposites with relational continuum and unity as well as provide analytical conditions for dualization of oppositeness in the same variety. Consider the fuzzy set of the universe of the form $\Omega=\{(\omega,\mu_\Omega(\omega))\mid\omega\in\Omega,\ \mu_\Omega(\omega)\in[0,1]\}$ and recall the structure of the *classical indicator function* with its sethood membership function of unit-uniform belonging of the form:

$$\mathcal{I}_{\mathrm{A}}(\omega_i)=\left\{\begin{matrix}1,\ \text{if}\ \omega_i\in\mathrm{A}=\{(\omega,\mu_{\mathrm{A}}(\omega_i))\mid\mu_{\mathrm{A}}(\omega)=1),\forall\omega\in\mathrm{A}\}\\ 0,\ \text{if}\ \omega_i\in\mathrm{A}^{\mathrm{C}}=\{(\omega,\mu_{\mathrm{A}^{\mathrm{C}}}(x)\mid\mu_{\mathrm{A}^{\mathrm{C}}}(\omega)=0),\forall\omega\notin\mathrm{A}\}\end{matrix}\right\},i\in\mathbb{I}_{\mathrm{A}},\mu_{\mathrm{A}}(\omega_i)\in\{0,1\},\mathrm{A}\subset\Omega=\mathrm{A}\cup\mathrm{A}^{\mathrm{C}},\mathrm{A}\cap\mathrm{A}^{\mathrm{C}}=\varnothing$$

Similarly, one may recall the *fuzzy indicator function* with non-uniform membership function of belonging at a *fixed level upper cut* of the form:

$$\mathcal{I}_{\tilde{\mathrm{A}}}(\omega_i)=\left\{\begin{matrix}1,\ \text{if}\ \omega_i\in\tilde{\mathrm{A}}=\{(\omega,\mu_{\tilde{\mathrm{A}}}(\omega_i))\mid\mu_{\tilde{\mathrm{A}}}(\omega_i)\geq\alpha\in[0,1])\}\\ 0,\ \text{if}\ \omega_i\in\tilde{\mathrm{A}}^{\mathrm{C}}=\{(\omega,\mu_{\tilde{\mathrm{A}}^{\mathrm{C}}}(\omega_i))\mid\mu_{\tilde{\mathrm{A}}^{\mathrm{C}}}(\omega_i)<\alpha\in[0,1])\}\end{matrix}\right\},i\in\mathbb{I}_{\tilde{\mathrm{A}}},\tilde{\mathrm{A}}\subset\Omega=\tilde{\mathrm{A}}\cup\tilde{\mathrm{A}}^{\mathrm{C}},\tilde{\mathrm{A}}\cap\tilde{\mathrm{A}}^{\mathrm{C}}\neq\varnothing$$

For the dualistic-polar analysis as needed in conditions for defining the general principle of opposites, the fixed level upper-cuts have corresponding *fixed level lower-cuts* of the form:

$$\mathcal{I}_{\tilde{\mathrm{A}}}(\omega_i)=\left\{\begin{matrix}1,\ \text{if}\ \omega_i\in\tilde{\mathrm{A}}=\{(\omega,\mu_{\tilde{\mathrm{A}}}(\omega_i))\mid\mu_{\tilde{\mathrm{A}}}(\omega_i)\leq\alpha\in[0,1])\}\\ 0,\ \text{if}\ \omega_i\in\tilde{\mathrm{A}}^{\mathrm{C}}=\{(\omega,\mu_{\tilde{\mathrm{A}}^{\mathrm{C}}}(\omega_i))\mid\mu_{\tilde{\mathrm{A}}^{\mathrm{C}}}(\omega_i)>\alpha\in[0,1])\}\end{matrix}\right\},i\in\mathbb{I}_{\tilde{\mathrm{A}}},\tilde{\mathrm{A}}\subset\Omega=\tilde{\mathrm{A}}\cup\tilde{\mathrm{A}}^{\mathrm{C}},\tilde{\mathrm{A}}\cap\tilde{\mathrm{A}}^{\mathrm{C}}\neq\varnothing$$

The fixed-level upper-cut and fix-level lower-cut are consistent with the conceptual system of dualities and polarities contained in the principle of opposites, where the duals and poles must be reconciled for a system of analytics over the space of identification-transformation dualities under static-dynamic processes.

The classical or the exact and the fuzzy or inexact indicator functions may be combined to establish the conditions of dividedness of the universal object space (Ω) into a family of classical or ordinary categories that constitutes a family of the areas of knowing or departments in the space of problem-solution dualities. Every classical or ordinary epistemic field is a derivative from a fuzzy epistemic field, where the membership functions of belonging are normalized to one. At the level of epistemological space, the epistemological object set with its information support as constructed from the signal dispositions under the space of acquaintance is considered as a fuzzy set or a fuzzy space

$$\Omega=\left\{\left(\omega,\mu_{\Omega}(\omega)\right) \mid \omega \in \Omega,\ \mu_{\Omega}(\omega) \in [0,1]\right\}=\left\{\bigvee_{\alpha_i \in [0,1]} \Omega_{\alpha_i}\right\}.$$

THEOREM 2.3.1.1: The Fuzzy Nestedness of the Universal Object Space (Ω) as Concentric Cones

Given a fuzzy set $\Omega=\left(x,\mu_{\Omega}(\omega)\in[0,1]\right)$ with increasing levels of presumption of the form $\alpha_1<\alpha_2<\cdots<\alpha_i<\cdots<\alpha_j<\cdots \mid i,j\in\mathbb{I}^{\infty}$ where $\mathbb{I}^{\infty}$ is an infinite index set of the varieties in (Ω) then $\Omega_{\alpha_j}\subset\cdots\subset\Omega_{\alpha_i}\subset\cdots\subset\Omega_{\alpha_2}\subset\Omega_{\alpha_1}, \forall i,j\in\mathbb{I}^{\infty}$ and $\Omega=\bigvee_{\{\alpha_i \mid i\in\mathbb{I}^{\infty}\}}\Omega_{(\geq\alpha_i)}$ for fixed-level upper-cuts $\Omega_{(\geq\alpha)}$ as an inverted cone and for a fixed-level lower-cuts $\Omega_{(\leq\alpha)}$, $\Omega_{\alpha_1}\subset\Omega_{\alpha_2}\subset\cdots\subset\Omega_{\alpha_i}\subset\cdots\subset\Omega_{\alpha_j}, \forall i,j\in\mathbb{I}^{\infty}$ and $\Omega=\bigvee_{\{\alpha_i \mid i\in\mathbb{I}^{\infty}\}}\Omega_{(\leq\alpha_i)}$ as a concentric cones that provide categorial nestedness.

The conditions of the theorem of the fuzzy nestedness lead to the theorem of *fuzzy decomposition* with fuzzy-stochastic conditionality for each fixed level cut because of neuro-decision-choice actions of cognitive agents under cognitive capacity limitations over the space of imagination-reality dualities. The fuzzy-stochastic conditionality at each fixed-level cut establishes the relational continuum and connectivity on the categories while the union establishes the unity under the principle of opposites with relational give-and-take sharing modes for mutual existence and destruction through the space of certainty-uncertainty dualities for all areas of knowing and information-knowledge development.

THEOREM 2.3.1.2: Fuzzy Decomposition of the Universal Object Space Ω into Categories

The fuzzy universal object set $\Omega = (\omega, \mu_\Omega(\omega))$ may be decomposed into categories by fixed-level (α-level) cuts into upper-level $\Omega_{(\geq \alpha)}$ nested sets or lower-level $\Omega_{(\leq \alpha)}$ nested sets such that given $\alpha_1 < \alpha_2 < \cdots < \alpha_i < \cdots < \alpha_j < \cdots | i, j \in \mathbb{I}_A$ where $\mathbb{I}^\infty$ is an infinite index set of either $\alpha - level$ upper-cuts or lower-cuts, then $\Omega = (\omega, \mu_\Omega(\omega)) = \underset{\{\alpha_i | i \in \mathbb{I}^\infty\}}{\text{Max}} \alpha_i \, \Omega_{(\geq \alpha_i)}$ for upper-level cut and $\Omega = (\omega, \mu_\Omega(\omega)) = \underset{\{\alpha_i | i \in \mathbb{I}^\infty\}}{\text{Max}} \alpha_i \, \Omega_{(\leq \alpha_i)}$ wheren $\alpha_i \in (0,1) \subseteq [0,1]$.

PROOF

The proof is obtained with the application of the conditions of dividedness from Lemma **2**.3.1 and fuzzy nestedness of Theorem 2.3.1 on the membership function $\mu_\Omega(\omega)$.

From the membership characteristic function and the conditions of dividedness, one can write for the upper level α-cuts where we have:

$$\mu_{\mathrm{A}}(\omega) = \begin{Bmatrix} 1.\ \mathrm{if}\, \mu_{\mathrm{A}}(\omega) \geq \alpha_i \\ 0,\ \mathrm{if}\ \mu_{\mathrm{A}}(x) \leq \alpha_{\mathrm{i}} \end{Bmatrix}, \mathrm{A} \subset \Omega;\ i \in \mathbb{I}^{\infty}$$

And for lower level α-cuts we have:

$$\mu_{\mathrm{A}}(\omega) = \begin{Bmatrix} 1.\ \mathrm{if}\, \mu_{\mathrm{A}}(\omega) \leq \alpha_i \\ 0,\ \mathrm{if}\ \mu_{\mathrm{A}}(x) \geq \alpha_{\mathrm{i}} \end{Bmatrix}, \mathrm{A} \subset \Omega;\ i \in \mathbb{I}^{\infty}$$

From the theorems of nestedness, one can write the union of the $(\alpha - levels)$ as:

$$\mu_{\Omega}(\omega) = \max_{\alpha_i}\left[\alpha_i \Omega_{\alpha_i}\right] = \max_{\{\alpha_i \leq \mu_{\Omega}(x_i)\}}\left[\alpha_i\right] = \mu_{\Omega}(\omega)$$

Similarly, for the lower level $(\alpha - cuts)$, one obtains the intersection of the $\alpha - levels$ as

$$\mu_{\Omega}(x) = \max_{\alpha_i}\left[\alpha_i \Omega_{\alpha_i}\right] = \max_{\{\alpha_i \geq \mu_{\Omega}(\omega_i)\}}\left[\alpha_i\right] = \mu_{\Omega}(\omega) \quad \square$$

The geometric structure of both the upper-fixed-level and lower-fixed-level cuts are shown with TFN (triangular fuzzy norm) in Figure 2.3.1. We must analytically notice that the upper and lower fixed-level cuts are consistent with the principle of opposites in dualistic-polar structures with relational continuum, unity and give-and-take sharing modes for dualistic-polar existence and destructions with intra-categorial and inter-categorial negative-positive dualities as well as cost-benefit dualities with their characteristic dispositions.

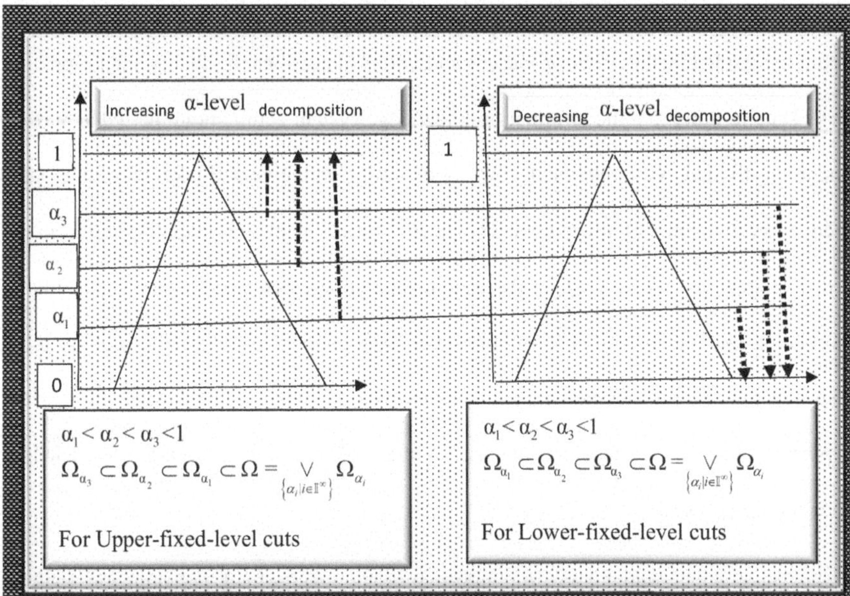

Figure 2.3.1: A Geometry of Upper Fixed-level and Lower fixed-level Fuzzy Decompositions into Categories and Dividedness with fuzzy-stochastic Conditionality

2.3.2 Reflections on the Lemma and Theorems on Categorial Dividedness and Nestedness

The important thing about the theorems of dividedness, nestedness and decomposition in the fuzzy information space is the theorems' relationships to the category formation, and the development of areas of knowing with departments of teaching, learning and research in relational continua as a unity of the general information-knowledge production and the continuous multidimensional search for the understanding of *what there was* (telescopic past), *what there is* (telescopic present) and *what would be* (telescopic future) under the *Sankofa-anoma* principle of the *time trinity* with neuro-decision-choice actions over the space of problem-solution dualities, where the neuro-decision-choice actions are guided by the *asantrofi-anoma conditions* of cost-benefit duality and rationality. Let us keep in mind that the information-knowledge system is time-point nested as well as generation-divided with relational continuum, unity and give-and-take sharing modes, where every generational information-knowledge stock is an evolutional offspring from the preceding generations as well as the parent and grandparent of the next generational information-knowledge stocks. Similarly, every generational paradigmatic toolbox is an offspring of previous generation paradigmatic toolboxes that have gone through refinements as parents and grandparents for the use in the knowing and information-knowledge-production processes.

These generational knowing, information-knowledge and paradigmatic toolbox developments are the foundations of the evolution of social philosophical consciencism as well as the social progress of nations and societies. The variety transformation dynamics is a time continuous process, where the past is indispensable from the present, and the past and present are the foundations of the future as seen over the space of identification-transformation problem-solution dualities, where the problems and solutions are studied over the epistemic field with corresponding relational and information fields which establish varieties and categorial connectivity under the α-algebra and γ-algebra which generate the interdependent areas of knowing in unity but not in disunity in such a way that the telescopic future is an evolutionary process from the telescopic present which is the result of the telescopic past with a system of correctional categorial converters or functors..

The categorial interdependencies of areas of knowing are not explicitly made available to students to enhance their academic and intellectual curiosities over the epistemic field in relation to neuro-decision-choice actions over the space of problem-solution dualities. For example, the theory of the decision-choice systems that are studied in economics and some areas in psychology are not linked to the general idea that all areas of human actions are governed by regimes of information-knowledge system and neuro-decision-choice actions in the space of fundamental-applied dualities where every category of applied varieties is also a category of fundamental dualities and vice versa. The categorial areas of teaching, learning and research are presented as they exist in the space of absoluteness rather than in the space of relativity of categorial interactions, intersectionality, and relationships in knowing and information knowledge production and that the varieties in the space of absoluteness are extremes to which the neuro-decision-choice agents work towards through thinking and reasoning over the space of inexact-exact dualities. The conflicts and collaborations between psychology and economics over the directions of the theories of decision and choice are healthy, giving rise to neuroeconomics, continual mutation of neuroscience and the possible development of a neuro-decision-choice school or a college within the universities for the study of neuro-decision-choice actions over the space of problem-solution dualities which is linked to a system of preference ordering over the goal-objective space under the principles of non-satiation and conditions of variety cost-benefit dualities with variety creative destruction, where there are continually infinite variety dualistic-polar destruction-replacement processes over the space of variety problem-solution dualities in all areas of human endeavor.

The past-present-future configuration in relation to the epistemic field is such that the argument for disunity of knowing and science is an artificial imposition, disregarding the essential point that every acquisition of knowledge is a neuro-decision-choice action and a problem-solution dynamic. Each α-cut reveals information about a defined state within the epistemic field which may be considered as a possible area of identification-transformation problem-solution dualities based on the conditions of the information field, the activities of which enhance the unity of knowing, information-knowledge accumulation and the nature of problem-solution disequilibrium dynamics over the space of production-consumption dualities. The importance of the lemma and theorems may be found over the path of creating an interdisciplinary (inter-categorial) process with the subjective phenomena of cognitive agents as the driving force of knowing, irrespective of the area of knowledge studies, where possibility-probability concepts and measurements (hard data) and non-measurements (soft data) find different expressions in the space of acquaintances. The sequence of category formation $\left\{\mathcal{C}_i, i \in \mathbb{I}^\infty\right\}$ with the fuzzy-stochastic indicator function of an upper-level cut will be represented as $\Omega_{(\alpha\geq)} = \left\{\mathcal{C}_i, i \in \mathbb{I}^\infty,\ \alpha \geq\right\}$ and with a lower-level cut as $\Omega_{(\leq\alpha)} = \left\{\mathcal{C}_i, i \in \mathbb{I}^\infty, \leq \alpha\right\}$.

2.4: THE CONCEPT OF UNITY-DISUNITY DUALITY IN KNOWING AND SCIENCE

The results of all the epistemic activities of cognitive agents are nothing but information processes under complete neuro-decision-choice actions over all areas of human endeavors with algorithms of thinking and the use of these algorithms over the space of problem-solution dualities. Nothing is knowable in the world of cognitive existence without information, and no problem is solvable without information-knowledge conditions in the space of problem-solution dualities. As it has been argued, information is the third dimension of universal existence. It is also a property of both matter and energy as the two other dimensions of universal existence in real and nominal states. Those who dispute information as a property of the mater-energy configuration must provide us with the meaning of information and the role of information in language development, communication over the space of source-destination dualities and many other phenomena in information-knowledge accumulation, storage and transmission, variety similarities, variety commonness and variety differences. A word is simply a word; it is a set containing characteristics as a disposition of identity representation in the space of varieties, where our language systems have built in the linguistic elasticity for the space of synonym-antonym dualities. In this epistemic framework, there is no experiential information without acquaintance at any level of existence and there is no knowledge without the experiential information structure which may be empirical, axiomatic, direct, and indirect in nature. The discussions of the meaning of information, the definition of information as a property of matter just like energy is provided in this monograph and in [R17.17] [R17.18] R21.5]. The classical theory of knowledge and information have no explicit definition of information and how information and knowledge are connected. The problem of this deficiency is simply, one is not sure of the inputs and outputs in the space of neuro-decision-choice actions. Again, without explicit definition of information, the concepts of misinformation, disinformation propaganda, lies, true false and related concepts and phenomena have little usefulness in communications over the spaces of source-destination dualities, analogue-digital dualities, continuous-discrete dualities and many such concepts and phenomena.

Adjectival qualifications of information make sense when the concept and phenomenon of information are defined and explicated.

2.4.1 Information Structures, Unity-Disunity Duality and Diversity in Knowing and Science

The universe presents itself to cognitive agents in terms of ontological space (Ω) with ontological information structure of the form $\left[\mathbb{Z}_{\Omega}=(\mathbb{X}_{\Omega} \otimes \mathbb{S}_{\Omega})\right]$ which is considered as the total information for ontological existence as the foundation in knowing involving the conditions of *what there was, what there is* and *what would be* in terms of identities and transformations. The $[\mathbb{Z}]$ is duality composed of characteristic disposition $(\mathbb{X})$ and signal disposition $(\mathbb{S})$. From the ontological space as the identity and the primary category of knowing, an epistemological space $(\mathfrak{E})$ with the information structure $\left[\mathbb{Z}_{\mathfrak{E}}=(\mathbb{X}_{\mathfrak{E}} \otimes \mathbb{S}_{\mathfrak{E}})\right]$ is created by cognitive agents with neuro-decision-choice processes over the acquaintance space. Corresponding to the ontological space is the space of ontological matter-energy varieties $(\mathbb{V}_{\Omega})$ with the information structure $\left[\mathbb{Z}_{\mathbb{V}_{\Omega}}=(\mathbb{X}_{\mathbb{V}_{\Omega}} \otimes \mathbb{S}_{\mathbb{V}_{\Omega}})\right]$ and corresponding to the epistemological space is also the space of epistemological matter-energy varieties $(\mathbb{V}_{\mathfrak{E}})$ over the space of real-nominal dualities with the information structure $\left[\mathbb{Z}_{\mathbb{V}_{\mathfrak{E}}}=(\mathbb{X}_{\mathbb{V}_{\mathfrak{E}}} \otimes \mathbb{S}_{\mathbb{V}_{\mathfrak{E}}})\right]$. The ontological varieties present ontic conditions in the ontological space, where the space of actual-potential dualities is defined, while the epistemological varieties present epistemic conditions in the epistemological space, where the space of imagination-reality dualities is defined in such a way that the space of the space of imagination-reality dualities is contained in the space of actual-potential dualities.

The ontic conditions present the characteristic dispositions of ontological varieties while the epistemic conditions present the characteristic dispositions of epistemological varieties as abstracted from the ontological signal dispositions at the acquaintance space under cognitive capacity limitations. In this respect, the ontic conditions are the primary category of knowing from the space of actual-potential dualities while the epistemic conditions are the derived category of knowing from the space of imagination-reality dualities as connected to the space of ontic-epistemic dualities. It is useful for the understanding of knowing and knowledge production leading to information-knowledge stock-flow disequilibrium dynamics to examine the foundations of the space of ontic-epistemic dualities and the space of imagination-reality dualities, where the connectivity is established by neuro-decision-choice actions of cognitive agents under cognitive capacity limitations.

THEOREM 2.4.1.1: Ontological-Epistemological Relationality

Given an epistemic field $(\mathscr{E})$ with an information support $(\mathbb{Z}_{\mathscr{E}})$ from the information field $(\mathscr{I})$ and a relational field $(\mathscr{R})$ also with an information support $(\mathbb{Z}_{\mathscr{R}})$ from the information field $(\mathscr{I})$, the relational conditions of the ontological and epistemological spaces are such that $(\Omega = \mathbb{V}_{\Omega})$, $\left[\mathbb{Z}_{\Omega}=\mathbb{Z}_{\mathbb{V}_{\Omega}} \Rightarrow (\mathbb{X}_{\Omega} \otimes \mathbb{S}_{\Omega}) = (\mathbb{X}_{\mathbb{V}_{\Omega}} \otimes \mathbb{S}_{\mathbb{V}_{\Omega}})\right]$ at the level of ontological space, and $(\mathfrak{E} = \mathbb{V}_{\mathfrak{e}})$, $\left[\mathbb{Z}_{\mathfrak{e}}=\mathbb{Z}_{\mathbb{V}_{\mathfrak{e}}} \Rightarrow (\mathbb{X}_{\mathfrak{e}} \otimes \mathbb{S}_{\mathfrak{e}}) = (\mathbb{X}_{\mathbb{V}_{\mathfrak{e}}} \otimes \mathbb{S}_{\mathbb{V}_{\mathfrak{e}}})\right]$ at the level of epistemological space, and at the levels of both ontological and epistemological spaces we have $(\mathfrak{E} \subset \Omega \Rightarrow \mathbb{V}_{\mathfrak{e}} \subset \mathbb{V}_{\Omega})$, and where $(\mathbb{Z}_{\mathfrak{e}} \subset \mathbb{Z}_{\Omega} \Rightarrow \mathbb{Z}_{\mathbb{V}_{\mathfrak{e}}} \subset \mathbb{Z}_{\mathbb{V}_{\Omega}})$ which also implies that $[(\mathbb{X}_{\mathfrak{e}} \otimes \mathbb{S}_{\mathfrak{e}}) \subset (\mathbb{X}_{\Omega} \otimes \mathbb{S}_{\Omega})]$ such that $[(\mathbb{X}_{\mathbb{V}_{\mathfrak{e}}} \otimes \mathbb{S}_{\mathbb{V}_{\mathfrak{e}}}) \subset (\mathbb{X}_{\mathbb{V}_{\Omega}} \otimes \mathbb{S}_{\mathbb{V}_{\Omega}})]$.

PROOF OF THEOREM 2.4.1.1

The proof begins with the condition that every matter-energy existence is a variety, the contents of which are identified by its characteristic disposition $(\mathbb{X})$ through a messaging system of signal disposition $(\mathbb{S})$. By the axiom of primary-derived duality, the epistemological space $(\mathfrak{E})$ is derived from the ontological space (Ω) under cognitive capacity limitations and hence is either equal or less than the ontological space (Ω). The existence of cognitive capacity limitation leads to a case where $(\mathfrak{E} \Subset \Omega)$. The condition where $(\mathfrak{E} \Subset \Omega \Rightarrow \mathbb{Z}_{\mathfrak{e}} \Subset \mathbb{Z}_{\Omega})$ implies that $(\mathbb{V}_{\mathfrak{e}} \Subset \mathbb{V}_{\Omega})$ and $[(\mathbb{X}_{\mathfrak{e}} \otimes \mathbb{S}_{\mathfrak{e}}) \Subset (\mathbb{X}_{\Omega} \otimes \mathbb{S}_{\Omega})]$. The conditions of characteristic-signal dispositions are such that $[\mathbb{Z}_{\mathbb{V}_{\mathfrak{e}}} \subset \mathbb{Z}_{\mathbb{V}_{\Omega}} \Rightarrow (\mathbb{X}_{\mathbb{V}_{\mathfrak{e}}} \otimes \mathbb{S}_{\mathbb{V}_{\mathfrak{e}}}) \subset (\mathbb{X}_{\mathbb{V}_{\Omega}} \otimes \mathbb{S}_{\mathbb{V}_{\Omega}})]$. The symbol $(\Subset)$ means a very small sub-set. But $(\mathbb{V}_{\Omega})$ is the collection of all matter-energy varieties in the ontological space and hence $(\mathbb{V}_{\Omega} = \Omega)$. But the ontological space and the space of ontological varieties contain the same information and hence $(\mathbb{Z}_{\Omega} = \mathbb{Z}_{\mathbb{V}_{\Omega}})$. Similarly, $[\mathfrak{E} = \mathbb{V}_{\mathfrak{e}} \Rightarrow \mathbb{Z}_{\mathfrak{e}} = \mathbb{Z}_{\mathbb{V}_{\mathfrak{e}}}]$ since the epistemological space contains all the epistemological varieties with the same information and hence $[(\mathbb{X}_{\mathbb{V}_{\mathfrak{e}}} \otimes \mathbb{S}_{\mathbb{V}_{\mathfrak{e}}}) = (\mathbb{X}_{\mathfrak{e}} \otimes \mathbb{S}_{\mathfrak{e}})]$ □

NOTE 2.4.1.1: On the Theorem 2.4.1.1 Ontological-Epistemological Duality

The containment of the epistemological space in the ontological space, the containment of the space of epistemological varieties in the space of ontological varieties and the containment of epistemological information in the ontological information simply show the cognitive capacity limitations of cognitive agents at all levels of knowing. Similarly, by an extension, the space of imagination-reality dualities is contained in the space of actual-potential dualities in that cognitive agents cannot know more than *what there was* (the past), *what there is* (the present) and *what there would be* (the future). The size, quantitative and qualitative exactness of the epistemological space is also constrained by the cognitive capacity limitation over the acquaintance space through the observational distribution of the variety-signal dispositions. The potential and the general acquaintance space is infinite. However, the actual acquaintance space opened to cognitive agents, at the space of individual-collective dualities, is very small and constrained by individual and collective cognitive capacity limitations and further constrained by geo-political capacity conditions such that the space of realities is very small compared to the space of the actuals. In general, the epistemological information structures for each variety, each category and the organic existence as encountered over the available space of acquaintance are vague and incomplete, creating *experiential information imperfection* which extends to the sequence of information derivatives of data, fact, knowledge, evidence, and evidential things as epistemic derivates from ontic states.

It is this information imperfection from the acquaintance space in the creation of knowledge by acquaintance that seems to have motivated Oskar Morgenstern's concern on the accuracy of economic observation [R22.13]. The problem of accuracy of observation is not limited to the economic information. It involves all areas of knowing and hence knowledge by acquaintance of all areas of science and non-science [R22.3] [R22.16] [R22.17][R2218] The complexity of this imperfection has been the source of intense disagreements in some variety identifications and transformations (e.g., environmental, and climatological varieties). The conditions of cognitive capacity limitations are related to defective information structures as cognitive agents examine what is possible and impossible information-knowledge variety and what is probable and improbable information-knowledge variety (quality-quantity in vagueness and volume). The conditions over the space of possibility-probability duality give rise to qualitative unsureness about the variety identities and hence uncertainty for anticipation and necessity. The conditions over the space of probability-improbability dualities give rise to quantitative unsureness and hence uncertainties expectation and freedom.

The defective information structure is amplified by misinformation, disinformation, propaganda and fake news to create a combination of *defective* and *deceptive* information structures to increase the space of the *penumbral region of decision* to affect the neuro-decision-choice activities over the space of problem-solution dualities with the intent to expand the zone of irreducible uncertainty and to increase the dualistic-polar dominance of failure over success or dominance of success over failure within the success-failure duality in conflicts for the games of the problem-solution dualities, where the reward is increasing benefits of the source and increasing costs for the receivers over the space of imagination-reality dualities. In this respect, two situations arise with cognitive agents in that if a neuro-decision-choice action over the space of problem-solution duality fails to achieve the desired goal, blaming and finger-pointing become powerful tools of morons in the irreducible zone of cognitive imbecility and do-nothingness, while thinking becomes a powerful instrument of the prudent in the expanding zone of wisdom and new search over the space of problem-solution dualities under the conditions of self-corrections to improve the quality of knowing as intellectual investment flows and information-knowledge accumulation as intellectual capital stocks. The space of imagination connects cognitive anticipations in the possibility space and expectations in the probability space to the potential space for continual self-correction neuro-decision-choice processes where uncertainties are generated by the ontic-epistemic differences. In this way the ontic variety identity is a constraint on the epistemic variety identity in all optimal claims of knowledge.

NOTE 2.4.1.2: On the Theorem 2.4.1.1, Complexities and Complications

The information imperfection leads to the existence of the *fuzzy-stochastic epistemological information.* The fuzzy-stochastic epistemological information creates the *fuzzy-stochastic knowledge by acquaintance* concerning all varieties, categories, areas of knowing and the total accumulated knowledge under stock-flow disequilibrium dynamics in cognitive relational complexities and neuro-decision-choice complications for general analytics and over the action space for all areas of knowing without an exception. The knowing and information-knowledge accumulations over all areas of knowledge must pass through the space of imagination containing the possibility and probability spaces where the possibility gives rise to the conception of possible and impossible worlds as well as probable and improbable worlds within the space of imagination. The degrees of complications and complexities in variety knowing vary over the space of variety quality-quantity dualities with differential distribution of quality-quantity relativities.

The relational conditions of the ontological and epistemological spaces generate interdependent information-knowledge stock-flow processes to support the information-knowledge disequilibrium dynamical system of never-ending organic processes with self-corrections, self-improving and self-motivation that constitutes essentially interdependent uniting forces of all the areas of knowing with a multiplicity of relational give-and-take sharing modes of complexities. There may be further partitioning of each intra-categorial structure with the use of different factors to facilitate decoding of deceptive information to enhance the neuro- decision-choice activities of thinking and reasoning within any relevant category of problem-solution dualities and the dynamic conditions of the relational and information fields. There are continually increasing variety differentiations, partition and divisions through categorial transformations in the space of social-natural dualities, which continually expand the information and relational fields associated with the space of varieties as cognitive agents exercise neuro-decision-choice actions over the space of the problem-solution dualities in destruction-replacement processes to achieve increasing preferred states under the psycho-economic principle of non-satiation over the space of varieties.

The fuzzy-stochastic knowledge by acquaintance as an experiential information structure becomes an input into the thinking-reasoning processes to create the knowledge by description which is, in fact, a fuzzy-stochastic knowledge by description accepted under fuzzy-stochastic conditionality with cognitive transversality conditions over the space of ignorance-knowledge dualities and polarities. The set of the attributes of the information-knowledge system in relation to the space of problem-solution dualities is an important uniting force of all areas of knowing whether any area or group of knowing is called science or non-science. Every set of attributes of the categorial information-knowledge structure belongs to the family of knowing and the family of information-knowledge accumulations. Let us keep in focus that every bit of information-knowledge acquisition is acquired by cognitive activity over the space of problem-solution dualities which is made up of varieties in dualistic-polar existence of negative-positive characteristic dispositions, where both the negative and positive duals reside in the space of cost-benefit dualistic-polar existence over the space of imagination-reality dualities, where the space of actions is defined. Similarly, both the negative and positive poles reside in the space of cost-benefit dualistic-polar existence over the space of imagination-reality dualities, where neuro-decision-choice actions take place.

It is analytically useful to see the intimate relationship between the negative-positive duality and the cost-benefit duality and the roles they play in understanding the unity and diversity of the evolving information-knowledge dynamics under neuro-decision-choice actions. The negative dual is not a cost dual, and the positive dual is not a benefit dual. The negative dual resides in cost-benefit duality within the space of imagination-reality dualities also containing the space of goals, objectives, and visions. Similarly, the positive dual resides in cost-benefit duality within the space of imagination-reality dualities also containing the space of goals, objectives, and visions. In this social process of knowing and information-knowledge accumulation, the activities of knowing and information-knowledge accumulation are undertaken for the maintenance of the existence of cognitive agents in the struggle within the zone of life-death dualities. Cognitive agents do not live to know and develop information-knowledge accumulation under the principle of non-satiation. Information-knowledge processes are undertaken for cognitive agents to function in the space of variety identification-transformation dualities under the conditions of non-satiation through neuro-decision-choice activities over the space of problem-solution dualities and polarities. The complexities and complications reveal themselves under the conditions, where a problem is a solution, and a solution is a problem relative to the elements in the space of goals, objectives, and visions in the journey from the space of relativity with imperfect information to the space of absoluteness with perfect information for varieties.

Problems are solved by variety substitution processes through creative destructions, where the best variety solutions are searched and then substituted for variety problems under cost-benefit rationality under the *asantrofi-anoma principle,* and where the variety solutions, as substitutions for variety problems, generate new variety problems in the information-knowledge stock-flow disequilibrium dynamics. It is this attribute of the problem-solution dynamics in the neuro-decision-choice processes that endows the general information-knowledge system with self-exiting, self-innovation and self-correcting power through the activities of knowing for a continual progress in variety identifications and transformations over the space of ontological-epistemological dualities for the understanding of the universal existence in accord with the time trinity (the *Sankofa-anoma principle* of past, present and future) in information-knowledge dynamics and decision-choice actions over the space of problem-solution dynamics.

The cognitive (neuro-decision-choice) activities proceed in two-forward steps under constructionism and information-knowledge creation, and two-backward steps under reductionism and the test of validity and information-knowledge acceptance. The two-forward steps involve acquaintance through cognitive observation, enhanced or unenhanced by experimentation and other supporting non-human facilities depending on the nature of the category (area of knowing) to create a memory of a list of experiences as the primary category of knowing at the epistemological space. The list of experiences is obtained from the ontological *signal dispositions* in relation to epistemological varieties and characteristic dispositions which form the *variety data* in terms of their repeated occurrences of the experiential information. The variety data of the experiential information become inputs to derive the *knowledge by description* from two-forward steps in the knowing over the imagination-reality dualities contained in the epistemological space.

Here, a special attention must be paid to the conceptual definitions of data, fact, and evidence in terms of nominal and real definitions as they relate to information. At the epistemological space, data is, a recollection and recording from the variety signal dispositions, and may be different from fact, knowledge, and evidence, all of which are sequential derivatives from variety characteristic-signal dispositions and claimed to be facts in the space of imagination-reality dualities. The two-backward steps involve initial acquaintance with the claim of knowledge, composed of knowledge by acquaintance and the knowledge by description as have been enhanced by some form of thinking, and other supporting human-decision-choice facilities to claim a list of variety knowledge, through the primary and derived categories of knowing, in relation to epistemological variety characteristic dispositions which form the variety data of knowledge by description and are taken as facts and evidence in support of neuro-decision-choice activities over the action space. The variety data of the primary and derived knowledge claims become inputs into modules of tests and acceptances of knowledge items through the two-backward steps of methodological reductionism.

2.4.2: Paradigmatic Paths in Support of either Unity or Disunity in Knowing and Science

The results of the activities of knowing over the epistemic field with the relational and information fields, through the epistemic activities over the space of acquaintance as the primary categories of epistemological knowledge and the corresponding knowledge by description as the derived categories of knowing, depend on the types of assumption that cognitive agents impose on the fuzzy-stochastic epistemological information which constitutes the primary category of knowing in the epistemological space, as a derived category from the ontological space. One information structure is designed by accepting the fuzzy-stochastic property of the epistemological information, which reflects human cognitive capacity limitations, as it is, and as representing the foundation of the inputs into neuro-decision-choice processes to derive the *variety knowledge by description*, without assumption, but with *fuzzy-stochastic conditionality* with possibilistic-probabilistic cognitive actions over the space of thinking-reasoning dualities. The alternative information structure is obtained by imposing the assumption of exactness to do away with vagueness, and inexactness, and to accept the observational incompleteness for the creation of *stochastic epistemological information* as inputs into the neuro-decision-choice process to derive the knowledge by description with *stochastic conditionality*, with probabilistic cognitive actions over the space of thinking-reasoning dualities.

The fuzzy-stochastic information with the fuzzy paradigm of thought leads to the construct and study of the theory of fuzzy-stochastic decisions in all areas of cognitive actions over the space of problem-solution dualities, where the fuzzy-stochastic information structure acknowledges the complete acceptance of an irreducible zone of cognitive capacity limitations in relation to observation, thinking and reasoning under the principle of opposites with relational continuum, unity and acceptance of contradictions. The classical-stochastic information with the classical paradigm of thought leads to the construct and study of classical-stochastic decisions in all areas of human actions over the space of problem-solution dualities, where the classical-stochastic information acknowledges the complete acceptance of exactness in relation to observation, thinking and reasoning under the principle of opposites with an excluded middle, relational separation, and non-acceptance of contradictions. The fuzzy-stochastic information shows the fuzzy part as capturing vagueness, qualitative dispositions, inexact dispositions, and subjective dispositions, while showing the stochastic part as capturing the volume-limitation and incompleteness of observation.

The classical-stochastic information shows the stochastics as capturing the volume-limitations and incompleteness of variety observation in the exact values in the space of quantity-quality dualities. In this respect, it is useful to note that *fuzzification* is to reduce the classical exact stochastic information to fuzzy-stochastic information and *defuzzification* is to convert the fuzzy analytical outputs to the classical equivalences with fuzzy-stochastic conditionality of the degrees of acceptance. The use of fuzzy membership functions is part of the fuzzification of the classical exact stochastic information, the use of fuzzy optimization is part of defuzzification while the use of hybrid numbers of fuzzy numbers and probabilistic numbers is a unification of qualitative and quantitative uncertainties in the space of inexact-exact dualities relative to the space of ontic-epistemic dualities where the ontic-epistemic differences in characteristic dispositions generate uncertainties over the space of qualitative-quantitative dualities.

All areas of varieties of knowing are related to the epistemological information and are united by the forward-backward steps in the knowing process over the space of problem-solution dualities irrespective of the category of belonging in the epistemic field, where every category in the epistemic field contains a set of problem-solution dualities. We must keep in mind that information-knowledge production involves cognitive activities in the space of problem-solution dualities and variety substitutions with cost-benefit rationality over the space of imagination-reality dualities under the activities of neuro-decision-choice actions, where every variety presents itself as a dualistic-polar entity. The stock-flow information-knowledge accumulation and the problem-solution processes are in infinitely disequilibrium dynamics, where problems generate solutions and solutions generate problems as foundation of the meaning of life and the manners of solutions as the essence of life. The results of the problem-solution processes generate the stock-flow infinitely information-knowledge disequilibrium processes, which become inputs into the dynamics of global spaces of problem-solution dualities and decision-choice actions [R17.18]. Let us keep in mind the meaning of dualistic-polar conditions of a variety, where the duality (dualistic condition) helps to solve the identification problem and polarity (polar condition) helps to solve the transformation problem in the dualistic-polar games of the identification-transformation processes in resolving the variety games of ignorance-knowledge conflicts with transversality conditions between variety zones of ignorance and zones of knowledge.

Given the two epistemological information structures of either the classical or fuzzy type, two different paths are opened to us in the analytical space for examining the conditions of the unity-disunity duality of knowing, science and non-science as we view the foundation of the information-knowledge construct about the universe through the cognitive actions over the space of problem-solution dualities, where the space of the problem-solution dualities is partitioned into categories under the principle of opposites to form epistemological and analytical spaces. The family of categories generates the family of actual and potential areas of knowing called the *epistemic field,* or an α-algebra, while the form of relations is called the *relational field* and the form of characteristic-signal dispositions is called the *information field.* When the epistemic field with the relational field and information field is transformed into characteristic-signal dispositions with paradigms of thought and neuro-decision-choice processes then we have the characteristic-signal field or γ-algebra from which hybrid algebras may be constructed as new problem-solution dualities emerge from neuro-decision-choice actions.

The nature of the conditions of unity-disunity duality in knowing and science finds expression in the information-knowledge space. The argument in support of unity is carried on with the fuzzy paradigm of thought when the fuzzy-stochastic epistemological information is the input of analysis and the epistemic field with embedding is constructed with a fuzzy-stochastic indicator function with an α-algebra on (Ω), and γ-algebra on $\mathbb{Z}_{\Omega} = (\mathbb{X}_{\Omega} \otimes \mathbb{S}_{\Omega})$, where the acceptance of each category is constrained by a *fuzzy-stochastic conditionality* as seen in terms of fuzzy-probabilistic reasoning under the principle of opposites with relational continuum, unity, a non-excluded middle and give-and-take-sharing modes. The argument in support of disunity is carried on with the classical paradigm of thought with relational separation, an excluded middle, no-give-and-take sharing mode and non-acceptance of contradiction, where the exact stochastic epistemological information is the input of the analysis and the epistemic field is constructed with the classical indicator function as a σ-algebra on (Ω), and each category is not constrained, but subject to an exact stochastic conditionality, as seen in terms of exact probabilistic reasoning, under the principle of opposites with relational separation and disunity. In this respect, what meaning should be attached to inexact probability or probability of probability and in what space should they be abstracted?

There are several concepts and representations that must be kept in mind to carry on the argument in support of either unity or disunity in knowing, and whether such knowing involves science or non-science in the epistemic field that is extended to epistemic-field imbedding. These concepts and representations are the ontological space(Ω) with ontological information structure$\left[\mathbb{Z}_{\Omega}=(\mathbb{X}_{\Omega}\otimes\mathbb{S}_{\Omega})\right]$, epistemological space$(\mathfrak{E})$ with epistemological information structure $\left[\mathbb{Z}_{\mathfrak{E}}=(\mathbb{X}_{\mathfrak{E}}\otimes\mathbb{S}_{\mathfrak{E}})\right]$, and the spaces of actuals$(\mathfrak{A})$ and possible$(\mathfrak{U})$ ontological varieties $(\mathbb{V}_{\Omega})$ as infinite space also with $\mathbb{Z}_{\mathbb{V}_{\Omega}}=\left(\mathbb{X}_{\mathbb{V}_{\Omega}}\otimes\mathbb{S}_{\mathbb{V}_{\Omega}}\right)$. We also have the space of epistemological varieties$(\mathbb{V}_{\mathfrak{E}})$ as infinite spaces also with characteristic-signal dispositions$\left[\mathbb{Z}_{\mathbb{V}_{\mathfrak{E}}}=\left(\mathbb{X}_{\mathbb{V}_{\mathfrak{E}}}\otimes\mathbb{S}_{\mathbb{V}_{\mathfrak{E}}}\right)\right]$. Additionally, it is useful to distinguish the family$(\mathscr{C})$ with a corresponding information structure$\left[\mathbb{Z}_{\mathscr{C}}=(\mathbb{X}_{\mathscr{C}}\otimes\mathbb{S}_{\mathscr{C}})\right]$ of categorial varieties of the form$(\mathcal{C})$ with information structure $\left[\mathbb{Z}_{\mathcal{C}}=(\mathbb{X}_{\mathcal{C}}\otimes\mathbb{S}_{\mathcal{C}})\right]$ from Ω, $\mathfrak{E}$ and $\mathbb{V}$ with their corresponding information structures as defined by their respective characteristic-signal dispositions from the information field $(\mathscr{I})$.

All areas of knowing as categories, whether science or non-science, within the epistemic field are identified by their characteristic-signal dispositions $\mathbb{Z}_{\mathcal{C}}=(\mathbb{X}_{\mathcal{C}}\otimes\mathbb{S}_{\mathcal{C}})$

which are the information structures from the information field $(\mathscr{I})$ needed to establish the areas' identities and their relationally categorial intersectionality and transformation. At the level of the epistemological space, the contents of the information structure as the characteristic disposition $(\mathbb{X}_{\mathcal{e}})$ for the identity of each variety $(\nu \in \mathbb{V})$ and the category $(\mathfrak{C}_{\nu})$ of its belonging are determined with neuro-decision-choice actions by cognitive agents operating either through the fuzzy paradigm of thought with the fuzzy-stochastic indicator function $\left[\mathcal{I}_{\tilde{\mathrm{A}}}(\omega_i)\right]$ or through the classical paradigm of thought with the classical indicator function $\left[\mathcal{I}_{\mathrm{A}}(\omega_i)\right]$, as specified in this chapter. In each case, the general set of characteristics as inputs for category formation must be made explicit, in the sense of establishing the information field required to construct the epistemic and relational fields as well as the development of epistemic field embedding. Each category is a representation of a set of problem-solution dualities for neuro-decision-choice actions.

2.4.3: Unity-Disunity Duality and Diversity-Oneness Duality in Knowing and Science

We have discussed the unity of knowing in relation to the unity of science and non-science in chapter one. It is useful now to discuss the unity-disunity duality over the epistemological space where there may be an argument in support of unity, disunity, diversity, oneness, or a distribution of the combinations of unity and disunity proportions in the unity-disunity duality. In all these epistemic discussions, it is useful to define the concepts of unity and disunity in knowing as seen over the epistemic field $(\mathscr{E})$ with a corresponding relational field $(\mathscr{R})$ and information field $(\mathscr{I})$ as may also be seen in the characteristic-signal field over the space of imagination-reality dualities contained in the space of actual-potential dualities.

DEFINITION 2.4.3.1 Unity and Disunity in Knowing and Science

Given the epistemic field $(\mathscr{C})$ with a relational field $(\mathscr{R})$, let $(\mathscr{U})$ be a relational categorial unity and $(\mathscr{E})$ be a *relational categorial disunity* (non-unity) such that $(\mathscr{U},\mathscr{E}\in\mathscr{R})$ is the set of available relations defined over the epistemic space with an information field $(\mathscr{I})$.There is a general categorial relational disunity $\mathscr{E}(\mathscr{C})$ if, $\left\{\mathcal{C}_i \neq \varnothing,\ \forall i, \in \mathbb{I}^\infty\right\} \subset \mathscr{C}$ and $\left(\bigcap_{i\in\mathbb{I}} \mathcal{C}_i = \varnothing, \left\{\mathcal{C}_i \cap \mathcal{C}_j = \varnothing, \mathcal{C}_i \nabla \mathcal{C}_j = \varnothing,\ \forall i \neq j \in \mathbb{I}^\infty\right\}, \left\{\bigcup_{i\in\mathbb{I}^\infty} \mathcal{C}_i \subseteq \mathscr{C},\ \forall i \in \mathbb{I}^\infty\right\}\right)$. It is said to have *relational categorial unity* $\mathscr{U}(\mathscr{C})$ if $\left\{\mathcal{C}_i \neq \varnothing,\ \forall i, \in \mathbb{I}^\infty\right\} \subset \mathscr{C}$ and additionally $\left(\bigcap_{i\in\mathbb{I}} \mathcal{C}_i \neq \varnothing\right) \subset \mathscr{C}$ with $\left\{\left\{\mathcal{C}_i \cap \mathcal{C}_j \neq \varnothing, \mathcal{C}_i \nabla \mathcal{C}_j \neq \varnothing, \forall i \neq j \in \mathbb{I}^\infty\right\}, \left(\bigcup_{i\in\mathbb{I}} \mathcal{C}_i\right) \subseteq \mathscr{C}\right\}$. If $\mathscr{U}(\mathscr{C})$ exists, then the categories are called diversities with connectivity in the space of give-and-take sharing modes, and their common elements are called non-zero inter-categories where their relations are called positive inter-sectionalities within the information field. If $\mathscr{E}(\mathscr{C})$ exists, then the categories are called individualities with complete non-connectivity in the space of no-give-and-take sharing modes, their common elements are called zero inter-categories and their relations are also called completely separable as disunity within the information field.

THEOREM 2.4.3.1: Categorial Relational Unity and Disunity in the Epistemic Field

If the epistemic field $(\mathscr{E})$ is constructed with a classical indicator function then there is the disunity relation $\mathscr{E}(\mathscr{E})$ under the conditions of the excluded middle and non-acceptance of contradictions, where opposites cannot exist in the same space and time within the classical paradigm of thought and the areas of knowing are mutually exact and distinct without mutual give-and-take sharing processes, where there are zero inter-categorial diversities. On the other hand, if the epistemic field $(\mathscr{E})$ is constructed with a fuzzy-stochastic indicator function then there is a unity relation $\mathscr{U}(\mathscr{E})$ with conditions of relational continua and unity under the conditions of acceptance of contradiction with the non-excluded middle, where opposites exist in the same space and time within the fuzzy paradigm of thought and the areas of knowing are inexact, mutually connected with give-and-take sharing modes and non-zero inter-categorial diversities.

PROOF OF THEOREM 2.4.3.1: Categorial Relational Unity and Disunity

The proof of the theorem is strait forward. With the fuzzy indicator function If $(\mathscr{E})$ is an *epistemic field* with a relational field $(\mathscr{R})$ and information field $(\mathscr{I})$ then there exists a fuzzy-stochastic indicator function:

$$\mathcal{I}_{\tilde{\mathcal{C}}}(\omega_i)=\begin{Bmatrix} 1\text{, if } \omega_i \in \tilde{\mathcal{C}} = \{(\omega_i,\mu_{\tilde{\mathcal{C}}}(\omega_i)) \mid \mu_{\tilde{\mathcal{C}}}(\omega_i) \geq \alpha^* \in [0,1])\}, \tilde{\mathcal{C}} \subset \Omega \\ 0\text{, if } \omega_i \in \tilde{\mathcal{C}}^C = \{(\omega_i,\mu_{\tilde{\mathcal{C}}}(\omega_i)) \mid \mu_{\tilde{\mathcal{C}}}(\omega_i) < \alpha^* \in [0,1])\}, \tilde{\mathcal{C}}^C \subset \Omega \end{Bmatrix}, i \in \mathbb{I}_{\tilde{\mathcal{C}}}, \Omega = \tilde{\mathcal{C}} \cup \tilde{\mathcal{C}}^C \text{ for}$$

the generation of the series $\{\mathcal{C}_i, i \in \mathbb{I}^\infty\}$, where the α^* is obtained as an optimal fuzzy value from a solution to a relationship to a dualistic-polar conflict in duality, and the epistemic field is called an (α-algebra) with $\{\mathbb{Z}_{\mathcal{C}_i} \cap \mathbb{Z}_{\mathcal{C}_j} = \varnothing, \forall i \neq j \in \mathbb{I}^\infty\}$ and $\{\mathbb{Z}_{\mathcal{C}_i} \nabla \mathbb{Z}_{\mathcal{C}_j} = \mathbb{Z}_{\mathcal{C}_i} \cup \mathbb{Z}_{\mathcal{C}_j} \subset \mathbb{Z}, \forall i \neq j \in \mathbb{I}^\infty\}$. In this case,

$$\left(\bigcap_{i\in\mathbb{I}} \mathcal{C}_i \neq \varnothing\right) \subset \mathscr{C} \text{ and} \left\{\{\mathcal{C}_i \nabla \mathcal{C}_j \neq \varnothing,\}, \left(\bigcup_{i\in\mathbb{I}} \mathcal{C} \;\forall i,j \in \mathbb{I}\right)\right\} \subset \mathscr{C}.$$

Alternatively, if the classical indicator function of the form:

$$\mathcal{I}_{\mathcal{C}}(\omega_i)=\begin{Bmatrix} 1\text{, if } \omega_i \in \mathcal{C} = \{(\omega,\mu_{\mathcal{C}}(\omega_i)) \mid \mu_{\mathcal{C}}(\omega_i) = 1), \forall \omega \in \mathcal{C} \subset \Omega\} \\ 0\text{, if } \omega_i \in \mathcal{C}^C = \{(\omega,\mu_{\mathcal{C}}(\omega)) \mid \mu_{\mathcal{C}}(\omega) = 0), \forall \omega \notin \mathcal{C} \subset \Omega\} \end{Bmatrix}, i \in \mathbb{I}_{\mathcal{C}}, \Omega = \mathcal{C} \cup \mathcal{C}^C \text{ is}$$

used to generate the series $\{\mathcal{C}_i, i \in \mathbb{I}^\infty\}$ where $\left(\bigcup_{i\in\mathbb{I}^\infty} \mathcal{C}_i \subset \mathscr{C}\right)$ and $\left(\bigcap_{i\in\mathbb{I}^\infty} \mathcal{C}_i = \varnothing\right) \subset \mathscr{C}$ such that. $\{\mathbb{Z}_{\mathcal{C}_i} \cap \mathbb{Z}_{\mathcal{C}_j} = \varnothing, \forall i \neq j \in \mathbb{I}^\infty\}$, and $\{\mathbb{Z}_{\mathcal{C}_i} \nabla \mathbb{Z}_{\mathcal{C}_j} = \mathbb{Z}_{\mathcal{C}_i} \cup \mathbb{Z}_{\mathcal{C}_j} \subset \mathbb{Z}, \forall i \neq j \in \mathbb{I}^\infty\}$, where the symbol ∇ represents a symmetric set difference, and thus there is a relational disunity $\mathscr{E}(\mathscr{C})$ with individual separations and no relational unity $\mathscr{U}(\mathscr{C})$ □.

NOTE 2.4.3.1: On the Theorem of Categorial Unity and Disunity

The categories of the epistemic field are actual and possible areas of knowing established and distinguished by defined characteristic dispositions from the space of imagination-reality dualities that indicate the nature of the distribution of socio-natural problem-solution dualities. For example, if $\mathfrak{C}_1$ = the area of science and $\mathfrak{C}_2$ = the area of non-science, then $\mathscr{C} = \{\varnothing, \Omega, \mathfrak{C}_1, \mathfrak{C}_2 \mid \mathscr{R}, \mathscr{I}\}$ given the relational and information fields. In this sense, classifications of areas of knowledge, the classificatory process and category formations are epistemic algebraic structures and epistemic processes over the space of identification-transformation dualities respectively, under the fuzzy arithmetic [R5.34], in solving the unity-disunity problem of conflicting relations in knowing and science relative to the conditions of variety ignorance-knowledge dualities in the information-knowledge development, area-specific languages, cross-categorial languages, common language of areas of knowing in their formation, communications and engineering of all forms. The nature of the epistemic structures and the inputs into the epistemic processes are the abstracted variety characteristic dispositions from the signal dispositions in the acquaintance space from decision-choice actions over the epistemological space to resolve conflicts in knowing over the space of ontic-epistemic dualities.

The upper-fixed-level cuts and the lower-fixed-level cuts present current information of the past-present dualities and future uncertainties in dualistic-polar conditions. Which of the fixed-level cuts may be appropriate in the information-knowledge analytics, just like which of the membership functions may be used in the algebraic analytics or big-small data analytics, will depend on the nature of the problem-solution duality at hand. It may be pointed out that there is a close and inseparable relation between fuzzy decomposition and the decomposition of areas of knowing. Within any category of decomposition, there is a subjectivity of dividedness. The subjectivity of dividedness is based on the fuzzy decomposition under the cognitive capacity limitations of human agents within the space of general certainty-uncertainty dualities in separation, intersectionality, and unification. The structure of unity-disunity duality is such that there is an argument in support of either unity or disunity but not both. The argument in support of simultaneous existence of unity and disunity over the epistemic field will always depend on the choice of the characteristic dispositions used to induce the category formation as well as the optimal process of the alpha-algebra (α-algebra), where the decision-choice actions of cognitive agents are internalized in the formation of the epistemic fields to the characteristic-signal fields to resolve the conflict over the space of oneness-dividedness dualities of the space of problem-solution dualities to deal with transformation-substitution processes over the space of variety ignorance-knowledge dualities.

The analytical process of the use of probabilities to establish dividedness or categories induces a circularity of reasoning, where categories are needed to establish probability distribution and where the probability distribution may be used to create the conditions for dividedness. The disunity relation on the epistemic field projects the existence of diversities with complete separations for individual purposes with no give-and-take-sharing relations relative to projected separations for individual purpose. The unity relation on the epistemic field projects the existence of diversities with relational continuum, interdependencies and give-and-take sharing relations for common and collective purpose. The unity in the epistemic field simply reveals the notion of different actions on different categories of problem-solution dualities as a contribution to the overall knowing and information-knowledge accumulation. In the unity relations of the epistemic field, one can speak of diversity in unity with cooperation and collaboration. In the disunity relations within the epistemic field, one speaks of diverse individuality with no cooperation and collaboration under the principle of the excluded middle.

In other words, one speaks of separate categorial existence and development without even any antagonistic relation. This kind of categorial disunity is not consistent with the variety existence in the universe, where categorial growth, development, peace and war, ecological diversity, biodiversity, the general sustainability of life, the food-chain, the input-output chain, the supply-demand chain, environmental-climatological relation, geomorphological-climatological existence, and socio-natural changes and stability are in categorial inter-dependencies for the unity of all the family of families of species (varieties) and the understanding of the celestial bodies, galaxies, family of celestial systems and the family of the galaxies, where there are continual creations and restructuring in terms of destruction-construction processes that give meaning to occurrences of earthquakes, volcanos, with similar activities taking place over other planets, planetary systems and galaxies. In this respect, restructuring educational systems, especially research universities with inter-categorial collaborations in teaching, learning and research will reduce individual and collective cognitive capacity limitations, improve individual and collective curiosity, raise the social cognitive capacity index, increase individual and social efficiencies of individual and collective decision-choice actions over the space of problem-solution dualities, speed up the dynamics of knowing and science and expand social visions under the psycho-economic principle of non-satiation that creates problem-solution disequilibrium dynamics within the space of imagination-reality dualities.

CHAPTER 3

THE PRINCIPLES OF DIVERSITY AND UNITY IN KNOWING AND SCIENCE

ABSTRACT: CHAPTER 3

This chapter deals with the conceptual and analytical foundations of the phenomena and conditions of diversity and unity in knowing and science through neuro-decision-choice actions over the space of problem-solution dualities and polarities, where the primary principle of unity of knowing and the primary principle of unity of science and non-science are established for identities and difference. This is followed by the development of the supporting derived principles. Categorial connectivity analytics are undertaken to show how all the knowing and the sciences are in relations to neuro-decision-choice actions under the social philosophical consciencism at generation points, where the dynamics of the social philosophical consciencism shape the dynamics of the generational preferences over the space of problem-solution dualities and polarities. From the space of problem-solution dualities, the dynamics of the social philosophical consciencism, methods and techniques of thinking and reasoning are introduced as meta-paradigms and distinguished from the organic paradigm of thought. All the neuro-decision-choice actions and the structure of variety problem-solution balances are defined over the space of cost-benefit dualities under the socio-psychological principle of non-satiation. Under these conditions, the knowledge by acquaintance as the primary category and the knowledge by description as the derived category are introduced and their relationships with the organic paradigm of thought and the space of meta-paradigms of thought are established. The chapter is ended with discussions on differences and similarities of information, knowledge, data, fact, and evidence.

K. K. Dompere, *The Theory of Epistemic Fields*, Studies in Systems, Decision and Control 498, https://doi.org/10.1007/978-3-031-42470-0_3

3.1 REFLECTIVE CONDITIONS ON THE DIVERSITY AND THE UNITY IN KNOWING

In the monograph entitled, *The Theory of Problem-Solution Duality and Polarity: Information-Decision-Choice foundations, the Unity of Knowing and the Unity of Science,* discussions were made to establish the relational structures among problem-solution dualities, problem-solution processes and information-knowledge processes and how these processes are relationally connected to the epistemic journeys of cognitive agents, with neuro-decision-choice actions as the vehicles, to travel between duals in the space of opposites, especially between the variety ignorance and the variety knowledge within the space of ignorance-knowledge dualities connected to the space of production-consumption dualities as a sub-space of the space of imagination-reality dualities[R8.16b]. In the discussions, several conceptual definitions were provided for distinctions and useful communications over the space of source-destination dualities. From the definitions of the concepts of the problem-solution duality and polarity, it is useful to state the conceptual principles of the diversity and the unity of science through the principles of diversity and unity of knowing as the general approach to the understanding of the universal variety connectedness and the oneness of the information-knowledge system in conceptualizing the conditions of oneness as an organized unity and the dividedness as an organized diversities with relational give-and-take sharing modes in the universal existence. There are two inter-supportive principles for the unity of knowing and the unity of science which constitute the intellectual capital stock-flow system over the spaces of variety input-output dualities variety ignorance-knowledge dualities and variety identification-transformation dualities.

The intellectual capital stock system is intimately connected to the spaces of production-consumption dualities, destruction-replacement dualities, and other important dualistic-polar structures within the space of imagination-reality dualities in which cognitive actions occur under all kinds of cognitive capacity limitations. The two inter-supportive principles are the *primary principle of existence* and the *derived principle of existence* based on the *general principle of opposites,* where the universe is seen as the matter-energy-information integrated form, containing varieties and categorial varieties on the basis of which all languages are formed and developed as the size of the information-knowledge accumulation expands under the matter-energy equilibrium, variety identifications and transformations, the problem-solution disequilibria and the information-knowledge disequilibria. These varieties and categorial varieties are seen in terms of diversities in categories with relational complexities of interdependence, continuum and unity that constitute the universal system of ontological existence under the principle of opposites from which the epistemological existence is a simple derivative by the collective neuro-decision-choice actions of cognitive agents over the space of imagination-reality dualities contained in the space of actual-potential dualities.

The space of actuals contains varieties of telescopic beginning and telescopic present, while the space of potentials contains varieties of telescopic future in the ontological space. The varieties of the telescopic beginning and past contain *what there was,* the varieties of the telescopic present contain *what there is;* and the varieties of the telescopic future contain *what there would be,* while the space of actual-potential dualities contains the games of natural variety transformations from actuals to potentials and from the potentials to the actuals. The space of natural variety transformations gives rise to the space of games of social variety identifications and transformations. The games involve social variety transformations from realities to potentials and to imaginations and from imaginations to realities with neuro-decision-choice actions over the space of theory-practice dualities. The space of realities contains varieties of *what is claimed to be known* as the variety realities from the telescopic past and the telescopic present of the space of the actuals. The space of imaginations contains varieties of *what would become known* or transformed from the space of potentials as telescopic future realities constrained by possibility-probability conditions. The space of imagination-reality dualities contains the game space of social neuro-decision-choice activities of variety transformations from imaginations to realities and to actuals, and from the identified realities to imaginations and then to the potentials.

The *natural game space* is defined between the variety actuals and the variety potentials with perfect information structures, where the existence of the space of actual-potential dualities is independent of the existence of any cognitive agent. The *social game space* is defined between the potential-imagination structures on one hand and actual-reality on the other hand with imperfect information conditions, where the existence of the space of imagination-reality dualities and its morphology at any time point depends on the existence of cognitive agents with the guidance of the ruling social philosophical consciencism at generational time point. The games in the natural game space are played between duals of dualities, poles of polarities and dualities and polarities, and are ontologically defined with ontological information, where the natural game space is the primary category to the social game space. The games in the social game space are played between duals of dualities, between poles of polarities and between dualities and polarities, and are epistemologically defined with epistemological information, where the social game space is composed of derived categories from the natural game space. The reward of every game is dualistic-polar identity dominance of the variety.

Let us keep in mind that the concepts of ontological and epistemological spaces are defined, explicated, and established by two characteristic dispositions in relational continuum and unity for identity and distinction [R8.16a]. The ontological space is the primary category of existence which is independent of cognitive existence, while the epistemological space is the derived category of existence that depends on the existence of the ontological space and cognitive agents in the creation and the establishment of its relational connectivity with the ontological space for knowing and decision-choice activities which find expressions in the space of imagination-reality dualities through neuro-decision-choice activities. The neuro-decision-choice activities include climatological and geomorphological dynamics in relation to the cognitive agents and general social activities in social production and consumption of varieties. The natural production is under natural variety transformations under natural dualistic-polar conditions. The social production and consumption cannot take place without the ontological existence, the varieties of which are indispensable inputs into the social production that generates variety outputs in the ontological-epistemological environments, where such outputs become inputs into ontological transformation processes in the space of ontological-epistemological dualities in the space of variety disequilibrium dynamics. It is the ontological-epistemological duality and its dualistic-polar conditions of variety interdependencies, relational continuum and relational unity in identifications and transformations that provide the analytical foundations for the study of climate science, geomorphological science, claims of climatological transformations caused by social productive consumption activities including actions with the geomorphological space over the space of problem-solution dualities in the epistemological space through neuro-decision-choice actions over the space of imagination-reality dualities. Natural variety transformations through natural productions are induced by natural technologies and processes, while social variety transformations through social productions are induced by social technologies which are learned from those of nature. The epistemological space is non-existence without the cognitive agents; the cognitive agents are non-existence without the ontological space.

The debate on the development of appropriate *environmental civilization* must not center around whether epistemological actions affect ontological processes and outcomes or not, because they do through the principles of relational interdependencies, connectedness, and universal oneness. The debate and the cognitive examination must center around the degree and direction of the effects since every variety outcome presents itself as a cost-benefit duality in the space of imagination-reality dualities under the *Asantrofi-anoma principle,* where one cannot enjoy the benefits of transformational outcomes without paying for the costs of the transformational outcomes over the spaces of uncertainty-certainty dualities and doubt-surety dualities constrained by the information-knowledge accumulation and cognitive capacity limitations in knowing and the acceptance of knowledge. Every epistemological action on the social production for benefits has ontological costs that require ontological restructuring. Over the ontological space, ontological varieties transform themselves over the space of quantity-quality dualities to maintain the ontological existence, ecological diversities, the bio-stability and the matter-energy equilibrium under the principle of destruction-replacement dynamics in the space of transformation-substitution dualities, where there are continual natural activities over the spaces of ontological production-consumption dualities and input-output dualities for a continual variety restructuring. The manner and efficiency of the ontological restructuring will depend on how social productive activities take place within the space of ontological-epistemological dualities as well as natural dualistic-polar conditions and their interactions with social dualistic-polar conditions.

In relations to the current debate on the natural environment and human actions, one must think of the ontological time of formations of fossil fuels and epistemological time of their destruction in social use as energy generation. The ontological processes take place as variety transformations in the space of input-output dualities without the space of cost-benefit dualities, where every variety input is transformed into variety output which is then transformed into variety input without a remainder as specified over the spaces of internalities and absoluteness to maintain the matter-energy equilibrium and the planetary stability, where all ontological input-output processes are internally recycled in perfect states without ontological externalities, where everything produced becomes an input into the ontological processes. The case of the epistemological processes is different. The epistemological processes under the epistemological time are defined over the spaces of input-output dualities, cost-benefit dualities, and internality-externality dualities in imperfect states, where the outcomes are established in the space of relativity to maintain the states of information-knowledge and ignorance-knowledge disequilibrium processes. The epistemological internalities form part of the epistemological input-output processes, where certain costs are excluded as external and not accounted in the cost-benefit decisions by cognitive agents. The epistemological variety externalities form part of the ontological input-output processes to maintain the ontological matter-energy equilibrium, where the epistemological externalities become ontological input for the production of ontological outputs in ontological time.

Importantly and cognitively, we always think of the value of the benefits of variety problem transformations into variety solutions not as approximation, but as exactness, where the costs of transformation are fully conceptualized to include the environmental costs. The human cognitive capacity limitations are such that the degrees of serious environmental impacts or non-serious impacts, in terms of climatological conditions, geomorphological changes, biodiversity and many other conditions are such that we must invoke the Gödel incompleteness theorem in using either the classical laws of thought or the fuzzy laws of thought and the Turing's conditions of the non-computability and the assigned probability of fifty-fifty to the outcomes of serious environmental impacts and non-serious environmental impacts. The fifty-fifty condition in the spaces of ignorance-knowledge dualities and uncertainty-certainty dualities is supported by the fuzzy optimization of dualistic-polar conflict resolutions in dualities and polarities, where the duals are mutually constraining as well as the poles are mutually constraining under a fuzzy-stochastic information structure in the games of variety transformations.

The optimal values of the fuzzy optimization of the dualistic-polar membership functions of mutual constraints lead to a point defining the principle of indifference or the principle of insufficient reasoning that separates dualistic-polar zones with a (50-50) dualistic-polar conditionality as well as indicate the line of the transversality conditions between the dualistic-polar zones where the acceptance of any of the zones is subjectively defined. Attached to the degrees of serious environmental changes are the conditions of irreversibility of neuro-decision-choice actions on serious environmental changes when they occur. Human neuro-decision-choice actions cannot completely reverse environmental damages through restoration or remediation in the zone of irreversibility. Human actions can stop the process of the environmental damages through the neuro-decision-choice actions over the space of problem-solution dualities with the techniques and methods used as well as limiting the desire for increasing varieties in the space of production-consumption dualities under the principle of non-satiation. The existence of irreversible outcomes in the neuro-decision-choice space points to the principle judicious subjectivity in human actions.

The relational interdependencies, continuum and unity of ontological varieties are such that our logical systems from laws of thought are seriously defective to capture the conditions of ontological complexities within the space of the ontological-epistemological dualities and the social cost-benefit system of production impacts on our socio-natural environment. There is no space of natural cost-benefit dualities in natural transformation processes, where nature transforms in accord with the natural laws of which there are no social understanding. The point consistent with the natural dualistic-polar conditions is that natural variety transformations take place on the principle of continual creations within the universal unity with continual variety differentiations with interdependence in the spaces of destruction-replacement dualities and input-output dualities under perfect natural information structure, where the natural laws are derived, in the last analysis, from the principle of opposites that generates energy and power for work .

In the information-knowledge support for the understanding and neuro-decision-actions in the space of imagination-reality dualities, substantial cognitive time is spent on *phantom problems* on degrees of social production impacts on the environment and the space of climatological-geomorphological dualities without the introduction of real cost-benefit conditions in social productions and consumptions, where social productions and consumptions are varieties, within the ontological productions with complexities of relational continua, unity and give-and-take sharing modes for mutual existence and destruction within the space of primary-derived dualities. Additionally, we do not discuss the impact of war and armaments on the environment including water, soil, vegetations and air that are essential supports of life. The notion that there are market solutions to the environmental problems is a phantom problem over the spaces of input-output dualities and cost-benefit dualities relative to the space of ontological-epistemological time dualities. The environmental deterioration is not caused by a stock of problems that can be solved one after the other to clean the toxic effects and have a good environment. It is caused by consistent supply of toxics from the neuro-decision-choice activities over the spaces of social production-consumption dualities and epistemological input-output dualities in static-dynamic processes. The solutions lie in the understanding of formations of hydrocarbons and fossil fuels over the space of information-knowledge dualities. The hydrocarbons and fossil fuels are buried under the earth with natural processes and technologies which have taken millions of years for their formations in the space of universal beginning-end duality, the outcomes of which contain different carbonic forms. The hydrocarbons and fossil fuels are ontological outputs of certain surface ontological output productions that become inputs into certain ontological storage processes to the protection of life and ecological balances and stability.

In this respect, a question arises as to what the neuro-decision-choice problem with the general environment and its dynamics is. The follow-up question is that can a solution be found to this problem if correctly identified and is this problem a phantom one? The other follow-up question is, can a solution be found to the problem through an epistemic process and technology in the space of input-output dualities to recycle the carbon externalities as inputs into the space of ontological-epistemological dualities or its resemblance within the neuro-decision-choice space in which the space of problem-solution dualities finds expressions in the space of relativity. Is the problem of finding neuro-decision-choice solutions to the problem of environmental decay from the interactive spaces of production-consumption dualities, input-output dualities, and cost-benefit dualities a phantom one over the space of ontological-epistemological dualities? The meaninglessness, phantomness, and integrity of all questions and answers on the environment decay and climatological changes must be related to the interrelated give-and-take sharing modes of input-output dualities, production-consumption dualities cost-benefit dualities, internality-externality dualities and neuro-decision-choice reversible-irreversible dualities over the space of ontological-epistemological dualities. The space of ontological-epistemological dualities contains the space of ontic-epistemic characteristic dispositions which contains the variety characteristic disposition that define the spaces of information-knowledge dualities and ignorance-knowledge dualities. The space of reversible-irreversible dualities relative to the spaces of gradual-abrupt dualities, digital-analogue dualities, and discrete-continuous dualities over the space of fuzzy decisions and actions offers epistemic directions to explain the tipping points for the analysis of climatological dynamics and the varieties in the space of controllability-non-controllability dualities. The conditions of tipping points must be related to the transversality conditions in the general epistemic analytics, as they relate to the epistemic journeys over the space of identification-transformation dualities relative to the space of quality-quantity dualities.

3.2: THE PRIMARY PRINCIPLE OF DIVERSITY AND UNITY OF KNOWING AND SCIENCE

The previous section was concerned with reflective conditions on the phenomena of diversity and unity. The reflective conditions of the diversity and the unity led us to examine the general connectivity of diversity and unity in knowing and science by stating the principle of unity in knowing and science. The focus here is to state the dualistic-polar conditions that help to define and establish the diversity and the unity in knowing, sciences and non-sciences as well as the nature of the universal information-knowledge system as the system of socio-economic productions and their developmental cognitive capacity limitations, and how these cognitive capacity limitations may be reduced by social actions. The dualistic-polar conditions are seen in terms of the conditions of primary-derived dualities in the relational continuum and unity of the diversity and the oneness of human efforts as collective neuro-decision-choice actions over the space of problem-solution dualities as defined over the epistemic and characteristic-signal fields. This is a non-traditional approach to understand the theory of knowledge through the theory of knowing as an intellectual economic production under the input-output processes, where the quality control of the traditional approach to the theory of knowledge must be seen in relation to the neuro-decision-choice supply-demand chain and input-output dynamics of social transformations changing the relationships between ignorance and knowledge. The conditions are stated as proposition.

PROPOSITION 3.2.1: The Primary Principle of the Unity of Knowing and the Unity of Science

The set of problem-solution dualistic-polar conditions of identifications and transformations of varieties defines *the primary principle of unity of knowing* and *the primary principle of the unity of science and non-science*, where every category of knowing has a set of problem-solution dualities and a corresponding set of question-answer dualities with inter categorial and intra-categorial give-and-take sharing relationships between the duals of a duality in the space of dualities, between the poles of a polarity in the space of polarities, among variety dualities, among polarities and among dualities and polarities in the process of knowing with continual relational give-and-take sharing modes over the epistemic field. The conditions of categories of problem-solution dualities constitute *the primary principles of diversity of knowing.*

NOTE 3.2.1: The Primary Principle of the Unity of Knowing and Science

All areas of knowing are about variety problem-solving and all areas of problem solving are about variety knowing under neuro-decision-choice actions to destroy the variety problems and establish a continual search for variety solutions in their replacements. The destruction-replacement processes are also information-knowledge production processes in the space of input-output dualities under a system of demand-supply chains in disequilibrium dynamics, where there are always the primary conditions and derived conditions. Similarly, all areas of teaching must be about the teaching of the development of skill set of thinking, and not simply about the teaching of the accumulated stock of information-knowledge stocks over the space of categorial problem-solution dualities. All areas of learning must be about the acquisition of skill set of thinking and not simply about learning the accumulated information-knowledge stock over the space of problem-solution dualities. The skill set of thinking helps in the development of further thinking skills for use as well as help in the utilization of the skill set for efficient actions over the space of problem-solution dualities with creativity and innovation through the system of neuro-decision-choice actions. The skill set of thinking influences the character of social philosophical consciencism which in turn is influenced by the contents of the social philosophical consciencism which is a product of neuro-decision-choice actions from the space of individual-collective dualities with conflicts and conflict resolutions, errors and error corrections, and tensions and changes over the variety space.

At the level of knowing, the use of books and accumulated information-knowledge structures in transmissions of cognitive skills establishes the initial conditions for the learners' developmental paths into the cognitive journey over the space of problem-solution dualities and polarities. The use of books is about the use of part of the stock of knowledge skills as have been generated by thinking for reasoning to assist neuro-decision-choice actions over the space of problem-solution dualities, where it is useful to note the conceptual unity and separation between thinking and reasoning and thinking skills and reasoning skills in the space of neuro-decision-choice actions over the space of problem-solution dualities. The thinking-skill toolbox is a collection of cognitive abilities for the development of framework of algorithms and theories for use by neuro-decision-choice agents for acquisition and processing of the relevant information-knowledge system on cost-benefit dualities of actual-potential varieties and the development of optimal neuro-decision-choice actions for the understanding of the imagination-reality replacement processes within the space of actual-potential dualities involving cognitive dynamics over the space of problem-solution dualities with an interdependence of thinking skill set and reasoning skill set as the only instruments for conceptualizing variety images over the space of variety imagination-reality dualities.

The reasoning skill set is the collections of abilities for the use of the thinking skill set to transform variety imagination to variety reality as well as transform variety reality to variety potential through variety imagination. The thinking skill set is the primary category of instruments of knowing and actions, while the reasoning skill set is the derived category of knowing and actions. The thinking skills are algorithms, the developments of which are independent of the users in that they may be used by any neuro-decision-choice agent in reasoning over the space of problem-solution dualities. An effective teaching in the space of knowing, therefore, must expose the learner to the concept of variety problem in relation to the concept of variety solution and their primary-derived processes, where every derived variety has a supporting primary variety which is also a derived variety from a primary variety in the past-present-future telescopic processes. Every variety imagination in the space of potentials is said to be a solution relative to a variety *reality problem* in the space of actuals if the potential benefit disposition of the variety solution dominates the reality cost disposition relative to the real cost-benefit conditions of the actual variety problem. In this respect, questions may always be asked: What are the duties of a teacher or supervisor of learning; what is the role of a learner; and what is the role of researcher in the whole enterprise of knowing and the information-knowledge production?

Every problem resides in the space of reality as a sub-space of actuals, but not every variety in the space of realities is a problem relative to the elements in the goal-objective space under individual and collective preference orderings over the space of varieties. Every potential variety is a solution to some variety problems through the space of imaginations and every variety solution is an imagination to be actualized through the space of realities relative to an element in the space of goals, objectives, and visions. Every actual variety in the space of realities has built-in potential characteristics to become a variety problem under dualistic-polar conditions relative to goal-objective varieties and information-knowledge requirements for creating increasing variety benefits relative to variety costs. The dynamics of the solution process are to always set the imaginative potential varieties considered as solutions against actual varieties considered as problems in the space of realities. Over the space of knowing and the information-capital accumulation, the dynamics of solution process are always to set a variety knowledge against a variety ignorance as one journeys through the spaces of uncertainty-certainty, doubt-surety, and decidability-undesirability dualities.

The identifications of variety problems and variety solutions are the problem-solution processes under neuro-decision-choice actions relative to individual and collective preferences. The problems require cognitive constructs of a *categorial convertors* that alter the terms of problem-solution relations of actual-potential varieties through the imagination-reality dualities irrespective of the areas of knowing under the real cost-benefit rationality. It must be kept in mind that there are inseparable and intimate relations among problem-solution dualities under a system of rules of give-and-take-sharing principle of real cost-benefit dualities and the structure of individual-collective dualities that begins with oneness, separates into diversities and combines into a relational unity to shape the paths and the directions of the general variety knowing, the selection of different problem varieties under cognitive actions to define the variety solutions over the path of enveloping of the information-knowledge structure and decision-choice activities that shapes the directions of human progress in the space of negative-positive dualities. It is important to keep cognizance of the powerful analytical roles of the concepts of cost-benefit duality, *asantrofi-anoma* principle and the cost-benefit rationality in all areas of knowing whether it is science or non-science, where every neuro-decision-choice action is guided by the conditions of the *asantrofi-anoma* principle and rationality over the space of identification-transformation dualities, where there is always a primary category and a derived category within the system of dualities and polarities to effect the paths of connectivity to knowing and the information-knowledge production over the space of problem-solution dualities that will lead to the examination of the development of methods and techniques in thinking and reasoning.

3.2.1 The Categorial Connectivity Analytics in Knowing and Science

The neuro-decision-choice activities over the space of problem-solution dualities or the problem-solving processes are not limited to one area of knowing whether the area is called science, engineering, non-science or planning under conditions of human existence, where the ultimate values of knowing and information-knowledge accumulations are their inputs into the neuro-decision-choice actions which then generate outputs to serve as inputs for further human actions over the space of input-output dualities. At the level of information-knowledge development in support of neuro-decision-choice actions, each area of knowing is a set of categorial problem-solution varieties that is of cognitive construct based on characteristic dispositions for study and understanding to abstract knowledge by acquaintance and knowledge by description that correspond to a specific category of knowing. The characteristic dispositions of variety problems and variety solutions present opportunities of enrichment of knowing and enhancement of efficiencies of the social management of the dynamics of the space of *ignorance-*

knowledge dualities as it relates to the space of socio-natural dualities through cognitive actions over the space of problem-solution dualities.

The whole development of this monograph and my views on knowing and the information-knowledge process of human existence, their growth and developments are based on the fundamental notion that the meaning of life finds expressions in the space of problem-solution processes and the essence of life finds expressions in the manner in which problem-solution dualities are inter-substituted in the space of social dynamics under dualistic-polar conditions, where all things exist in dualities and polarities. In this respect and in the universal space of dualities and polarities, the acts of thinking and reasoning are social balancing actions between the forces of opposites in dualities and in polarities over the space of individual-collective dualities. The balancing of the thinking-reasoning actions takes place in dualistic-polar game space. The universal system of knowing is such that successful neuro-decision-choice outcomes of problem-solution activities in one area of knowing increase the possibility and probability of successes as *knowing spillover effects* in other areas of the spaces of problem-solution dualities, knowing and information-knowledge developments through the space of imagination-reality dualities, where the information-knowledge stocks are transformed into flows of intellectual capital-service inputs into further knowing activities. The *knowing spillover effects* are always in complexities of intra-categorial and inter-categorial variety relationality that unifies all areas of knowing with give-and-take sharing modes over the space of problem-solution dualities contained in the space of imagination-reality dualities which include science, non-science, social science, and non-social science in the space of justified-non-justified dualities.

In this respect, there are intra-categorial diversity and unity and inter-categorial diversity and unity in the space of knowing and the space of information-knowledge disequilibrium processes. Under this analytical structure, the inter-categorial and inter-categorial diversities and the unities of knowing and knowledge may be seen as a dynamic special fruit-cocktail tree in complexity with a permanent organic stem, many branches, many different leaves, many different fruits, and many different seeds with capacities for growth and development representing the space of actual-potential dualities. The organic stem is the general knowledge, the roots are categorial convertors or functors for categorial transformations, the branches are the categories of information-knowledge areas, the fruits are the categories of outputs of knowing and information-knowledge processes, the seeds are inputs into the growth of knowing and the information-knowledge accumulation, where the discoveries of the contents of the tree are defined by the space of problem-solution dualities. The neuro-decision-choice actions are to discover the hidden roots, branches, leaves fruits and seeds to update the information-knowledge stocks to expand the horizon of search processes for the elements of the actual-potential roots, branches, leaves, fruits, and seeds as the fruit-cocktail tree grows in the space of actual-potential dualities. The discovery processes of actual-potential roots, branches, leaves, fruits, and seeds are presented as problem-solution dualities and question-answer dualities for the actions of neuro-decision choice agents.

It is the *knowing spillover effects* from the dynamics of the space of problem-solution dualities mapped onto the space of question-answer dualities that cognitively unite all areas of knowing as interdependent system of categories, especially social and non-social sciences in give-and-take sharing modes to ensure the interactive processes of the categorial and organic growths of the information-knowledge stocks and the discovery of new elements of needed toolboxes. The discovery of new methods and techniques provides pathways for the discovery of other methods and techniques for increasing knowing to maintain the stock-flow conditions of information-knowledge disequilibrium dynamic effects that provide continual processes of neuro-decision-choice actions over the space of problem-solution dualities in any socio-natural setting contained in the space of ontological-epistemological dualities. The existence of problem-solution dualities engulf all areas of knowing and are distinguished by the differences in characteristic dispositions that point to differential needs of techniques and methods required for the understanding of other cognitive activities over the spaces of the problem-solution dualities and the question-answer dualities under the control of neuro-decision-choice actions and information-knowledge developments in social sciences for efficient organization, management and control of the space of neuro-decision-choice activities in the information-knowledge search under the principles of variety input-output dualities and variety identification-transformation dualities in the space of destruction-replacement dualities under the accumulated information-knowledge conditions from social sciences and their connectivity to non-social sciences. The most important space of knowing and development of information-knowledge stocks is the space of problem-solution dualities, where necessity and freedom find expressions and over the space of question-answer dualities where curiosity and creativity find expressions and meanings.

It is this important idea of the role that social science plays in its own development and in relation to the development of non-social sciences and all areas of knowing that led to the development and discussions in the monograph [R3.26] [R17.15] in that the progress of knowing whether in science, non-science, social science, non-social science, engineering, non-engineering and input-output science depends on the progress of knowing efficiency of social sciences, the institutions that they help to create and their effects on the evolution, command, management and decision-control systems in all areas of knowing. The progress of knowing as the production of intellectual investment flows and the information-knowledge accumulation as the production of intellectual capital stocks depends on the social sciences in organizing and understanding of human behaviors that demand the needs in institutional casting for the development of other areas of knowing over the space of problem-solution dualities and allocations in the space of resources. The dependency of other areas of knowing and neuro-decision-choice actions over the spaces of problem-solution dualities and question-answer dualities is usually taking for granted. We speak of the progress of science without speaking of the roles of social science, economics, politics and legal conditions that make scientific progress or increasing knowledge accumulation or successes and failures in neuro-decision-choice actions possible over the space of imagination-reality dualities, where techniques and methods of reasoning impose demands on thinking and reasoning which impose demands on the spaces of question-answer dualities and ignorance-knowledge dualities. . Let us examine the roles of techniques and methods of thinking and reasoning relative to diversities and unities of knowing and science under the principle of connectivity through categorial give-and-take-sharing modes.

3.2.2: Techniques, Methods, Diversity and Unity in Knowing and Science

The developments of techniques and methods for working in different areas of knowing are in relational continua in uses to push forward the progress of knowing and science. No area of knowing, whether called science or non-science, can be claimed to be superior in the space of knowing. All areas of knowing present themselves in a *horizontal degree of knowing relevance* and not in *vertical degrees of knowing importance* due to the dualistic-polar principles of interdependency in the space of cost-benefit dualities. In other words, by the principle of interdependency in the space of give-and-take-sharing modes, the areas of knowing are in horizontal integration and unity in dynamics over the space of the problem-solution dualities. The general epistemic process is such that there are intra-area vertical integrations and intra-area horizontal integrations over the area of problem-solution dualities in operational continua and unity to effect the general give-and-take sharing principle required for feedbacks and self-correcting processes to maintain the social information-knowledge disequilibrium dynamics.

Every area of knowing resides in the space of relativity to affirm relational continua and unity of diversities. Every relational knowing does not reside in the space of absoluteness to justify relational separation and disunity of diversities in knowing and science. No important progress can be made in the information-knowledge stock-flow disequilibrium dynamics without the information-knowledge structure of social sciences relative to the efficiencies of the organization of neuro-decision-choice actions which affects all areas of human activities over the spaces of unity-disunity and oneness-dividedness dualities. It is here that organizational decomposition and oneness analytics of neuro-decision-choice actions become critically important in the space of problem-solution dualities in all areas of knowing and information-knowledge production relative to methods and techniques of acquaintance to construct an experiential information or the knowledge by acquaintance over the acquaintance space relative to the development of the space of paradigms of thought and its utilization in thinking and reasoning within the neuro-decision-choice processes, where each dualistic-polar category may require a different set of techniques and methods to deal with the specificities od the problem-solution contents and the question-answer situations.

3.2.3: Dualistic-Polar Conditions, Techniques, and Methods for Thinking and Reasoning

In the state of dualistic-polar conditions, every polarity $(\mathbf{p} \in \mathbb{P})$ is made up of *problem pole* $(\mathbf{p}_{\mathfrak{p}})$ and a *solution pole* $(\mathbf{p}_{\mathfrak{s}})$ in the form $\left(\mathbf{p}_{\lambda} = \left\{\left(\mathbf{p}_{\mathfrak{p}} \bullet \mathbf{p}_{\mathfrak{s}}\right) \mid \mathfrak{p} \in \mathfrak{P}, \mathfrak{s} \in \mathfrak{S}, \lambda \in \mathfrak{R}\right\}\right)$, where the space of the problem-solution polarities may be defined as $\left(\mathbf{p} \in \mathbf{P} = (\mathfrak{P} \otimes \mathfrak{S})\right)$ .The problem pole $\left(\mathbf{p}_{\mathfrak{p}}\right)$ has a residing problem-solution duality, $\lambda \in \mathfrak{R}$, where the problem disposition is greater than the solution disposition $(\mathfrak{p} > \mathfrak{s})$ in the sense that the real cost disposition is greater than the real benefit disposition $(C > B)$ for the problem-solution duality in the problem pole. Similarly, the solution pole $(\mathbf{p}_{\mathfrak{s}})$ has a residing problem-solution duality $(\lambda \in \mathfrak{R})$, where the solution disposition is greater than the problem disposition $(\mathfrak{s} > \mathfrak{p})$ in the sense that the real benefit disposition is greater than the real cost disposition $(B>C)$ with $\left[\lambda \in \mathfrak{R} = \left\{\lambda = (\mathfrak{s},\mathfrak{p}) \mid \mathfrak{s} \in \mathfrak{S}_{(BUC)}, \mathfrak{p} \in \mathfrak{P}_{(BUC)} \text{ with} \mathfrak{R} = \left(\left(\mathfrak{S}_{(BUC)}\right) \otimes \left(\mathfrak{P}_{(BUC)}\right)\right)\right\}\right]$ as the Cartesian product of the solution space $\left(\mathfrak{S}_{(BUC)}\right)$ and the problem space $\mathfrak{P}_{(BUC)}$.

The implication here is that every variety simultaneously contains benefit disposition and cost disposition in the same space and time. This dualistic-polar variety conditions form the foundations of the theory of cost-benefit analysis as the general theory of decision-choice systems, where costs and benefits are mutually constraining in all optimizations of decision-choice problems such that every objective in an optimization is a benefit (cost) constrained by a cost (benefit). The multi-criterial decision-choice problem is a multi-benefit (multi-cost) decision-choice problem, where the system of constraints is a system of costs (benefits) in the neuro-decision-choice processes [R15.12] [R5.14] [R17.13] [R17.14]. All neuro-decision-choice analyses are based on dualistic-polar conditions of the general principle of opposites, where the duals of dualities are mutually constraining, the poles of polarities are mutually constraining and the dualities and polarities of the space of opposites are mutually constraining in all representations of neuro-decision-choice problems in the space of mathematic-non-mathematic dualities. This principle of opposites is not made explicit in all studies of decision theories and theories of optimization [R6.3] [R6.8].

The concepts of variety problem and variety solution are defined in terms of cost-benefit relations as assessed by cognitive agents under the principle of opposites in relation to preferences over the space of relativity relative to a variety in the space of goals, objectives, and visions. Every variety exists as a union of a cost disposition that constitutes the cost dual of the internal duality and a benefit disposition that constitutes the benefit dual of the internal duality. The real cost-benefit relations are induced by cognitive preferences of the individuals for individual neuro-decision-choice actions and the cognitive preferences for collective neuro-decision-choice actions in the space of relativity, where comparative analytics can be performed on the cost and the benefit dispositions. The differences in individual preferences over the space of varieties induce differences in individual assessments of what constitute variety problems and variety solutions which are translated into the conflict zone of agreeably collective assessments of problem-solution dualities in the variety problem-solution of destruction-replacement-decision processes under the principle of opposites in all areas of knowing due to differential real cost-benefit assessments of variety problems and variety solutions. It is this conflict zone of problem-solution dynamics that generates the forces of self-corrections in the knowing system which is constrained by cognitive capacity limitations of cognitive agents over the space of imagination-reality dualities. It is also these variety dualistic-polar conflicts which create a system of self-corrections of the individual and collective neuro-decision-choice actions over the spaces of varieties and problem-solution dualities and polarities as new information-knowledge items are encountered in the space of cost-benefit dualities and affect the contents of the social philosophical consciencism which contains all justified-belief-unjustified-belief dualities, proven-unproven knowledge dualities, and other dualities involving misinformation, disinformation, lies, deceptions. propaganda and other dualistic-polar information-knowledge structures to affect the general conditions of neuro-decision-choice actions over the space of imaginations and to shape the elements in the space of realities.

The problem-solution analytics and question-answer analytics are to develop methods of thinking and strategies of reasoning to change the relational terms of problem-solution configuration of any problem-solution polarity in favor of solution such that the relational real cost-benefit condition of the actualized variety is of the form, where the real benefit dominates the real cost in relational terms $\left(B > C \mid \frac{B-C}{B} > 0, \text{ or } \frac{B-C}{C} > 0 \right)$ as the neuro-decision-choice action demands, where the cost-benefit units are to defined [R5,13] [R5.14]. The actualized variety is said to be *unintended* if the real benefit of the actualized variety solution is less than the real benefit of the problem variety $\left(B < C \mid \frac{B-C}{B} < 0, \text{ or } \frac{B-C}{C} < 0 \right)$. Unintended consequences are not neuro-decision-choice paradoxes. They are external outputs which are explainable by the conditions of cognitive capacity limitations about correct construct of the variety real cost information and the variety real benefit information in the space of real cost-benefit dualities, where positive-negative externalities are not accounted for in the cost-benefit information as an input into neuro-decision-choice process.

All the dualistic-polar conditions of the space of opposites in understanding the static problem-solution existence of identifications and the dynamic problem-solution existence of transformations of varieties are not unique to one area of knowing. They are the uniting forces of all areas of knowing structured in the identification of *what there was, what there is* and *what there would be* in the backward-forward telescopic opposites in dualistic-polar structures which define the continual production and the disequilibrium dynamics of information-knowledge structures as inputs for continual neuro-decision-choice actions , where there are the relational interdependencies of give-and-take-sharing modes of all areas of knowing, and where any solution generates new problem-solution dualistic-polar structures and new information-knowledge outputs.

These conditions of the space of problem-solution dualities and polarities with the corresponding real cost-benefit implications of neuro-decision-choice actions constitute the *primary category of principles of diversity and unity of knowing* which is supported by a *derived category of the principles of diversity and unity* of knowing for completeness. The negative-positive variety dualities are defined in the socio-natural existence. The cost-benefit variety dualities are defined in the social existence. In the social state, the negative actual dual is $\left(\mathbb{X}_{v_{\mathfrak{a}}}^{N\mathfrak{a}}\right)$ and $\left(\mathbb{X}_{v_{\mathfrak{a}}}^{P\mathfrak{a}}\right)$ is the positive actual dual, where $\left(\mathbb{X}_{v_{\mathfrak{a}}}^{\mathfrak{a}} = \mathbb{X}_{v_{\mathfrak{a}}}^{N\mathfrak{a}} \cup \mathbb{X}_{v_{\mathfrak{a}}}^{p\mathfrak{a}}\right)$ with $\left(\mathbb{X}_{v_{\mathfrak{a}}}^{N\mathfrak{a}} = \mathrm{B}_{v_{\mathfrak{a}}}^{\mathfrak{a}} \text{ or } \mathrm{C}_{v_{\mathfrak{a}}}^{\mathfrak{a}}\right)$ and $\left(\mathbb{X}_{v_{\mathfrak{a}}}^{P\mathfrak{a}} = \mathrm{C}_{v_{\mathfrak{a}}}^{\mathfrak{a}} \text{ or } \mathrm{B}_{v_{\mathfrak{a}}}^{\mathfrak{a}}\right)$, where $\left(\mathrm{B}_{v_{\mathfrak{a}}}^{\mathfrak{a}}\right)$ is the internal real benefit dual and $\left(\mathrm{C}_{v_{\mathfrak{a}}}^{\mathfrak{a}}\right)$ is the internal real cost dual abstracted in the space of imagination-reality duality, where the variety real problem-solution dualities are defined in the neuro-decision-choice processes. The $\left(\mathbb{X}_{v_{\mathfrak{a}}}^{\mathfrak{a}} = \left(\mathbb{X}_{v_{\mathfrak{a}}}^{N\mathfrak{a}} \cup \mathbb{X}_{v_{\mathfrak{a}}}^{p\mathfrak{a}}\right) = \left(\mathbb{X}_{v_{\mathfrak{a}}}^{N\mathfrak{a}} \otimes \mathbb{X}_{v_{\mathfrak{a}}}^{p\mathfrak{a}}\right)\right)$ is the actual negative-positive duality, where $\left(\mathbb{X}_{v_{\mathfrak{a}}}^{N\mathfrak{a}}\right)$ is the actual negative dual and $\left(\mathbb{X}_{v_{\mathfrak{a}}}^{p\mathfrak{a}}\right)$ is the actual positive dual. The epistemic point to keep in mind is that the negative dual may be either a real cost or a real benefit relative to the positive dual which may also be a real benefit or a real cost as assessed over the neuro-decision-choice space in relation to a goal-objective element or a set of goal-objective elements by cognitive agents and preference orderings that are held over the space of varieties as either the external benefit dual or the external cost dual that defines their cost-benefit rankings among varieties, where each variety has a residing internal cost-benefit duality which defines its internal identity and external cost-benefit duality which defines its external comparative relativity for decision-choice actions.

3.3 THE DERIVED PRINCIPLE OF DIVERSITY IN THE UNITY OF KNOWING AND SCIENCE

It is the differential distribution of the information conditions of the space of problem-solution dualities and polarities that establishes the unifying forces of all areas of knowing including science, non-science, exact sciences in the classical paradigmatic tradition and inexact sciences in the fuzzy paradigmatic tradition. Information is defined as characteristic-signal dispositions of varieties, where the characteristic dispositions are the content dispositions, and the signal dispositions are the message dispositions. The conditions of the *spaces of problem-solution dualities and polarities* constitute *the primary category of diversity and unity of knowing* without exemption of any area of knowing. The *primary category of diversity and unity* is amplified by the *derived category of diversity and unity of knowing*, where the spaces of problem-solution dualities and polarities present the conditions of static and dynamic variety information-knowledge processes for all areas of knowing. The derived set of conditions of the category of diversity and unity of knowing involves the conditions of common paradigmatic tools of epistemic processes of knowing in all areas of knowledge production where information constitutes the input.

In the development of the theory of knowing, we must consider the static conditions and the dynamic conditions of the knowing processes and what are the common factors that unite the search for knowledge of varieties and categorial varieties. In this respect, and among some important elements of knowledge acquisition, the theory of knowing is divided into the sub-theories of info-statics and info-dynamics [R17.17] [R17.18]. The sub-theory of info-static deals with variety *identification problem-solution dualities* to unravel the identity of *what there is* and *what there was*. The sub-theory of info-dynamics deals with variety *transformation problem-solution dualities* to unravel the identities of *what would be* relative to *what there is* and *what there was*. Each of the theories of info-statics and info-dynamics is further divided into *knowledge by acquaintance* and *knowledge by description*. It has been pointed out that the past-present-future time trinity of *Sankofa-anoma* tradition in the telescopic past, telescopic present and telescopic future involves forecasting from the known past information and discounting from the known future information to obtain information-knowledge inputs in additions to the current information for current decision in order to set a variety solution against a variety problem over the space of problem-solution dualities to maintain the information-decision-choice connectivity of the past-present-future structures involving past-present dualities, present-future dualities and past-future dualities of the knowing process and the information-knowledge disequilibrium dynamics, where the disequilibrium path is an enveloping of temporary variety transformation equilibria.

The knowing system involves all varieties with quantitative and qualitative characteristic dispositions in the universal system under the time trinity of the past-present-future configuration of the universal existence. The set of conditions of the time trinity links all areas of knowing, where every variety presents a different path of knowing journey through the space of problem-solution dualities of variety identifications and transformations on the basis of which variety information structures are abstracted, real and normal languages are constructed to initialize the dynamics of knowing, and variety discoveries are claimed with knowledge constructed, justified, falsified, verified and corroborated with characteristic dispositions from the acquaintance space through the spaces of true-false and acceptance-rejection dualities on the basis of paradigmatically collective rationality as a mode of quality control. The general conditions for defining *knowledge by acquaintance* and the construction of *knowledge by description* provide the framework to abstract the *derived category of the principle of diversity and unity* of knowing and the diversity and unity of science in support of the *primary category* of the principle of knowing and information-knowledge production.

The theory of knowing points to the idea that there is the knowledge by acquaintance and the knowledge by description. The knowledge by acquaintance and knowledge by description simultaneously involve the interactive conditions of problem-solution dualities of variety identifications in terms of info-statics and problem-solution dualities of variety transformations in terms of info-dynamics over the space of ignorance-knowledge dualities, where the theory of info-statics helps to identify the variety ignorance by establishing conditions of necessity and hence initializing required conditions for change while the theory of info-dynamics helps to initialize the conditions of freedom and hence initiate a process to set variety knowledge against ignorance. The necessity and initial conditions of knowing reside in the *categorial conversion* and the freedom and sufficiency of knowing reside in the *social philosophical consciencism* that has taken hold in a particular society within a defined generation time. In this respect, the theory of knowledge is concerned with the knowledge definition and quality control of items that enter the intellectual accumulation and information-knowledge stocks. The theory of knowing involves identification of variety identities and their transformations. The variety identifications and transformations include natural processes and technologies connected to natural variety changes.

The Primary category of the principle of opposites is defined by the conditions that all areas of knowing involve activities over the problem-solution dualities from which the derived category of diversity and unity of knowing may be abstracted by understanding the similarities and differences from the theory of info-statics and the theory of info-dynamics as the foundations of the theory of problem-solution dynamics in knowing and information-knowledge development, viewed in terms of variety problem-solution replacement dynamics at both static and dynamic states, where the union of the space of identification problem-solution dualities and the space of transformation problem-solution dualities constitutes the organic space of problem-solution dualities. In a simple statement, every problem-solution duality and every neuro-decision-choice action are about either *variety identification*, *variety transformation* or both to continually generate information-knowledge varieties for acceptance into the intellectual capital stocks. The information-knowledge acceptance process must be related to quality controls over the space of intellectual input-output dynamics for neuro-decision-choice processes under the *general theory of entropy* and the cultural dynamics, where the cultural dynamics, containing thinking and reasoning skills defined over the space of individual-collective dualities , must be related to the social Philosophical Consciencism that has taken hold of the society which then produces a set of systemic constraints for cognitive acceptance decisions which are valued in the space of dynamics of real cost-benefit dualities of varieties. In this respect, knowledge by description is a logical refinement of knowledge by acquaintance (experiential information) through *methodological constructionism* with a paradigm of thought and its thinking-reasoning toolbox. The corresponding knowledge by description is examined for acceptance through *methodological reductionism* with the same paradigm of thought and an appropriate toolbox in the space of methodological constructionism-reductionism dualities [R21.5], where the methodological constructionism and reductionism are supplemented by methodological nominalism through linguistic systems.

3.3.1 Similarities and Differences of the roles of Info-Static and Info-Dynamic Conditions in the Diversity and Unity of Knowing and Diversity and Unity of Science

Any problem of knowing presents info-static conditions of variety identification and any solution presents info-dynamic conditions for variety transformation, where the info-dynamic conditions presenting freedom and sufficiency conditions that reside in the info-static conditions which present necessity and initial conditions as well as reside in info-dynamics through the space of imagination-reality dualities as containing the actual-potential real cost-benefit dualities and mutual replacements. The concept and the phenomenon of knowing are defined in the epistemological space as a sub-space of the space of ontological-epistemological dualities. The concept and the phenomenon of transformation are defined in both the ontological space and the epistemological space as sub-spaces of the space of ontological-epistemological dualities, while the sub-spaces are defined by the principle of dualistic-polar dominance of characteristic dispositions under the general principle of opposites. The info-static conditions present the primary category, while the info-dynamic conditions present the derived category in the space of the problem-solution dualities relative to the spaces of imagination-reality and identification-transformation dualities, where the space of realities houses problems and the space of imaginations houses solutions under the general dualistic-polar principle in the space of varieties.

The structure of the epistemic field to deal with individual dualities and polarities within the principle of opposites in all variety identifications and variety transformations requires the use of the concepts of *primary-derived duality* and the space of variety primary-derived dualities in the dynamics of problem-solution polarities within the universal system of varieties in the space of knowing and information-knowledge production. The epistemic derivatives of secondary factors of diversity and unity of knowing must connect all areas of information-knowledge search over the epistemological space containing the space of acquaintance through the neuro-decision-choice actions over the space of imagination-reality dualities, the elements of which are the problem-solution varieties, where each variety exists as internal duality and constitutes as external dual and hence a pole of a polarity for transformation dynamics as illustrated in the monograph [R8.16]. The dualistic-polar approach is an important and powerful analytical method to explain changes and transformations within societies, and between societies, between imperialism and neocolonialism, and between incumbencies and the fringes in the study of oppression and freedom, (0-1) dualities, nothingness-somethingness dualities absoluteness-relativity dualities, new-old dualities, life-death dualities and other such dualities with information from the spaces of input-output dualities and construction-destruction dualities. Let us keep in mind that the meanings of concepts and phenomena make sense in the space of relativity with conditions of relational continua and unities under the principle of acceptance of contradictions, where opposites can exist in the same space and time, and not in the space of absoluteness with the conditions of relational separation and disunity under the principle of the excluded middle with non-acceptance of contradictions, where opposites cannot exist in the same space and time.

The derived conditions which constitute the secondary category of the principle of diversity and unity of knowing of sciences in the space of exact-inexact dualities must first be abstracted from the space of acquaintance that generates *experiential information structures* or knowledge by acquaintances consistent within particular areas of knowing and the space of *certainty-uncertainty dualities* with relational continua and unity in the process of knowing through the neuro-decision-choice activities within the space of variety problem-solution dualities and polarities defined over the space of imagination-reality dualities contained in the space of actual-potential dualities over the epistemological space. The experiential information structures and the certainty-uncertainty dualities link all areas of production of knowledge by description without an exception into a unified epistemic structure. All areas of knowing must account for elements of uncertainty, doubt, unsureness, anticipation, and expectations in relation to substitution-transformation processes within the ignorance-knowledge dualistic-polar space.

The experiential information structure is named as knowledge by acquaintance by Russell and in philosophical discussions of the theory of knowledge [R22.11] [R22.14] [R22.18] [R22.19]. In these discussions, there is a clear distinction between the theory of knowing and the theory of knowledge even though they are both examined in the space of imagination-reality dualities as a sequential inter supportive structures in the actions of knowing and information-knowledge production processes. The knowledge by acquaintance is what has been referred to in this research sequence as the experiential information structure or epistemological information as the first step and as the primary category in knowing and information-knowledge production processes. Each variety experiential information structure is an input into the development of knowledge by description about that variety. The methods of acquaintance will vary over matter-energy varieties and categories due to the differences in the distribution of characteristic dispositions of varieties. The methods of acquaintance are derived from the set of all senses over the space of primary-derived dualities which may be enhanced by other technological know-hows. No other type of knowledge can be produced without an experiential information (knowledge by acquaintance), where the acquaintance spaces, the methods and techniques of acquaintance are broadly defined to include the currently known and unknown boundaries and elements.

The nature of certainty-uncertainty dualities enhanced by the conditions of exact-inexact dualities and vagueness-clarity dualities was central in the Russell-Brouwer debate on mathematical representation of information and paradigmatic derivations of knowledge by description about all varieties [R3.6] [R3.13] [R3.15] [R3.86] [R3.88]. The same set of conditions of the nature of certainty-uncertainty duality amplified by the conditions of the nature of exact-inexact duality is also central in the current debate between *the classical paradigmatic thinking and reasoning* with the excluded middle, relational disunity with non-acceptance of contradiction as a truth value in the decision-choice space in all variety information-knowledge developments and the *fuzzy paradigmatic thinking and reasoning* with the relational continuum, unity, non-excluded middle and acceptance of contradictions as a truth values in the neuro-decision-choice space in all variety information-knowledge developments.

The classical paradigm of thought deals with the knowledge by acquaintance under stochastic and not under fuzzy dispositions while the fuzzy paradigm of thought deals with the knowledge by acquaintance with fuzzy-stochastic dispositions. The epistemic nature of the variety experiential information will vary over different areas of knowing since different areas of knowing are generated by different categorial varieties from different characteristic dispositions as projected by the different signal dispositions in the epistemological space, where the instruments of acquaintance may vary in the production of experiential information or knowledge by acquaintance that will serve as inputs into the production of knowledge by description. At the basic level of knowing, variety identities are examined by the theory of info-statics and at the level problem-solution dualities, the changes in relation and properties are examined by the theory of info-dynamics, where knowing and the information-knowledge production must pass through the spaces of exactness-inexactness dualities, uncertainty-certainty dualities, doubt-surety dualities, and decidability-undecidability dualities before entering into the space of further variety production of information-knowledge refinement over the space of error-correction dualities.

Given the space of certainty-uncertainty dualities as derived from the space of ontic-epistemic dualities, the variety experiential information is then processed by an acceptable *paradigm of thought* as an element in the derived category of diversity and unity of knowing and complemented by the *methodological constructionism-reductionism duality,* where every variety duality and every variety polarity exist in the relational continua and unity such that every duality contains its duals, every pole of a polarity houses a duality, every duality has an opposite duality and every polarity has an opposite polarity . From the epistemic process of problem-solution replacement activities, we state the conditions of the derived category of diversity and unity of knowing and diversity and unity of science. The primary category of the principle of diversity and unity of knowing is composed of cognitive activities over the *space of problem-solution dualities*. In other words, all areas of cognitive activities are firstly involved in neuro-decision-choice activities over the space of variety problem-solution dualities irrespective of the sizes and difficulties. The derived category of the principle of diversity and unity of knowing contains the common elements of the information-knowledge construction and the information-knowledge certainty in relation to acceptable paradigms of thought under the social Philosophical Consciencism that has taken hold on the society, where the evolving social philosophical Consciencism establishes the expanding constrained boundaries of effective thinking, reasoning, relations, and neuro-decision-choice actions. The process is that the theory of info-statics relates to the identification problem-solution dualities, the establishment of categories with necessity and initial conditions for categorial conversion and the theory of info-dynamics relates to the transformation problem-solution dualities with freedom and conditions of sufficiency in knowing and the information-knowledge production over the space of ignorance-knowledge dualities with categorial convertor and functor actions.

PROPOSITION 3.3.1.1: The Derived Category of the Principle of Diversity and Unity of Knowing and Unity of Science

The set of the derived conditions of the principle of diversity and unity of knowing is the derived category from the space of problem-solution dualities which is made up of the *experiential information structures,* the *certainty-uncertainty dualities, paradigms of thought* and the *methodological constructionism-reductionism dualities* which incorporate the test conditions of doubt-surety dualities, computable-incomputable dualities, and acceptance-unacceptance dualities, all of which are driven by neuro-decision-choice actions of cognitive agents with cognitive capacity limitations over the space of individual-collective dualities relative to the space of variety cost-benefit dualities.

3.3.2: Analytics of the Primary and Derived Categories of the Principles of the Diversity and Unity of Knowing and the Diversity and Unity of Science

All areas of knowing and information-knowledge production whether they are grouped as science or non-science, derive their existence, successes and failures in diversity and unity from the space of problem-solving activities as the primary category under neuro-decision-choice actions. These unified problem-solving activities, as a set of principal categories, are amplified by experiential information structures, conditions of inexactness-exactness, vagueness-clarity and certainty-uncertainty dualities, paradigms of thought and conditions of the methodological constructionism-reductionism duality for the derivations and verifications of the results in the continual interactive processes over the space of doubt-surety dualities under the conditions of the space of ignorance-knowledge dualities. The complexity of knowing within any variety problem-solution duality which induces an actual-potential variety replacement takes place through cognitive information-decision-choice interactive processes within the space of imagination-reality dualities relative to some elements in the space of goals, objectives and visions , where every human activity is seen in terms of a variety problem-solution duality under neuro-decision-choice actions, while replacements are effected through the dualistic-polar outcomes in the space of cost-benefit dualities under the principle of non-satiation in the cognitive existence over the space of imagination-reality dualities which is derived from the space of actual-potential dualities. Let us keep in mind that a variety problem is a variety solution relative to goal-objective elements as defined in relative cost-benefit dominance of characteristic dispositions as assessed by cognitive agents.

The common set of conditions of the variety problems from the problem space constitutes the *primary category of diversity and unity* of knowing and the diversity and unity of science. The common set of the conditions of the construct and the acceptance of the variety solutions from the solution space constitutes the *derived category of diversity and unity* of knowing and the diversity and unity of science. In other words, the diversity and unity of knowing, which is composed of science and non-science, over the space of exact-inexact dualities find expressions in the primary and the derived categories of cognitive activities over the space of problem-solution dualities. The *primary category of the diversity and unity of knowing* is about neuro-decision-choice activities over the *space of problem-solution dualities*, while the *derived category of diversity and unity of knowing* is about the space of paradigms of thought within the space of construction-reduction dualities in relation to the system of quality controls of information-knowledge claims. It must be noted that the development of paradigms of thought is shaped by the general conditions of the space of problem-solution dualities as identified by their corresponding characteristic-signal dispositions with continual influences from the social philosophical consciencism which also expands with the information-knowledge disequilibrium dynamics at every generation time depending on the nature of the inter-generational information-knowledge transmission processes. The social philosophical consciencism is always in disequilibrium dynamics incorporating new, proven, and unproven ideas, false and true claims, socially justified and unjustified norms, and all dimensions of the evolving culture including knowing and the information-knowledge system.

The epistemic structures of knowing and information-knowing production, through the space of opposites and the logic of the principle of opposites, are such that every pole of a polarity has a residing duality which is opposite to each other as well as generates dualistic-polar transformations to set the variety solution against the variety problem based on the cost-benefit rationality in the space of imagination-reality dualities. The analytical framework is such that there is the negative pole defined by a negative duality and there is the positive pole defined by a positive duality, all of which are in relational continuum and unity as defined by the residing characteristic dispositions. The negative pole has the residing negative duality while the positive pole has the residing positive duality as have been defined and discussed in the previous chapters of this monograph, where the negative polarity is one with dominating negative duality and the positive polarity is one with dominating positive duality under the principle of opposites with relational continua and unity.

The identity and character of very duality are established by the dominant dual in such a way that the negative dual establishes the negative duality if it dominates the positive dual and vice versa. Every dual of a duality also has a residing cost-benefit duality relative to an element in the goal-objective space which is a sub-space of the space of imaginations contained in the potential space. The epistemic structure of knowing must always be related to the space of problem-solution dualities contained in the space of imagination-reality dualities, where every problem is a variety that has a residing cost-benefit duality, and every solution is a variety that has a residing cost-benefit duality in relative differentials for a variety neuro-decision-choice identity, discrimination, and ranking. Analytically, the space of problem-solution dualities is a conformal mapping onto the space of question-answer dualities from the space of imagination-reality dualities, where the associated space of cost-benefit dualities activates neuro-decision-choice activities over the action space to bring about changes in variety relationships among variety imaginations and variety realities. At the space of imagination-reality dualities, there are conformal mappings among the space of problem-solution dualities, the question-answer dualities, cost-benefit dualities, and negative-solution dualities such that the mappings preserve the angular transformations of problem-solution, question-answer and cost-benefit enveloping curves in transformations in the neuro-decision-choice space, where the mapping is carried on by elements from the space categorial convertors or functors. The conformality of mapping in the space of categorial convertors or functors include orientation-reversing and non-orientation reversing mappings to allow for reversible and irreversible neuro-decision-choice actions.

It is the presence of irreversible decisions in the space of social actions that points to the possible existence of the problem phantomness to finding a solution to the reversibility of environmental decay, destruction of ecological balance, variety extension as generated by human actions in the space of ontological-epistemological dualities. Our known social technologies and processes are not sufficiently good to replace the natural elements in the space of natural input-output dualities, where the actions of natural technologies and processes take ontological time that resides in the fuzzy epistemological space. For example, our currently known social process-technological structures do not help in bringing dead people to life when we transform them from variety life to variety death in the space of social input-output dualities. Similarly, our known technologies and processes cannot stop earthquake and volcanoes from accruing. The cognitive capacity limitations point to the general concept and phenomena of the space of knowability-unknowability dualities which will be discussed in the monograph epistemic anomalies. Let us keep in mind that all epistemological negative externalities become inputs into ontological transformation decisions if such production-consumption externalities do not enter as inputs into the space of social input-output processes. In the space of ontological-epistemological duality, we have the space of the ontological input-output dualities as the primary category of transformations with supply-demand chain dynamics. Corresponding to it is the space of epistemological input-output dualities as the derived space of space of category of transformations with epistemological supply-demand chain dynamics. The space of ontological transformations has a corresponding space of perfect and efficient ontological processes and technologies as the primary instruments to effect the ontological transformations with information-knowledge equality. Similarly, the space of epistemological transformations has a corresponding space of imperfect and inefficient epistemological processes and technologies as the derived instruments to effect the epistemological transformations with information-knowledge inequality where such inequality gives us the direction to explain and understand anomalies in the space of thinking-reasoning dualities.

Every cost disposition is simultaneously a benefit disposition relative to an element in the goal-objective space defined in terms of opportunity cost. Similarly, every benefit disposition is simultaneously a cost disposition relative to an element in the goal-objective space. Every problem is a variety reality in the space of the actuals waiting to be replaced by a variety imagination in the space of variety potentials, and every variety solution is a variety imagination from the space of variety potentials waiting to replace a variety reality in the space of variety actuals. It is here that the economic concept of *opportunity cost* finds an analytical justification and significance in all areas of neuro-decision-choice activities including those over the space of ignorance-knowledge dualities of science and non-science, where the principle of opposites presents the general conditions for the understanding of the universal internal and external dynamics for continual variety transformations in the multiplicity of socio-natural existence, such as climatological and geomorphological varieties, and social developmental stages in the space of ontological-epistemological dualities.

In this respect, the variety reality problem-solution duality is the real opportunity cost relative to the variety imagination problem-solution dualities. Similarly, the variety actual problem-solution duality is the *real opportunity cost* for the variety potential problem-solution duality in terms the *asantrofi-anoma conditions* and *rationality* over the space of variety cost-benefit dualities and corresponding to the asantrofi-anoma problem-solution duality is the *asantrofi-anoma principle* as the general principle for providing variety ranking and choice under the universal cost-benefit conditions relative to the elements in the space of imagination-reality dualities. The *asantrofi-anoma* problem-solution dualities, as has been discussed in the previous chapters, states that cost and benefit dispositions simultaneously and inseparably reside in every variety over the space of imagination-reality dualities as contained in the space of actual-potential dualities such that a neuro-decision-choice action must relate to varieties in balancing the simultaneous existence of real cost disposition and real benefit disposition whether measurable or non-measurable or computable or non-computable, where in general one cannot select the real benefit disposition of a variety and leave the real cost disposition which is intimately attached to the real benefit under the dualistic-polar give-and-take sharing modes [R5.13[[R5.14][R8.16].

The asantrofi-anoma principle is the epistemic foundation of the theory of real cost-benefit analytics as the general approach to the understanding of neuro-decision-choice processes over the space of problem-solution dualities, where the neuro-decision-choice actions are continually influenced by the social philosophical consciencism, while the results of the neuro-decision-choice actions influence the contents of the philosophical consciencism at every generational times, where each generation has its social philosophical consciencism inherited from the previous generation as its primary philosophical category of support and as a primary social philosophical support for the next generation's social philosophical consciencism, and each generation has a set of problem-solution dualities within the general space of problem-solution dualities relative to the space of goals, objectives and visions [R17.15] [R17.16] [R17.17] [R17.18].

It is useful to note that in the problem-solution dynamics, every *derived variety* (a solution) has a *primary variety* (a problem) in the transformation space in terms of *offspring-parent path* analytics that allow the establishment of *telescopic backward-forward dualities* within one of either past-present duality or present-future duality over the space of problem-solution dualities, where the problem-solution connectivity is a duality that finds expressions in the relational continua and unity with variety give-and-take sharing modes. The relational continua and unity of dualities and polarities, as essential properties of the space of opposites and the principle of opposites, present an analytical situation, where the duals of any duality exist in relational *give-and-take-sharing* structures over the spaces of antagonistic-cooperation dualities, conflict-resolution dualities, and tension-change dualities without which the duals and the duality have no definable existence. Similarly, the poles of any polarity exist in relational *give-and-take-sharing* structures through their residing dualities without which the poles and the polarity have no definable existence, and where every pole of a polarity has a residing duality that generates the *polar internal energy* for variety transformations in terms of problem-solution relational dynamics. The relational structures of the varieties in diversity and unity are such that there is a relationally inseparable connectivity among polarities and dualities to maintain the general unity of existence which is then transformed into diversity and unity of knowing, and diversity and unity of science. It is always important to understand the interplays of dualities and polarities relative to the dynamics of the elements of the space of problem-solution dualities and ignorance-knowledge dualities as they relate to the diversities and unity of knowing and information-knowledge stock-flow processes, where knowing is the production of *intellectual investment flows* and information-knowledge accumulation is the production of *intellectual capital stocks* in disequilibrium dynamics on the infinite path to unknowing destination of information-knowledge stocks due to individual and collective cognitive capacity limitations. All the socio-natural transformation under the principle of opposite present themselves as dualistic-polar games with dualistic polar transformation for changes in *properties* and *relations* with the maintenance of the variety relational unity under different relational continua.

The field of human conditions for curiosity of knowing is infinite with infinite varieties, where human and non-human lives demand never-ending decision-choice activities over the space of problem-solution dualities which links all stages of knowing, production-consumption processes and input-output processes in inseparably complex interdependent modes, where identifications of variety problems relative to goal-objective elements and variety solutions to variety problems are information-knowledge productions that enhance the accumulated information-knowledge stocks as well as increase the conditions of possibility and probability of problem solving of other varieties from the *knowledge by acquaintance* to the *knowledge by description* under the methodological constructionism-reductionism duality, where the methodological constructionism involves the development of knowledge by description and methodological reductionism involves in quality control through the development of conditions of knowledge by description over the space of justified-unjustified dualities.

The diversities and the unity of knowing and the diversity and the unity of science consist of human cognitive activities over the *space of problem-solution dualities* under acquaintance with experiential information structure, The experiential information structure is connected to the certainty-uncertainty dualities with doubt-surety, exact-inexact paradigm of thought and the methodological constructionism-reductionism dualities and not directly with the methods and techniques. The methods and techniques of neuro decision-choice actions in an area of knowing are imposed by the nature of the variety characteristic dispositions from the space of acquaintance such that the set of tools for variety solutions may vary with inter-dependencies over all areas of knowing. The basic problem-solution structure in relation to certainty-uncertainty, doubt-surety, cost-benefit dualities, and others may be represented in an epistemic geometry as in Figure 3.3.2.1.

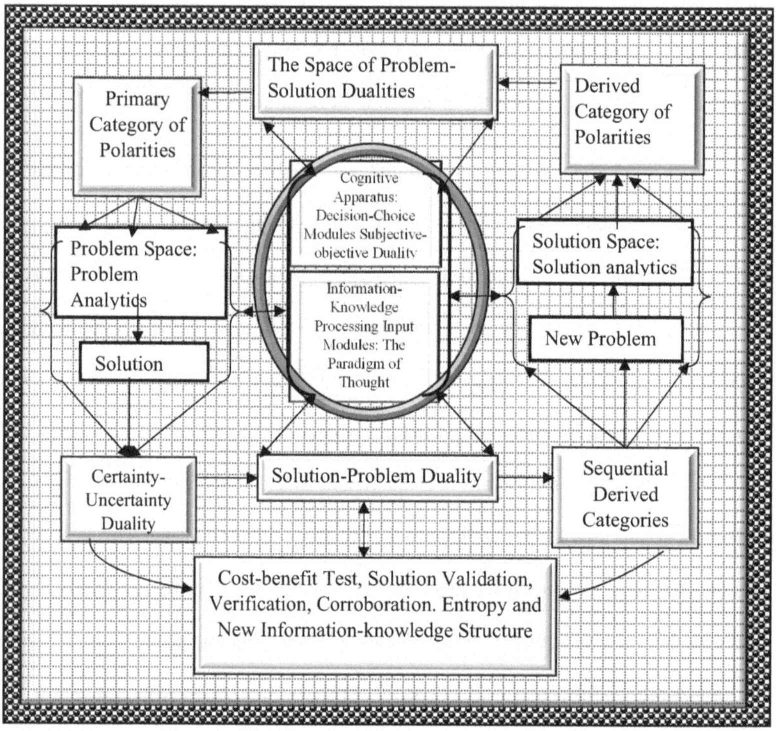

Figure 3.3.2.1: A Cognitive Geometry of Dynamics of Problem-Solution Duality within Certainty-Uncertainty Duality Under the Space of Sequential Primary-Derived Dualities with Real Cost-Benefit conditions and Validation of Variety Solution as defining the Diversity and Unity Principle of Knowing, Science and Non-science.

The essential elements of the cognitive geometry of the problem-solution dualities with interactive conditions of certainty-uncertainty dualities are that these dualities are defined in the epistemological space that accommodates all variety realities and variety imaginations in the areas of knowing as cognitive activities over the space of variety problem-solution dualities relative to the space of actual-potential dualities. The space of the problem-solution dualities is also an infinite collection of dynamic processes where one can speak of the general *problem space* and the general *solution space* with cognitive modules that establish a set of solution-search algorithms called *thinking* with a paradigm of thought and the use of these algorithms called *reasoning* to balance cost-benefit conditions in relations to the elements in the goal-objective space under a paradigm of thought for neuro-decision-choice ranking in the epistemic field with the relational and the information fields. The essential activities of thinking over the space of problem-solution dualities are to map variety solutions from space of imaginations as a sub-space of the space of potentials onto the variety problems in the space of realities as a sub-space of the space of the actuals through information-knowledge processes to resolve conflicts in preferences over varieties in the variety space under cost-benefit conditions and rationality over the general dualistic-game space.

It is through the conditions of thinking-reasoning process that the Newell-Simon works [R8.41][R8.42][R8.60][R12.47] on the human problem solving make an important epistemic and analytical contributions to the understanding of cognitive activities in the static and dynamic spaces of knowing and help the development of the theory of knowing under methodological constructionism-reductionism duality, where *knowledge by acquaintance* is obtained by *methodological experientialism* and recorded as the *observational data* from variety *signal dispositions* and crystalized as *fact* and taken as *evidence* in judgmental encounters, and the *knowledge by description* is derived by *methodological constructionism* from the observational data or the evidence by an approved and selected *paradigm of thought*, and information-knowledge certainties are verified, falsified or corroborated by *methodological reductionism* under the same paradigm of thought, working with statistics in the domain of general entropy [R8.41][R8.42][R8.60]. The knowing process with its output involves two separate but interdependent complexities and a system of sequential inputs of primary and derived varieties, and a system of sequential output processes of primary and derived nature. The complexity of knowing and the development of information-knowledge stock-flow disequilibrium may be geometrically presented as in Figure 3.3.2.2.

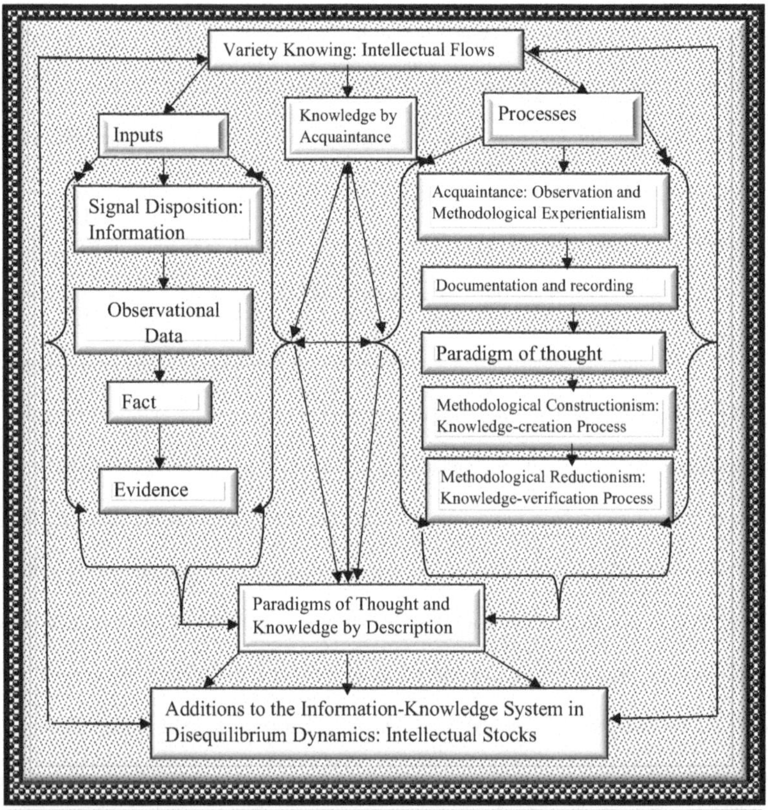

Figure 3.3.2.2 A Cognitive Geometry of Input-Process of Information-Knowledge Dynamics from the knowledge by acquaintance to knowledge by description under the Space of Methodological Construction-Reduction Dualities from Intellectual Investments to Intellectual Information-Knowledge Capital Accumulation in Closed and open loop over the space of Internal-External Dualities.

It must be understood that all algorithms as seen from mathematical logic as a member of the family of abstract languages (FAL) and non-mathematical logic as a member of the family of ordinary languages (FOL) are analytical tools to help neuro-decision-choice actions over the space of problem-solutions dualities that may pass through the spaces of uncertainty-certainty dualities doubt-surety dualities, acceptance-nonacceptance dualities in a continual interdependent complexity of diversity and unity in the space of imagination-reality dualities. This is how the different areas of mathematics as a language must be viewed as algorithmic representation of information-knowledge structures under nominalism and as computable systems to process and abstract knowledge as input into the neuro-decision-choice system. The mathematical representations are constrained representation of derived information-knowledge system to assist in processing conditions in the space of acquaintance-description dualities. These algorithms are only valid, generate paradoxes, incompleteness, and ill-posed presentations of problems within the paradigm of their creations. Every language is an accessory and facilitator to knowing and information-knowledge structures and not simply an information-communication structure within the space of source-destination dualities. The complexity of meaning of information and its relationships to data, fact, evidence and knowledge and language and mathematical representation has been discussed in the monograph of the theory of info-statics [R17.15][R17.17], however a short reflection in terms of diversity and unity of knowing and science may be useful in understanding some details of neuro-decision-choice actions over the space of problem-solution dualities as contained in the space of imagination-reality dualities as a sub-space of the space-actual-potential dualities. The space of imagination is a sub-space of the space of potentials; the space of realities is a sub-space of the space of actuals, the space of imagination-reality dualities is a subset of the space of actual-potential dualities which is contained by the space of ontological-epistemological dualities. The spaces of actual-potential and imagination-reality dualities are always in disequilibrium states within the matter-energy equilibrium states, where the space of actual-potential dualities is of nature and the space of imagination-reality dualities is of social.

3.3.3 Relational Complexity of Information, Data, Fact, Evidence and Knowledge in Decision-Choice Actions, Diversity and Unity of Knowing and Science.

The relationality of information, data, fact, evidence, knowledge, and evidential things is defined by cognitive agents over the space of imagination-reality dualities. Such a relationality is complex presenting these concepts in distinction and unity of thought. It is the complexity of the relationships among the concepts of information, data, fact, evidence, evidential things and knowledge as inputs in neuro-decision-choice actions that makes it difficult to establish diversity and unity of knowledge and science with the use of methods and techniques of information-knowledge search, where distinctions are not made among them and are interchangeably used in the (FOL) and (FAL) as they are used in communication over the source-destination dualities. This is the *input complexity* in the space of neuro-decision-choice activities over the action space. What is information and what are the conceptual relationships among information, data, fact, evidence, and knowledge, and which of these enter as inputs into the neuro-decision-choice processes and what are the outputs? This *inputs complexity* also leads to *output complexity*, in the sense of a question. Is the output of any neuro-decision-choice action an information, data, fact, evidence, or knowledge? Is information an output of input or both? There is another analytical complexity involving the way information inputs are transformed into knowledge outputs. This is the *processing complexity*. The input complexity is such that there is always confusion in verbal and written source-destination processes, where information is viewed as either data, fact, evidence, or knowledge. Information is neither data, fact, evidence nor knowledge even though they are intimately related and derivable from information as the primary category of input where data, fact, evidence, and knowledge are derived categories. This confusion creates further confusion as what is data analytics, big data analytics, information technology, science of evidence and statistical science, data science and others. Is big data analytics the same as big fact analytics, big evidence analytics and many others in the sequential structure within the space of primary-derived dualities?

The processing complexity amplifies the confusion in the input complexity such that some important theoretical and application difficulties arise in the use of any paradigm of thought such as the classical and fuzzy paradigms in the input-processing toward the information-knowledge stocks. Why there are confusions of these concepts in the family of ordinary languages, they cannot be accepted in critical analytical works in information-knowledge constructs to generate input-output structures for the space of neuro-decision-choice actions that affects the production of fundamental and applied information-knowledge systems. Here, one must keep in mind the effects of information-knowledge inputs on neuro-decision-choice actions that affect the backward-forward telescopic path of historical outcomes relative to progress in the space of negative-positive dualities under the Sankofa-anoma conditions of past-present-future connectivity. It is within these conceptual difficulties that defective information, misinformation, and disinformation may be weaponized by cognitive agents in communications over the space of source-destination dualities to affect the outcomes of neuro-decision-choice actions within the space of dualistic-polar transformation games as well as shape and distort the paths of the cultural history and civilization in the past-present-future telescopic processes.

The enveloping of the dynamics of the space of human problem-solution dualities is infinite that relates to the spaces of general identification problem-solution dualities and social transformation problem-solution dualities with relational continua and unity. Every identified problem and every solved problem are generations of information and possible contributions to knowledge and every contribution to knowledge is a contribution to ignorance reduction and every knowledge about new variety is a contribution to a decrease in ignorance. Every problem-solution duality is either a primary or a derived category in the human problem-solution enveloping processes just as the knowledge by acquaintance is primary category of knowing while every knowledge by description is a derived category of knowing irrespective of whether one is considering scientific or non-scientific groups of varieties. Every problem in the space of problem-solution dualities defines conditions of existence of reality couched in an indestructible information set, while its solution in the same duality destroys the existing variety but retains the information conditions, and simultaneously defines the conditions of variety transformations which then establish new conditions of reality couched in an indestructible information set to generate a new problem-solution reality. Each problem-solution duality is under forces of continual mutation to increase the number of problem-solution dualities over the infinite space. The evolving space of variety problem-solution dualities provides information-knowledge conditions for explaining the meaning and essence of life, where the meaning of life finds expressions in the neuro-decision-choice activities over the space of problem-solution dualities and the essence of life finds expressions in the manners in which variety solutions are abstracted from the dynamics of the solution space under dualistic-polar game conditions.

The solution to the initial *variety problem-solution duality* is transformed into a *variety solution-problem duality*, where the conditions of the new variety reality create a new problem-solution duality as a derived category in the problem-solution sequence. The initial problem-solution duality is the primary category in the transformation process of *parent-offspring process* of the information-knowledge accumulation in all disequilibrium dynamics, where infinity is the upper bound of the knowing process as the production of intellectual investment flows and the information-knowledge accumulation is the production of intellectual capital stocks as inputs into the space of social productions to generate outputs in the space of input-output dualities. The dynamics of the primary-derived-duality structures of the parent-offspring processes are such that the derived categories of problem-solution dualities are obtained from the primary category of a problem-solution duality by methodological constructionism. The derived problem-solution dualities are reducible by methodological reductionism to the primary problem-solution dualities with continual stock-flow information-knowledge disequilibrium processes under the principle of methodological constructionism-reductionism dualities. The nature of the derived problem-solution duality will always depend on the type and manner of a variety solution found for the primary problem-solution duality. The journeys within the space of constructionism-reductionism dualities are facilitated by statical methods, entropic analyses, and evidential analytics. The derived problem-solution duality (variety offspring) is always different from the primary problem-solution duality (parent) as well as shares with it some of its elements of the characteristic disposition to provide us with information conditions of backward-forward generic tracing and the development of ancestral categories in interdependent modes over the space of destruction-construction dualities. It is this categorial inheritance of characteristic dispositions that allows backward-forward generic tracing to establish the knowing process over the space of telescopic past-present dualities as well as projections over the space of telescopic present-future dualities, where every present variety has a past variety as its parent which then constitutes a parent of variety future and establishes variety DNA sequence and inheritances, where the space of knowability-unknowability dualities creates the space of cognitive anomalies, such as phantom problems.

Every derived category (current problem) has an intimate connection to the primary category (primary problem) through a backward-telescopic process while every current category (current problem) serves as a primary category (parent) for a future derived category by a forward-telescopic process in a never-ending continuous process generating the information-knowledge dynamics in continual disequilibrium states. The forward-backward telescopic process of neuro-decision-choice activities in the dynamics over the space of problem-solution dualities is not limited to some areas of knowing. It is the core of the meaning of human life and existence in the spaces of ignorance-knowledge dualities and the problem-solution dualities. Life stops when its meaning and essence are not neuro-decision-choice defined over the space of problem-solution dualities with information-knowledge stock-flow disequilibrium dynamics, where the social problem-solution processes are also on the path of disequilibrium dynamics. The equilibrium of the problem-solution processes implies information-knowledge equilibrium states which imply that the flow equilibrium in knowing as an intellectual production and the stock equilibrium in the information-knowledge accumulation is the production of intellectual capital stocks which will imply the ultimate state of no change and no life as we know it. The Information-knowledge stocks are essential as inputs into neuro-decision-choice actions which are essential over the space of problem-solution dualities which is generated by cognitive needs for increasing net benefits over the space of individual-collective dualities under the socio-economic principle of non-satiation which stopes at the problem-solution and information-knowledge stock-flow equilibria.

Let us keep in mind that every variety problem and every variety solution are identified by their characteristic dispositions as abstracted by cognitive agents from the ontological variety signal dispositions over the space of acquaintance within the epistemological space to obtain epistemic characteristic dispositions that represent the ontic characteristic dispositions. The abstracted variety epistemic characteristic dispositions from the ontological signal variety dispositions may have cross-sectional and time-series differences depending on the individual and collective cognitive capacity limitations and the source-destination transformation quality. One may have different variety classifications into categorial varieties with common characteristic disposition, where this characteristic disposition may be seen as the set of methods and techniques of knowledge-finding activities over the space of problem-solution dualities. The techniques and methods in dealing with variety problem-solution replacement dynamics are not efficient factors to establish categorial diversities and the unity of knowing and the unity of science in the epistemic fields. In fact, the development of organizations of the enterprise of knowing and the enterprise of science finds expressions over the space of problem-solution dualities, where the efficiency of cognitive activities over the space of problem-solution dualities is continuously governed by the internal balances of variety cost-benefit dualities in the dualistic-polar space. Areas of knowing, learning, teaching and research in fundamental and applied information-knowledge systems are not established by common methods and techniques but by common problem-solution dualities.

In a sense, the demarcation of science and nonscience by methods and techniques of information-knowledge process is rejected. Such demarcation must be established by similarities and differences of variety characteristic dispositions over the space of problem-solution dualities in such a way that the set of common problem-solution dualities of sciences establishes the area of science with the corresponding department, where each department is charged with a set of problem-solution dualities of science; and similarly for non-science and the corresponding departments, where each department is also charged with a set of problem-solution dualities of non-science. In the process of knowing as the production of intellectual investment flows and the process of information-knowledge accumulation as the production of intellectual capital stocks, some department of sciences and non-sciences may be combined to create multidisciplinary departments for a class of problem-solution dualities relative to the dynamics of research universities and institutions. The rise of multidisciplinary departments and institutions are not formed by the commonness of methods and techniques but by the commonness of the set of problem-solution dualities. The dynamics of departmental and institutional creation will be dictated by the dynamics of the space of problem-solution dualities and their interactions relative to the space of parent-offspring dualities in the space of backward-forward dualities. The conditions of the categories of problem-solution dualities in the epistemic fields must form the frameworks for the redesigning and development of the evolving research universities and their controls and managements.

In the search for common attributes for establishing the principles of the diversities and the unity of knowing and the diversity and the unity of science in the space of oneness-dividedness dualities, it is the case that we are not simply looking to a reduction of areas of knowing to one primary category of knowing, where all other categories are sequential derivatives. What is essentially important, here, is the existence and presence of the relational continua of diversities of areas of knowing that allow for the principle of give-and-take-sharing processes in the mutual existence of categorial areas of information-knowledge stock-flow systems in relationally categorial interdependencies over the penumbral areas of the space of acquaintance-description dualities that contains the space of vagueness-clarity dualities. The understanding of the dependencies and penumbral zones of knowing are accomplished by the development of the epistemic field supported by the relational and the information fields, where the epistemic field contains multiplicities of categories to capture the problem-solution dualities and their envelopes. As it has been argued, the diversity and unity of knowing and the diversity and unity of science find expressions in the shared purpose of variety identification-transformation activities over the space of imagination-reality dualities contained in the space of actual-potential problem-solution dualities with continual the actual-potential variety destruction-replacement processes to effect the information-knowledge disequilibrium dynamics and the mutual give-and-take sharing modes of the results of success-failure outcomes of cognitive actions under the neuro-decision-choice activities over the action space. where the success-failure outcomes may be explained by cognitive capacity limitations in the spaces of vagueness-clarity and exact-inexact dualities which contain the spaces of success-failure dualities, error-correction dualities, anticipation-expectation dualities, and necessity-freedom dualities.

The problem-solution dualistic-polar approach to the understanding of the diversities and the unity of knowing and the diversity and the unity of science points to the general and specific interdependencies of areas of knowing and the continual mutation of categorial, institutional and departmental varieties that establish and maintain information-knowledge disequilibrium dynamics. Here, there is the grammar of knowing and information-knowledge production that is linked to the neuro-decision-choice actions and the space of problem-solution dualities, where the language of the enterprise of knowing and the enterprise of neuro-decision-choice activities over the space of problem-solution dualities forms a unity of communications among categorial varieties in a horizontal line of degrees of importance in the shared epistemic environment of the continual expansion of the space of problem-solution dualities and the information-knowledge system.

From the parent (primary category) of a problem-solution duality more than one offspring (derived category) of problem-solution dualities may emerge from the variety solution depending on how the solution is abstracted. In this respect, it may be kept in mind that different varieties have different names in relation to methodological nominalism for distinctions and communications that correspond to their identities and the categories of their belonging. The continual expansion of the space of problem-solution dualities and the universal information-knowledge system cannot be used to justify the current claims of information-knowledge about an expanding universe. The claim of expanding universe must be reconciled with matter-energy universal equilibrium where matter and energy cannot be destroyed, and the existence of a universal vacuum for matter-energy expansion seen in the space of nothingness-somethingness dualities.

The space of nothingness-somethingness dualities is ontological and hence the question of "why there is something rather than nothing" is a phantom question with a corresponding phantom problem. It is just like trying to find out an answer to the question of why there is cost rather than benefit or why there is earthquake rather than no earthquake or there is sky and no sky. The question and the corresponding answer find expression in the space of ontic-epistemic dualities contained in the space of ontological-epistemological duality that is contained in the space of beginning-end duality and the space of nothingness-somethingness dualities that houses the space of knowability-unknowability dualities (for further explanation see [R8.16b], where anomalies are discussed). Different planetary system may expand within the universe as seen in terms variety destruction-construction and transformation-replacement processes, where stars and other planets may be born under the general principle of opposites. The universe is in the-matter-energy stock-flow equilibrium such that there is a matter-energy flow equilibrium, matter-energy stock equilibrium and matter-energy variety disequilibrium. The concept of expanding universe must be examined in the space of ontological-epistemological dualities. The concept and the phenomenon of an expanding universe find residence in the conditions of the ontological space contained by the spaces of the beginning-end duality and the nothingness-somethingness duality that house the space of knowable-unknowable dualities as well as the space of ontological variety disequilibrium dynamics. The information-knowledge conditions on the concept and phenomenon of expanding universe find residence in the conditions of epistemological space contained by the spaces of past-present dualities and fullness-emptiness dualities that house the space of known-unknown dualities as well as the space of epistemological variety disequilibrium dynamics. The epistemological space is contained in the ontological space. The expansion of epistemological space in terms of information-knowledge about the universe does not imply the expansion of the ontological space in terms of its universal existence. The space of expansion-non-expansion duality and the dynamics of the space of problem-solution dualities point to the notion that the more we know the less we know of the universe.

At the level of ontology, there is variety matter-energy disequilibrium states, where the number of varieties operate in the space of expansion-contraction dualities. The variety mater-energy disequilibrium dynamics over the space of actual-potential dualities generate variety characteristic-disposition disequilibrium dynamics to expand the space of variety information and the space of information-knowledge dualities. At the level of epistemology, each neuro-decision-choice action over the space of problem-solution dualities generates new information such that the information-knowledge stock-flow processes operate in disequilibrium states such that the problem-solution dynamics also operate in disequilibrium states. The expanding information-knowledge system about the universe, however, may be used to justify the expansion of the space of variety ignorance-knowledge dualities as new solutions generate new problems requiring new solutions, where some destruction-replacement processes may generate more difficult variety problem-solution dualities with increasing difficulties over the space of cost-benefit dualities in relational intentionality and externalities defined over the space of negative-positive dualities.

Let us keep in mind that the conditions of a variety problem in the space of the realities contained in the space of actuals initialize the conditions of many variety solutions in the space of imaginations contained in the space of the potentials from which an optimal variety solution may be abstracted, while the conditions of optimal variety solution to the previous variety problem initialize conditions of potential variety problem to be actualized in an infinite sequence of ignorance-knowledge identification-transformation disequilibrium dynamics. The relational complexity of information, data, fact, evidence, evidential things and knowledge, and their interchangeable usages over the space of source-destination dualities help to create an intentional system of propaganda, misinformation, disinformation, lies and fake news with deception and info-wars for the multiplicity of interpretations under the *social philosophical consciencism* over the zones of penumbral confusion that affect neuro-decision-choice actions and social administrative inefficiencies as well as form a system of increasing constraints on knowing and information-knowledge stock-flow processes such as the critical race theory, critical legal theory, theories of critical history or other theories in all areas of knowing to shape the cost-benefit distribution over the space of individual-collective dualities.

CHAPTER 4

THINKING AND REASONING OVER THE SPACE OF IGNORANCE-KNOWLEDGE DUALITIES AND POLARITIES IN KNOWING AND SCIENCE

ABSTRACT: CHAPTER 4

Chapter 4 of this monograph presents two essential elements for the construct of the principles of diversity and unity of knowing as well as the principles of diversity and unity of science, where the latter is seen in relation to the former which is the parent in a parent-offspring process in the epistemic field and characteristic-signal field. Knowing is a primary foundation in all human actions over the space of production-consumption dualities, where the space of consumption-production dualities is intimately connected to the space of neuro-decision-choice actions in all areas of human existence, where any category of knowledge as intellectual stock is a derivative from category of knowing as an intellectual investment flow. The construct of the diversity and unity principles of knowing and science is based on the fundamental notion that all human activities are neuro-decision-choice actions to solve *problems* in relation to *things,* analytically called varieties with characteristic dispositions that establish variety identities, commonness and differences called *information-knowledge conditions*. This foundational framework allows for the development of an epistemic field as a mathematical and philosophical entity to examine the thinking-reasoning path to knowing, the diversity and unity of knowing and the partition of the space of knowing into categories of knowing in terms of disciplines and inter-disciplines. This chapter examines extra factors in support of the primary and derived principles of diversities and unity in knowing and then questions, the idea of disunity of intellectual

K. K. Dompere, *The Theory of Epistemic Fields*, Studies in Systems, Decision and Control 498, https://doi.org/10.1007/978-3-031-42470-0_4

investments and accumulation of information-knowledge stocks. It set the stage for the examination of the organization of knowing as *intellectual investment flows* and conditions of variety information-knowledge claims as *intellectual capital stocks* over the space of ignorance-knowledge dualities.

4.1 REFLECTIONS ON THE RELATIONAL CONNECTIVITY OF THE MIND, THINKING, REASONING, AND PROBLEM-SOLUTION DUALITY

The principle of unity of knowing, based on cognitive activities over the space of problem-solution dualities, is claimed to unite all categories of knowing including engineering and socio-economic planning as seen in explanatory and prescriptive sciences with the corresponding explanatory and prescriptive theories and practices, where knowing is a journey over the space of ignorance-knowledge dualities traversing from the zones of ignorance to the zones of knowledge, where the zonal crossings are defined by epistemic transversality conditions. The fuels of these journeys are information and its derived sequential varieties such as data, facts, knowledge, and evidence of which discussions were made at the ending of the chapter three of this monograph. The vehicles of the journeys are the minds through thinking and reasoning of which we turn our attentions, where we seek to discusses the meanings of thinking and reasoning and their differences in knowing and information-knowledge accumulations as social productions.

In the knowing and information-knowledge production processes over the epistemic field, every variety rests in the space of negative-positive dualities, where the negative and the positive duals are defined by their characteristic dispositions that place them in appropriate categories. The initial characteristic dispositions are ontic in the sense of *what there was* (past telescopic), *what there is* (present telescopic) and *what there would be* (future telescopic). For each variety, we start with an ignorance dual as the problem dual and journey toward the knowledge dual as the solution dual where the vehicle to make the journey is neuro-decision-choice module acting through thinking and reasoning from the space of acquaintance to the space of description. Acquisitions of information-knowledge items of varieties are based on interpretive conditions of ontological variety signal dispositions over the space of acquaintance to produce experiential information or knowledge by acquaintance and for the epistemic processing of the experiential information to obtain the knowledge by description. These essential elements in all categories of diversity and unity of knowing and information-knowledge production are grouped into the *primary categories* that find meaning in the space of problem-solution dualities and the *derived categories* that find meaning in the logical processes for epistemic maneuvering over the space of problem-solution dualities under the control and the management of neuro-decision-choice activities to effect variety problem-solution and destruction-replacement processes, the results of which go to update the universal information-knowledge stock-flow *disequilibrium dynamics*. The nature and structure of information-knowledge disequilibrium dynamics are discussed in relational comparison of matter-energy conditions of *equilibrium* in the space of static-dynamic dualities in [R17.17] [R17.18], while the conditions of information-knowledge acceptance into the intellectual capital stocks are discussed in the space of certainty-uncertainty dualities with varying degrees of distribution over the neuro-decision-choice space. The space of certainty-uncertainty dualities is derived from the space of ontic-epistemic dualities to give rise to imagination containing the space of anticipation-expectation dualities.

It must be understood that the elements in the space of success-failure dualities are the varieties inducing the information-knowledge identification conditions of varieties and variety stock-flow disequilibrium dynamics over the epistemological space. It is also the information-knowledge disequilibrium processes that provides initial conditions for variety transformation-substitution and destruction-replacement processes. These initial conditions relating to necessity and possibility are discussed extensively in the theory of categorial conversion [R17,15] and linked to freedom and probability in the theory of Philosophical Consciencism [R17.16]. The initial conditions involve the information-knowledge states of variety solutions to the variety-identification problems, where there is enough knowing of the variety characteristic dispositions to reveal the variety identities, initial conditions, necessity, and possibility as induced by the conditions of categorial conversion for neuro-decision-choice activities over the action space. When the problem of the variety identification is solved, the variety identity is known from its characteristic disposition to establish its initial static conditions to examine the changing conditions of variety transformations and changes either as internal or external dynamic conditions.

The dynamic conditions involve the acquisition of information-knowledge states of solutions to the variety-transformation problems which involve the understanding of the changing nature of variety dispositions, where the transformations are linked to conditions of freedom and probability as induced by conditions of social philosophical consciencism that have taken hold of the social conscience to cognitively simplify and guide individual and collective preferences over the socio-natural space for neuro-decision-choice actions. The social philosophical consciencism also helps to simplify cognitive agents' perceptions of the nature of varieties over the space of quality-quantity dualities, where such perceptions induce socially systemic discriminations and risks of the neuro-decision-choice actions. Let us keep in mind that the social philosophical consciencism contains justified and unjustified beliefs, legal and political precepts, and economic conditions over the space of production-consumption dualities. The space of production-consumption dualities resides in the spaces of justified-unjustified dualities, peace-war dualities, justice-injustice dualities, kindness-wickedness dualities, bias-unbiased dualities and many other such dualistic-polar structures in reference to the general distributional conditions over the social space of real cost-benefit dualities in the production and real production of conditions of existence over the space of individual-collective dualities.

The information-knowledge accumulation process involves the space of success-failure dynamics with self-correction processes of the individual and collective neuro-decision-choice actions over the space of the problem-solution dualities. The space of success-failure dualities is intimately connected to the space of problem-solution dualities under the guidance of neuro-decision-choice actions on information-knowledge processes defined over the space of uncertainty-certainty dualities. Given the apparatuses of the brain and the mind, the success-failure processes over the space of problem-solution dualities are intimately connected to *thinking, reasoning* and *neuro-decision-choice actions* in the space of certainty-uncertainty dualities under a real cost-benefit rationality in all areas of knowing. The need and curiosity to understand the interactive conditions among problem-solution dynamics and success-failure processes, in the space of certainty-uncertainty dualities, are essential for understanding human progress and the dynamics of different types of civilizations from the point of view of knowing, understanding and defining the conditions from the neuro-decision-choice space and their connections to the conditions of the space of problem-solution dualities. What are viewed as essentials in the neuro-decision-choice space over the space of individual-collective dualities are cost-benefit-defined and what generate conflicts and tensions over the action space are also cost-benefit-motivated, where the utility of variety for a choice is net-benefit defined and the disutility of variety for choice is net-cost-defined.

The need and curiosity to understand the interactive, intersectional and interdependent conditions among problem-solution dynamics, success-failure processes, the brain, mind, thinking, reasoning and neuro-decision-choice actions in the space of certainty-uncertainty dualities define the conditions for the development of a powerful knowing area of *neuro-decision-choice theory* to help in understanding the concepts of thinking and reasoning in decision-choice space, and their connections to the space of variety preferences, the space of problem-solution dualities, social transformations and developments over economic, political and legal spaces of the basic social organization and the social philosophical consciencism at generational points. The neuro-decision-choice theory may be seen as making up of the *neuro-economic*- decision-choice theory that deals with the varieties in the space of economic problem-solution dualities and related varieties of production, cost-benefit distribution and consumption of intellectual investment and intellectual capital accumulation of information-knowledge system and a supporting system of economic institutions ; the *neuro-political* decision-choice theory that deals with the varieties in the space of political problem-solution dualities and related social power, system of institutional arrangements for governess and social management of the social cost-benefit distribution system, and the *neuro-legal* decision-choice theory that deals with varieties in the space of legal problem-solution dualities, corresponding legal institutions and related social control systems of rules with restrictions, judgments, enforcements and punishments in the space of incentive-disincentive dualities.

The neuro-decision-choice theory, therefore, deals with the organicity of the social neuro-decision-choice theory for identifying and understanding varieties in the space of sociological problem-solution dualities. The social system is composed of three structure of economic structure that contain the space of economic problem-solution dualities, political structure that contains the space of political problem-solution dualities, and legal structure that contains the space of legal problem-solution dualities in which some social problem-solution dualities may belong. The economic structure is expansive containing scientific and non-scientific sub-spaces of production-consumption dualities including the political, legal structures and itself. The neuro-economic decision-choice performances that generate resources in the social system are affected by the behavioral performances of other areas of neuro-decision-choice behaviors and the corresponding institutional arrangements. Each set of the structural institutions provides a vehicle through which neuro-decision-choice actions are carried on over the corresponding space of problem-solution dualities to set variety solutions against variety problems over the path of parent-offspring dynamics. The sets of the structural institutions exist in the state of mutual interdependencies, intersectionality and in relational give-and-take-sharing modes in a unified complexity, where the categories of problem-solution dualities are mutually defined and the inputs into the space of neuro-decision-choice actions are information-knowledge items and the outputs from the space of neuro-decision-choice actions are information-knowledge items defined in the space of input-output dualities. The epistemic paths of relational structures of social set-up for problem-solution processes and information-knowledge generation are shown as a cognitive geometry in Figure 4.1.1.

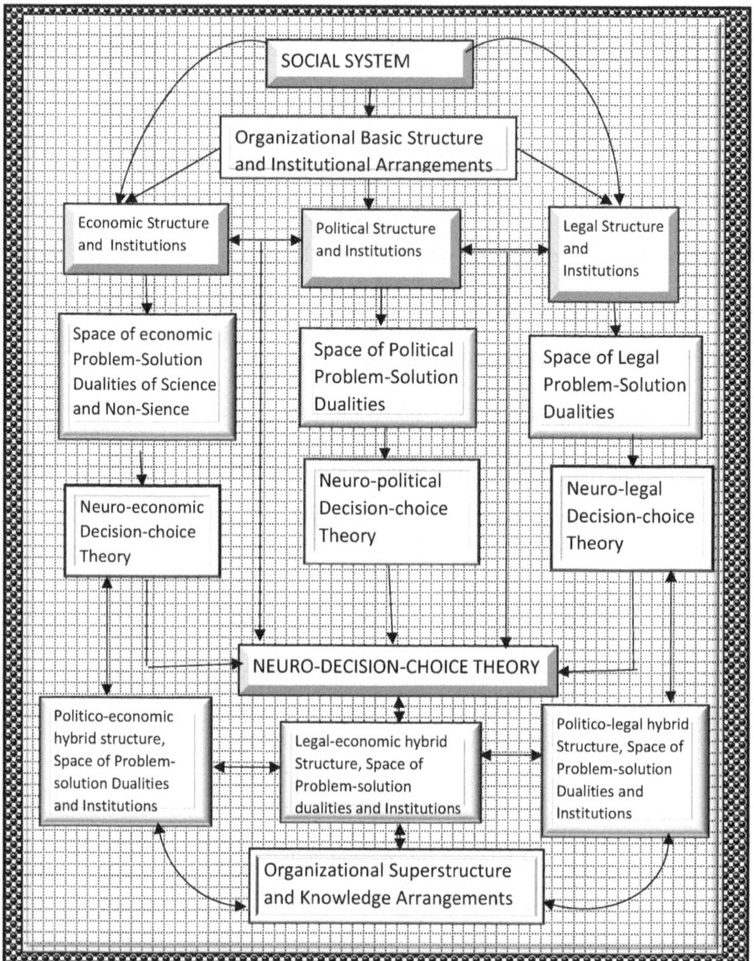

Fig 4.1.1 **A Geometric Paremiology of Neuro-decision-choice Theory and its Components for the Development of the Fundamental and Applied Information-Knowledge Systems**

The spaces of variety problem-solution dualities and polarities have been defined in the previous chapters and in [R8.16a]. These spaces of problem-solution dualities involve all areas of human and non-human existence including the totality of cognitive existence and dualistic-polar conditions of all varieties. Each variety in the space of being exists simultaneously as an internal problem-solution duality and as an external pole which takes the form either as a problem pole or a solution pole which together define a problem-solution polarity. The epistemic structure in understanding and justifying the unity and diversity principles in knowing and science is such that there are variety problem-solution dualities and variety problem-solution polarities in mutually interdependent relational give-and-take sharing modes to explain variety destruction-replacement processes in knowing and information-knowledge production system in the space of internal-external dualities. The concepts of problem and solution have also been defined and explicated for distinction from their uses in ordinary languages, scientific works, and other analytical works in the space of thinking and reasoning. These definitions are structured in relation to principle of opposites in the spaces of relativity and cost-benefit dualities in relative dualistic-polar dominances. Some questions arise as to the meaning of thinking and reason over the space of cognitive activities. Let us first consider the concept and the phenomenon of thinking, then consider the concept and phenomenon of reasoning, then consider their interdependent give-and-take-sharing relations within the social philosophical consciencism of social set-ups.

4.1.1: The Concept of Thinking in Knowing, Information-Knowledge Production and Problem-Solution Dualities

It is useful now to give a workable definition of *thinking* involving the cognitive activities over the spaces of problem-solution dualities, question-answer dualities, knowing and information-knowledge production. All these activities are neuro-decision-choice defined and determined given the concept and the facility of the mind relative to the space of variety identification-transformation dualities in knowing as the production of the intellectual investment flows and the information-knowledge accumulation as production of intellectual capital stocks. It is with the presence of space of identification-transformation dualities that the concept and phenomenon of thinking find residence through the system of neuro-decision-choice actions, where the neuro-decision-choice actions are about variety identification, variety transformation or both over the space of cost-benefit dualities, while thinking is about the abstraction of the best variety cost-benefit configuration in relation to goals, objectives and visions defined over the space of individual-collective dualities. In this frame, let us keep in mind that when one says that a solution is worse than the problem, one has in mind the idea that the net cost-benefit of the variety solution is such that cost outstrips the net cost-benefit of the problem due to increasing cost externalities attached to the variety solution and perhaps there are either defective information, defective thinking or both. To relate thinking to neuro-decision-choice systems over the spaces of problem-solution dualities, question-answer dualities, knowing, information-knowledge production and action space a working definition of thinking is required.

DEFINITION 4.1.1.1: Thinking

Thinking is an individually self-induced activity from the faculty of the mind using elements in social philosophical consciencism and experiential information to epistemically develop algorithmic steps for internal self-following in reasoning to reconcile real cost-benefit conflicts in individual variety-selection processes with a given preference order in the process of cognitive activities over the spaces of problem-solution dualities, question-answer dualities, ignorance-knowledge dualities and identification-transformation dualities under neuro-decision-choice activities over the action space of the general spaces of production-consumption dualities and input-output dualities. It is the act of the mind to produce a system of epistemic structures and varieties in the space of imaginations in the development of variety transversality conditions of epistemic journeys within the space of ignorance-knowledge dualities relative to the space of decidability-undecidability dualities.

Let us remind ourselves of, or at least be familiar with, the goals and objectives of the theory of definition and the construction of dictionaries in both the family of ordinary languages (FOL) and the family of abstract languages (FAL). The discussions on the theory of definition may be found in [R3.26] R3.33] [R3.82]. Every concept or word is a linguistic variety with a defined characteristic disposition that presents its identity and identification in the linguistic system of vocabularies and concepts, where every defined or undefined word has its identity in a system of verbal dualities over communication actions through the space of source-destination dualities, where every concept, word or term rests in the spaces of vagueness-clarity and exact-inexact dualities relative the space of fix-point mapping. It is these linguistic characteristic dispositions of the general linguistic systems that present dualistic-polar structures over the definitional space of real-nominal dualities, where there are synonyms and antonyms to establish dualistic-polar conditions in knowing and information-knowledge production in the space of relativity and not in the space of absoluteness. The space of relativity allows comparative analytics in differences and similarities which then provides the conditions for ranking and neuro-decision-choice actions over the space of intra-categorial varieties and inter-categorial varieties, where the conditions are created for categorial conversions and transformations.

The concept of thinking, based on the demands of the theory of definitions, must be defined in a way that is contained in the space of synonym-antonym dualities with fuzzy conditionality for linguistic substitutions and non-substitutions in communicational processes over the space of source-destination dualities relative to the space of vagueness-clarity dualities. It is this fuzzy conditionality that ranks the degrees of meanings of words and terms, and belonging in definitional similarities, commonness, and differences of categories of synonyms and antonyms in linguistic identities in the space of substitutions under the principle of opposites. In this framework of theories of definition, measurement of phenomena, concepts and usages, the characteristic disposition of the concept of thinking must be established to give it meaning and essence before any adjective may be attached. As an individual subjective activity from the faculty of the mind, the output will differ from cognitive agent to agent even though the inputs may be the same, where the input relevance may also vary over cognitive agents, ideas and needs over the space of source-destination dualities in the general space of cognitive actions. Thinking is an action to produce outputs from inputs. The inputs are information-knowledge set from the variety source signal dispositions and the outputs are information-knowledge set as characteristic dispositions from the destinations, where the characteristic dispositions may come out as variety thoughts, ideas, sounds, designs in all forms, which we shall call epistemic structures with multiplicity of dimensions.

The variety thoughts are epistemic algorithmic steps as outputs for explanation, understanding and formation of ideas about variety identities, behavior as well as for prescriptions for variety transformations over the space of problem-solution dualities relative to the space of decidability-undecidability dualities. The question as to what the thinking is behand a neuro-decision-choice is simply appealing to know the epistemic algorithms behind the decision leading to neuro-decision-choice action and conclusion. The statement that one must do some thinking to see how best to act on a variety problem-solution duality in the space of problem-solution dualities is simply stating that, one must develop epistemic algorithm to set the best solution against a problem. The statement that this thinking is outdated is simply stating that a set of epistemic algorithms is outdate relative to a variety problem-solution duality in the space of problem-solution dualities relative to the social philosophical consciencism at that generational time point. The statement that new information has changed thinking about economic development strategies as well as variety transformation processes means that the new information that has expanded the contents of the social philosophical consciencism to provide extra inputs into the thinking process has brought about the revision of the set of epistemic algorithms relative to the understanding of variety transformations and economic developmental process.

The space of thinking in relation to the space of problem-solution dualities and polarities contains different paths and forms of the development of epistemic algorithms depending on categories of cognitive agents, preference schemes, goal-objective structures, visions, and the categories of problem-solution dualities from either the epistemic field or the characteristic-signal field. Over the space of quality-quantity dualities, there are different kinds of thinking of qualitative and quantitative nature including critical thinking, analytical thinking, abstract thinking, creative thinking, innovative thinking, concrete thinking, convergent thinking and divergent thinking with their identities established by their characteristic dispositions for the corresponding sets of algorithms in relation to the varieties in the space of problem-solution dualities contained in the space of imagination-reality dualities. All thinking, epistemic algorithms and variety solutions are defined in the space of imagination, relative to goals, objectives and visions, also reside in the space of imaginations, where behind every neuro-decision-choice action is a set of epistemic algorithms developed from thinking to guide actions over the space of identification-transformation dualities to find and understand *what there was* (the past) and *what there is* (the present) in the space of actuals, with anticipations in the space of possibilities and expectations in the space of probabilities of *what would be* (the future) in the space of imaginations contained in the space of potentials within the *sankofa-anoma* tradition of time trinity on which information trinity is imposed and time structure is only meaningful relative to information structure. We must always keep in mind that the concepts of *what there was* (the telescopic past), *what there is* (the telescopic present) and *what there would be* (the telescopic future) are surrogate representations of information on past varieties, present varieties and future realities defined in terms of characteristic-signal dispositions, where the characteristic dispositions are the contents of the information and signal dispositions are the messages carrying the contents of the information to reveal the variety identities.

4.1.2: The Concept of Thinking, Paradigms of thought and Social Philosophical Consciencism

The act of thinking produces two system of organic paradigms of thought, which then guide further thinking when the organic paradigms are socially and collectively accepted as rules of thought. There is also two system of micro-paradigms of thought. Each organic paradigm of thought contains a system of micro-paradigms of thought. Each organic paradigm of thought produces organic laws of thought as a set of organic algorithms that is used by individuals and the collective in all neuro-decision-choice actions over all areas of human endeavor of knowing and information-knowledge production in relation to the spaces of acceptance-rejection, doubt-surety, and ignorance-knowledge dualities. The micro-paradigms of thought are the epistemic micro-algorithms under the guidance of macro-paradigms of thought, some aspects of which have been discussed in [R4.7] [R4,8] [R8.16] in relation to the classical paradigm of thought with exact and clear information structures and the fuzzy paradigm of thought with inexact and vague information structures. In this respect, there is the organic thinking and micro-thinking over the information space relative to the space of neuro-decision-choice actions. The macro-paradigms of thought contain the set of micro-algorithms as the set of methods and techniques developed from within the macro-paradigms. The set of the micro-paradigms are under the guidance of the macro-paradigms and consistent with their laws of thought, logic and mathematics to be used over the space of problem-solution dualities under the constraints of on information structures to produce a guidance to the individual and collective thinking over the space of inexact-exact dualities relative to the space of vagueness-clarity dualities as contained in the spaces of beginning-conclusion and imagination-reality dualities. The organic thinking types as the macro-decision-choice algorithms are accepted by all cognitive agents in the development of the micro-algorithms that may be used by cognitive agents for specific variety problem-solution duality or specific category of problem-solution dualities.

Given the space of imagination-reality dualities, thinking presents itself in the space of individual-collective dualities. Within the space of individual-collective dualities, there is the system of individual thinking types that are individual specific with individual cognitive capacity limitations which constrain individual thinking over the space of efficient-inefficient dualities. There is also the system of collective thinking that is collective-specific with collective cognitive capacity limitations which also constrain collective thinking over the space of efficiency-inefficiency dualities, where the collective thinking involves conflicts and individual aggregations in the system of individual thinking types. The aggregations and conflicts reduce the space of individual cognitive capacity limitations in thinking through increasing inputs and learning with multiplicity of relational give-and-take-sharing mode, while the individual conflicts and collaborations reduce the space of collective capacity limitations in collective thinking through individual and collective reexaminations of the individual experiential characteristic dispositions with give-and-take-sharing modes and the collective organic paradigm of thought, where the organic paradigm of thought is part of the contents of the social philosophical consciencism.

All types of thinking reveal themselves in all areas of neuro-decision-choice actions individually and in combinations over the space of problem-solution dualities defined over the space of fundamental-applied dualities. The characteristic disposition of *creative thinking* is the cognitive ability to conceive new and visionary ideas in the space of imaginations to contribute to the established explanatory and prescriptive theories on varieties at the fundamental level of knowing. At the applied level of cognition, the creative thinking is about the development of new micro-paradigms of thought within the macro-paradigms of thought for analysis to claim conclusions in the space of identification-transformation dualities. At the fundamental level of knowing, *analytical thinking* is a process defined over the space of deductive-inductive dualities, whereby the method deduction, the oneness is decomposed into parts to examine the relational continuum and by the method of induction and the decomposed parts are examined for relational oneness, especially over the space of problem-solution dualities to produce new intellectual investment flows and increased intellectual capital stocks. At the applied level of knowing, analytical thinking is a process that helps neuro-decision-choice actions to reconcile deductive-inductive transformation processes over the space of variety input-output dualities. Within the space of fundamental-applied dualities, *critical thinking* is a cognitive process that helps neuro-decision-choice agents to efficiently act over the spaces of variety identification problem-solution dualities and quality-control duality for validity, accuracy, and correctness. Similarly, *concrete thinking* is a neuro-decision-choice process exercised over the space of fundamental-applied dualities in knowing to help neuro-decision-choice agents to journey from the space of imaginations to the space realities. It is, therefore, the practice of the fundamental knowing in the space of variety imaginations to the space of variety realities through the applications which may proceed through learning by doing or apprenticeship systems under the guidance of the prevailing social philosophical consciencism. In the general process of knowing and information-knowledge production, learning how to think is also learning how to use thinking to learn how to think and to know to produce intellectual investment flows. It is here, that teaching how to think is central to the enterprise of knowing and information-knowledge production as established over the space of problem-solution dualities.

4.1.2.1: Types of Thinking, Paradigms of thought and Social Philosophical Consciencism in Knowing and Information-Knowledge Production

Given the definition and explication of the phenomenon and the concept of thinking, we may now examine variations of thinking as types of thinking with differential characteristics over the spaces of problem-solution, ignorance-knowledge, input-output, cost-benefit, and production-consumption dualities. We shall briefly discuss the abstract thinking, convergent thinking, divergent thinking, and innovative thinking in the space neuro-decision-choice actions of human activities. Each of these types of thinking are neuro-decision-choice-generate under conditions appropriate to enhance benefit over cost. Let us keep in mind that thinking is related to the development of algorithms for the use in reasoning over the space of problem-solution dualities. There are two types of the algorithms. There is the macro-algorithm which is the paradigm of thought. There is also the set of micro-algorithms which is developed under the established rules of thought of the macro-paradigm within which different types of thinking are developed and adjustments are made to both the macro-algorithm and micro-algorithms as needs arise.

4.1.2.1A: DEFINITION: Abstract Thinking

An abstract thinking is a cognitive process using some elements outside the social philosophical consciencism and defined over the space of problem-solution dualities in the space of variety fundamentals with interlocking penumbral zones of confusions characterized with difficult and hidden parts in relational complexities, meanings, identities of connectivity of variety characteristic dispositions over the space of imagination-reality dualities such that variety imaginations may be indistinguishable from variety realities which may produce elusions and delusions of knowing and information-knowledge items over the space ignorance-knowledge dualities. It relates to conditions of fundamental varieties, the results of which provide the necessary framework for applications and linkages to conditions of applied and critical thinking in the space of variety production-consumption dualities under the principle of non-satiation over the space of imagination-reality dualities abstracted from the space of actual-potential dualities to abstract the essential conditions of cognitive transversality conditions of epistemic journey from variety ignorance to variety knowledge to provide conditions of actions in the space of decidability-undecidability dualities.

4.1.2.1B: DEFINITION: Convergent Thinking

Convergent thinking is a cognitive process over the space of imagination-knowledge dualities using a finite set of elements from the social philosophical consciencism to guide neuro-decision-choice actions over the spaces of problem-solution dualities and ignorance-knowledge dualities as contained in the space of imagination-reality dualities which is also a sub-space of actual-potential dualities. Convergent thinking is to reduce the number of relevant pathway combinations between a variety problem and a variety solution as well as varieties in the space of ignorance-knowledge dualities to reveal the epistemic transversality conditions for optimizing the relative uncertainty-certainty degrees of cognitive journey and actions over the space of decidability-undecidability dualities.

4.1.2.1C: DEFINITION: Divergent Thinking

Divergent thinking is a cognitive process over the space of imagination-reality dualities using complex explorations of infinite set of elements from the social philosophical consciencism to guide neuro-decision-choice actions over the space of problem-solution dualities as contained in the space of imagination-reality dualities which is also a sub-space of actual-potential dualities. Divergent thinking is to employ all possible irrelevant and relevant pathway combinations between variety problems and variety solutions in order to find the most effective variety solutions in replacement of variety problems or variety knowledge in replacement of variety ignorance to discover the epistemic transversality conditions of cognitive journeys within the space of variety ignorance-knowledge dualities minimizing the degree of uncertainty of error and maximizing the degree of certainty of correctness to justify the action over the decidability-undecidability dualities.

4.1.2.1D: DEFINITION: Applied Thinking

Applied thinking is a cognitive process using an experiential information, the results of abstract thinking and selected elements in the social philosophical consciencism and defined to specify the relevant problem-solution set abstracted from the space of applied problem-solution dualities in the space of variety transformations to create varieties with greater benefit dispositions over cost dispositions for use in human pursuit of happiness under the principle of non-satiation over the space of imagination-reality dualities such that variety imaginations may be transformable to variety realities of knowing and information-knowledge items over the space of ignorance-knowledge dualities. Applied thinking relates to conditions of basic varieties, the results of which provide the necessary framework for judging the validity and usefulness of abstract thinking over the spaces of variety production-consumption dualities contained in the space of imagination-reality dualities as abstracted from the space of actual-potential dualities, where the results of applied thinking produce the variety test conditions of cognitive transversality conditions of both abstract and applied epistemic journeys from the variety ignorance to the variety knowledge to maintain conditions of actions in the space of decidability-undecidability dualities.

4.1.2.1D: DEFINITION: Critical Thinking

Critical thinking is a cognitive process using some of the available experiential information composed of available data, facts, evidence and evidential things as analytical inputs into a neuro-decision-choice action with the help of a macro-paradigm of thought and its contents of micro-algorithms to create results from the analysis which are needed to examine the validity of the transversality conditions over the space of imagination-reality dualities under cognitive journeys over the space of ignorance-knowledge dualities which includes journeys from the space of variety knowledge by acquaintance to the space of variety knowledge by description, where within the center of critical thinking are the practices of principle doubt, practices of awareness of taxonomy of competing alternatives, the practice of avoidance of disinformation, misinformation, propaganda and the practices of awareness of cognitive capacity limitations over the space of individual-collective dualities for claims of justified framework in support of epistemic transversality conditions for claims over the spaces of ignorance-knowledge dualities and problem-solution dualities, where there is adherence to the practices self-direction, self-motivation and self-correction with pursuance of rigorous adherence to the goal of epistemic excellence through overcoming the phenomena of egocentrism and national exceptionalism contained in the social philosophical consciencism at generational time points.

4.1.2.1E: DEFINITION: Innovative Thinking

Innovative thinking is a cognitive process over the space of imagination-reality duality using search explorations of finite set of elements from the social philosophical consciencism to guide neuro-decision-choice actions to make relevant improvements of the methods and techniques of the existence algorithms of neuro-decision-choice actions on knowing and information-knowledge production over the space of problem-solution dualities as contained in the space of imagination-reality dualities which is also a sub-space of actual-potential dualities. Innovative thinking is to explore qualitative and quantitative improvements of the known pathway combinations between variety problems and variety solutions over the space of efficient-inefficient dualities relative to the macro-algorithm (macro-paradigm of thought) and micro-algorithms (micro paradigms of thought) within the macro-algorithm (macro-paradigm of thought), where innovative thinking is from within the existing social philosophical consciencism with intra-categorial changes, and the abstract thinking is from without the existing social philosophical consciencism with inter-categorial changes.

4.1.3: The Concept of Reasoning in Knowing and Information-Knowledge Production

Given the discussions about the concepts and phenomena of thinking and the types of thinking with their corresponding characteristic dispositions to establish their identities and being, let us turn an attention to the concept and phenomenon of reasoning relative to the concept and phenomenon of thinking. The examination of the concept and phenomenon of reasoning is necessary because various levels of confusion tend to arise in interchangeability of thinking and reasoning in knowing as the production of intellectual investment flows and information-knowledge accumulations as the production of intellectual capital stocks. Reasoning is not thinking and vice versa even though they are closely related in conceptual interdependencies in the space of give-and-take sharing modes. Let us define and explicate the concept and phenomenon of reasoning and examine its decomposition to types. It must be understood that organization involves decomposition, composition, disunity, unity, diversity, oneness, dividedness, nothingness and somethingness contains in disorganization over the space of input-output dualities relative to the space of ontological-epistemological dualities.

DEFINITION 4.1.3.1: Reasoning

Reasoning is an individually self-induced activity of the mind as a thought process in neuro-decision-choice actions in using the variety characteristic dispositions as inputs to follow the epistemic algorithms developed by thinking to arrive at the variety selection over the space of variety problem-solution dualities in the sense that thinking is the product of any mind as an epistemic guide to reasoning as a cognitive process to exercise neuro-decision-choice actions over the spaces of problem-solution and question-answer dualities, where thinking is the *primary category* of mental action and reasoning is a *derived category* of mental action of neuro-decision-choice activities over the spaces of problem-solution and question-answer dualities contained in the space of imagination-reality dualities as derived from the space of actual-potential dualities contained in the space of ontological-epistemological duality.

4.1.3.1: The Concept of Reasoning, Paradigms of thought and Social Philosophical Consciencism

Given the macro-paradigm of thought with the corresponding organic algorithm and micro-paradigms of thought with the corresponding micro-algorithms from the concept and phenomenon of thinking in the space of imagination-reality dualities, it is useful, now, to examine the relationships of reasoning to both macro-paradigm with macro-algorithm and the set of micro-paradigms with the set of micro-algorithms. Both the macro-paradigm of thought and the set of micro-paradigms are the product of thinking. Similarly, both the phenomena of thinking and reasoning are products of the mind. Each constructed organic paradigm of thought not only contains a system of micro-paradigms of thought but it provides the organic rules and steps for general reasoning, while the micro-paradigms provide steps within the organic paradigm for specific category of problem-solution dualities to move from a variety ignorance to a variety knowledge within categories and between categories abstracting the needed epistemic transversality conditions for the cognitive journeys over the space of ignorance-knowledge dualities. The organic algorithm provides a set of organic rules for neuro-decision-choice actions over the space of acquaintance-description dualities in addition to providing rules of quality control over the space of acceptance-rejection dualities using reasoning with the paradigms of thought applied to the experiential information as inputs and to abstract knowledge by description as outputs over the space of neuro-decision-choice actions. The organic paradigm of thought may be associated with the organic thinking that provides a set of general rules to guide the development of micro-paradigms and methods for reasoning over the spaces of problem-solution dualities and ignorance knowledge dualities.

There are two organic paradigms of thought which are of interest in these discussions over the knowing process and the information-knowledge processes as cognitive activities in production and consumption of information. The natures of the two organic paradigms are related to the structure of information as input into the thinking processing facility. They are the classical paradigm of thought and the fuzzy paradigm of thought. The organic classical paradigm of thought, with its set of rules of thinking, provides the organic classical pathway of reasoning and the development of a dynamic set of classical micro-algorithmic pathways in processing exact and clear information structures over the space of absoluteness, where reasoning of acceptance of contradictions in thought are not allowed under the principle of excluded middle, and opposites are not allowed to exist simultaneously in the same space and time in relational separation, diversity and disunity under the principle of opposites . The organic fuzzy paradigm of thought, with its set of rules of thinking, provides the organic fuzzy pathway of reasoning and the development of fuzzy micro-algorithmic pathways in processing inexact and vague information structures over the space of relativity, where reasoning in acceptance of contradictions in thought are allowed under the principle of relational continuum, and unity and opposites with diversity, disunity and decomposition are allowed to exist simultaneously in the same space and time under the principle of opposites.

Just as there is the organic thinking and micro-thinking, there is also an organic reasoning and a micro-reasoning corresponding to the organic thinking. The organic reasoning is governed by the rules of organic thinking which contains the macro-rules of thought. Similarly, the micro-reasoning is governed by the rules of the micro-thinking which holds the micro-paradigms of the methods and techniques of reasoning within the organic paradigm of thought as exercised over the spaces of problem-solution dualities and ignorance-knowledge dualities. Let us keep in mind that the space of question-answer dualities has conformal mappings onto the space of problem-solution dualities which then has conformal mappings onto the space of neuro-decision-choice actions. The legitimacy of the rules of micro-reasoning depends on the set of rules provided by the organic paradigm of thought. Given the organic paradigm, there is a continual development of micro-paradigms as new problems are encountered new epistemic anomalies are encountered and old epistemic anomalies are revisited and examined over the spaces of problem-solution and question-answer dualities in all areas of knowing and information-knowledge production as new variety conditions emerge in the general space of human actions generating information-knowledge disequilibrium system corresponding to problem-solution and question-answer disequilibrium dynamics. In this epistemic structure, the organic paradigm is in equilibrium within the development of the set of laws of thought while the set of micro-paradigms of thought is in disequilibrium with the development of the micro-algorithms over the spaces of problem-solution, question-answer and ignorance-knowledge dualities under continual disequilibrium states to provide the information-knowledge system with a continual self-correction properties, where the disequilibrium states are envelopes of temporary equilibrium states. The rules of quality control of knowing as the production of intellectual investment flows and information-knowledge accumulations as the production of intellectual capital stocks are always contained in the organic paradigm of thought that accounts for deficiencies from the space of possibility-probability dualities contained in the space of imagination-reality dualities.

Both the concepts and phenomena of thinking and reasoning find meaning and residence in the space of imagination-reality dualities with different roles over the space of variety identification problem-solution dualities to establish the information conditions of variety identities, and variety transformation problem-solution dualities to establish the information conditions of variety characteristic changes. Given the space of imagination-reality dualities, reasoning, like thinking, presents itself in the space of individual-collective dualities. Within the space of individual-collective dualities, there is the system of individual reasoning types which are individual specific with individual cognitive capacity limitations derived from the contents of the existing social philosophical consciencism. These cognitive capacity limitations constrain individual uses of thinking algorithms for reasoning over the space of efficient-inefficient dualities relative to the space of problem-solution dualities, even under the same information input with the same organic paradigm of thought to discover the variety transversality conditions over the space of variety ignorance-knowledge dualities. Thinking is used to create the vehicles for reasoning to take variety imaginations in the space of imaginations contained in the space of potentials to variety realities in the space of varieties contained in the space of actuals. The thinking and reasoning processes maintain the relational unity between the space of imagination-reality dualities and the space of actual-potential dualities in backward-forward logical movements over the space of methodological constructionism-reductionism dualities, where every variety reality claimed by the thinking-reasoning processes may or may not be a member of in the space of actuals. It may belong to the space of impossible world. Every variety in the space of imaginations belongs to the space of potential and every variety in the space of potentials is not necessarily in the space of imaginations, even though it has a degree of capacity to belong to it; however, every variety in the space of imaginations belongs to the space of potentials with dualistic-polar conditions.

Similarly, within the space of individual-collective dualities, there is also the system of collective reasonings that is collective-specific with collective cognitive capacity limitations which also constrain collective reasoning over the space of efficiency-inefficiency dualities relative to the space of problem-solution dualities that is defined over the individual-collective dualities. The collective reasoning, like the collective thinking, involves conflicts and aggregations of individual neuro-decision-choice reasoning outcomes in the system of individual reasoning types, where these reasoning types follow the thinking types with the same signal dispositions and perhaps different decoding. The aggregation and conflicts of the individual reasoning in the collective reduce the space of individual cognitive capacity limitations in reasoning through increasing inputs and inter-individual learning with a multiplicity of relational give-and-take-sharing modes. The individual conflicts and collaborations reduce the space of collective capacity limitations in the collective reasoning through the individual give-and-take-sharing feedback system for continual corrections with individual reasoning constrained and enhanced by collective reasoning which is also constrained by the system of individual reasoning, where all elements of thinking and reasoning are drawn from the contents of the prevailing social philosophical consciencism which provides the foundations for national character in the space of common mistakes and system of beliefs. Let us examine the different types of reasoning relative to the types of thinking.

4.1.3.2: Types of Reasoning, Paradigms of thought and Social Philosophical Consciencism

Within the space of reasoning processes using the algorithmic and non-algorithmic varieties from the space of thinking algorithms over the space of individual-collective dualities, reasoning may take the form of inductive, deductive, abductive, and intuitive types. The *inductive* reasoning and *deductive* reasoning are directly connected to algorithmic varieties, while the *abductive* reasoning and *intuitive* reasoning are indirectly connected to algorithmic varieties through the social philosophical consciencism at any generation time that provides a general collective framework in the space of thinking-unthinking dualities. In the simple definitional structures, *deductive reasoning* is a logical procedure in using the rules of micro-algorithmic paradigms within the rules of macro-paradigm of thought to execute a neuro-decision-choice action over the space of identification-transformation dualities, where the processing of information support is derived from the variety general to variety particular. It is a reasoning using a paradigm of thought from the organic information-knowledge conditions of the system to ascertain information-knowledge conditions about the variety knowing on the unknown individual varieties. Let us keep in mind that variety knowledge (ignorance) is different from knowledge (ignorance) variety in these discussions in knowing as the production of intellectual investment flows and in information-knowledge accumulation as the production of intellectual capital stocks. Variety knowledge is an epistemic variable in the space of knowing, while the knowledge variety belongs to the information-knowledge stocks both of which are defined in the fuzzy space with a fuzzy conditionality relative to the general principle of opposites.

Using any of the varieties in the system of thinking algorithms under the organic paradigm of thought, the *inductive reasoning* processes several individual variety information-knowledge conditions to ascertain the information-knowledge conditions about the general system in the knowing and the information-knowledge production in the intellectual capital stock-flow disequilibrium dynamics. The *abductive reasoning* is a set of cognitively processes of justified and unjustified current variety characteristic dispositions from the space of acquaintance to abstract conclusions or arrive at neuro-decision-choice actions without a micro-paradigm or algorithm.

Indirectly, the abductive reason in the production of intellectual capital stock-flow disequilibrium dynamics is a non-micro-algorithmic based. It is based on perceptive analytics of available experiential information and a form of wisdom from ontological variety signal dispositions, defined from the conditions of characteristic dispositions, to arrive at the knowledge by description. The *intuitive reasoning* is also a cognitive capacity for a set of unestablished analytical processes within or outside an organic paradigm of thought to immediately comprehend the variety characteristic dispositions from the signal dispositions in the space of identification-transformation dualities without direct reference to any micro-paradigm of thought that presents methods and techniques of thinking algorithms to process the experiential information in order to abstract correct characteristic dispositions as the knowledge by description of varieties. Alternatively stated, intuitive reasoning is a process of arriving at knowledge by description from knowledge by acquaintance within an organic paradigm of thought without a direct use of a micro-paradigm of thought over the space of ignorance-knowledge dualities, where both the macro-paradigm and the micro-paradigms are contained in the social philosophical consciencism at a generation time.

The system of deductive, inductive, abductive and intuitive reasoning types and the system of thinking types are governed by the logical conditions of an organic paradigm in which deduction and induction are conscious processes under micro-paradigms of thought and abduction and intuition are unconscious processes without any micro-paradigm of thought, where both move from knowledge by acquaintance to knowledge by description, where the outcomes from them must satisfy the set of rules of the organic paradigm of thought over the space of quality-quantity duality for quality control and the management of the information-knowledge stock-flow disequilibrium processes over the epistemic field with relational and information fields, where behind every neuro-decision-choice action lies a thinking-reasoning dualities. In the space of problem-solution dualities, learning to reason with thinking is learning to think and learning to think is learning to use reasoning to produce thinking over the spaces of problem-solution, question-answer, and ignorance-knowledge dualities in the space of principle of opposites.

PROPOSITION 4.1.3.2.1: The Roles of Thinking and Reasoning in the Epistemic Field

Under the principle of opposites with relational continuum and unity in knowing and information-knowledge production over the space of problem-solution dualities and polarities within the epistemic field, the role of thinking is used to develop an organic paradigm of thought and micro-algorithms within the organic relevant to specific categories contained in the epistemic field with corresponding relational and information fields for examining and understanding the dualistic-polar conflicting forces between the duals of dualities and the poles of polarities. The reasoning is to use the thinking algorithms in the space of macro-micro dualities to solve specific problems of conflict and control processes to bring about the desired outcomes over the space of goals, objective, and visions. The combined thinking-reasoning actions are parts of the contents of the philosophical consciencism for balancing actions between the forces of opposites to manage the dualistic-polar games in human actions, where decision is about problem-solving processes which are about changes of property, relation, or both.

4.2: SEMANTIC ANALYTICS OF THINKING AND REASONING IN KNOWING AND INFORMATION-KNOWLEDGE ACCUMULATION

The system of reasoning of deductive, inductive, abductive and intuitive types, and the system of thinking types are governed by the logical conditions of an organic paradigm in which deduction and induction are conscious processes under micro-paradigms of thought and abduction and intuition are unconscious processes without any micro-paradigm of thought, where the thinking and reasoning types establish experiential information from ontological variety signal dispositions and process the knowledge by acquaintance to acquire knowledge by description, where the combined outcomes from them must satisfy the set of rules of the organic paradigm of thought over the space of quality-quantity dualities for quality control and the management of the information-knowledge stock-flow disequilibrium dynamics over the spaces of production-consumption and input output dualities in the epistemic field with relational and information fields. The thinking and the reasoning systems are under the guidance of the social philosophical consciencism as the organic vehicle containing all relevant and irrelevant varieties of information-thought conditions with approved non-approved thought situations about all conditions in the epistemological space for use in the space of imagination-reality dualities contained in the space of actual-potential dualities.

In the general usages over all areas of knowing, we speak of algorithm as step-by-stem or defined rules to be followed to solve problems or to find answers to questions without relating it to thinking, reasoning and decision. Here, the algorithms, as products of thinking, are seen in terms of mathematical and non-mathematical programs in the system of principle of opposites, where the duals and poles are mutually constraining in their existence to abstract solutions to problems with goals and objectives which constrain the goals, just as costs and benefits as duals are mutually constraining in all neuro-decision-choice systems over the variety space. In the theories of knowing and knowledge production in the development of intellectual capital stocks, it is useful to relate algorithms to thinking and decision as a set of steps to be used for reasoning by any neuro-decision-choice agent over the space of problem-solution dualities relative to the space of imagination-reality dualities to move and acquire the logical conditions in dualistic-polar structures with mutual constraints in dualistic-polar constraints. The above definitions suggest that it is useful to note that thinking precedes reasoning and that, thinking and reasoning together define the neuro decision-choice processes over the spaces of problem-solution dualities, question-answer dualities and ignorance-knowledge dualities relative to the space of identification-transformation dualities under variety curiosities, variety creativity and variety motivations over the space of incentive-disincentive dualities translated from the space of cost-benefit dualities. Thinking is an individual process in the development of algorithms, while reasoning is a collective process in the use of algorithms.

The point here is that the epistemic connectivity of the mind, thinking and reasoning is an element that helps the logical process to understand the principles of diversity and unity of knowing, as well as diversity and unity of science and non-science, all are viewed as oneness in the multiplicity of journey over the space of ignorance-knowledge dualities through the spaces of *certainty-uncertainty* dualities, doubt-surety dualities and other relevant dualistic-polar structures under the disequilibrium dynamics of intellectual capital stock-flow processes over the space of fundamental-applied dualities, where the fundamental dual relates to abstract theoretical constructs and the applied dual relates to applicational constructs. Similarly, it is *the mind, thinking and reasoning* in collaborative actions through the neuro-decision-choice actions *that examine elements* in the spaces of true-false, acceptance-rejection and decidability-undecidability dualities and make selections for applications in the action space, where the action space depends on elements from the space of applied thinking-reasoning dualities relative to the space of identification-transformation dualities.

The driving force in the thinking activity in the development of epistemic algorithms is the conflicts in variety real cost-benefit dualities as they relate to variety problem-solution dualities and question-answer dualities over the organic space of identification-transformation dualities. It is epistemically useful to point out that the concept of real cost-benefit analytics in preference orderings and its role in decision-choice actions has not been well articulated and given its generality in economic theories of decision and choice as foundations of general neuro-decision-choice theory that applies to all areas of human existence. The analytical force of cost-benefit conditions and rationality is not well understood and incorporated in the development of psychological theory of decision and choice including neuroeconomic theory and in the criticism against economic theory as well as the psychological theory of decision and choice. The cost-benefit analytics have not received attention in the development of theory of knowing and the theory of knowledge. Some of the fundamental discussions about the positions involving agreements and disagreements of economic and psychological theories of decision-choice actions are discussed in the monographs [R3.26] [R4.8] [R4.9] [R24.5].

It must be understood that the development of neuro-decision-choice theory must be general to encompass the interactive conditions of thinking and reasoning over all human conditions that lead to decision, choice, and selection over the space of variety dualities under different information processing capabilities and psychology of preferences in relation to the principle of non-satiation in an infinite horizon. It must also be understood that neuro-decision-choice actions require the existence of varieties and ranking capacity among varieties from their identities defined by their characteristic dispositions as abstracted from the ontological signal dispositions. Neuro-decision-choice actions do not exist in the state of variety commonness. The existence of varieties differences and capacity for ranking them require a criterion space for variety rankings and the techniques and methods for ranking. One must ask oneself what the criterion of ranking is. The position, here, in the principles of diversity and unity of knowing and science in the construct of fundamental and applied information-knowledge system, is that every element in the criterion space is reduceable to real cost-benefit characteristic disposition, where neuro-decision-choice actions are undertaken with real cost-benefit rationality and where a variety problem is the real opportunity cost for variety solution as a real benefit. Epistemically, the real variety cost-benefit dispositions for neuro-decision-choice actions in all neuro-decision-choice situations are such that the real cost dispositions, as the cost duals or poles, constrain the real benefit dispositions as the benefit duals or poles, and the real benefit dispositions as benefit duals or poles constrain the real cost dispositions as the cost duals or poles in dualistic-polar settings, where neuro-decision-choice outcomes are always at the mercy of information-knowledge structures in the space of quality-quantity dualities, and certainty-uncertainty dualities are experiential-characteristic defined. The concept and phenomena of costs and benefits are extensively discussed in the monograph [R5.13], and the process of aggregation is discussed in the monograph [R5.14]. The use of the concept of relative cost-benefit disposition to define the concept and phenomena of problem-solution situations are discussed in the monograph [R5.14], where decision-choice situations are related to conflicts in variety relations and variety properties relative to cost-benefit conditions . The conditions are such that every decision is a dualistic-polar transformation of either variety property disposition, variety relational disposition or both

[R8.16b], where the neuro-decision-choice actions are the resolutions in the space of dualistic-polar games.

It may also be pointed out that all theories and variety ideas are *thinking frameworks* developed for reasoning practices over the spaces of problem-solution dualities, question-answer dualities ignorance-knowledge dualities within categories and inter-categories contained in the epistemic field with the corresponding relational and information fields irrespective of the size of the duality from the space of imagination-reality dualities in a sense, the theories of computable systems, algorithms explanations, prescriptions, analytical frameworks and mathematical representations are thinking frameworks for reasoning in the space of cost-benefit dualistic-polar games to abstract the best variety as seen by the cognitive agent or cognitive agents relative to some varieties in the space of goals, objective and vision defined in the space of imaginations contained in the space of potentials. The thinking algorithms as defining the framework of reasoning are guided by an organic *paradigm of thought* in the information-knowledge processing over the spaces of problem-solution and question-answer dualities under the general principle of opposites with relational continuum and unity. Under the general principle of opposites, we have the classical paradigm of thought as an organic paradigm from which a set of classical micro-paradigms are developed for classical reasoning with exact and clear information structure to obtain knowledge as an input into the space of classical neuro-decision-choice actions. There is also the fuzzy paradigm of thought as an organic paradigm from which a set of fuzzy micro-paradigms are developed for fuzzy (approximate) reasoning with inexact and vague information structure to obtain knowledge as an input into the space of fuzzy neuro-decision-choice actions.

4.2.1 The Principle of Opposites, and Thinking-Reasoning Processes in Knowing and Science

Let us now examine the roles of thinking and reasoning in all spaces of dualities and polarities under the principle of opposites. The structure and forms of duals of duality, poles of polarity and dualities of polarities have been defined, explicated, and explained. The dualistic-polar conditions in the last analysis relate to the varieties in the problem-solution dualities and question-answer dualities as seen from the organic space of identification-transformation dualities in relation to the space of imagination-reality dualities. From the viewpoint of the principle of opposites, every computable system developed with a micro-algorithm from a micro-paradigm of thought has a corresponding incomputable system in relational continuum and unity such that there are spaces of computability-uncompatibility, explainability-unexplainability, predictability-unpredictability, decidability-undecidability dualities and polarities and many such dualistic-polar phenomena in the space of knowing and neuro decision-choice actions in all categories of human knowing, actions and information-knowledge production. The space of identification problem-solution dualities provides us with the relevant information about variety identities and static behavior to establish a necessity, initial conditions, and anticipations relevant for actions by functors or categorial convertors. The space of transformation problem-solution dualities provides us with information about the dynamics of variety identities, behaviors, freedom of actions and sufficient conditions to effect transformation.

It is the dualistic-polar conditions in the search for variety identities and transformations that create zones of epistemic conflict of individual and collective acceptances of variety identities, claims of truth and claims of knowledge items in the economic, political, and legal spaces of neuro-decision-choice actions where the values of these conflicts are assessed over the space of cost-benefit dualities in relation to relative social cost-benefit distributions. At the space of acquaintance, observations, and abstractions of variety characteristic dispositions from the variety signal-dispositions may vary from an individual to an individual creating disparities of variety experiential information or knowledge by acquaintance in relation to claims of information-knowledge items over the space of doubt-surety dualities. It is also at the presence of these dualistic-polar conflicts in the space of acceptance-rejection dualities that creates epistemic frameworks of conditions for variety changes and endows the information-knowledge system with the properties of self-excitement and self-correction for improving the elements in the space of doubt-surety dualities. The dualistic-polar phenomena of decidability-undecidability, computability-incompatibility, explainability-unexplainability, predictability-unpredictability dualities find expressions over the neuro-decision-choice space under the conditions in the space of uncertainty-certainty and doubt-surety dualities. The conditions and behaviors of these variety dualities and polarities will be examined in another chapter of this monograph or another supporting monograph in detail. These conditions belong to the space of problem-solution dualities involving intellectual investment capital stock-flow process in knowing and information-knowledge systems to create epistemic zones over the space of agreement-disagreement dualities.

PROPOSITION 4.2.1.1: Thinking and Reasoning under the Principle of Opposites

Thinking and reasoning capacities are interdependent compositive cognitive actions derived from the existing social philosophical consciencism, given cognitive abilities, to manage and administer dualistic-polar competitive forces under the principle of opposites with relational continuum and unity to create preferred cost-benefit results by neuro-decision-choice actions over the spaces of production-consumption and input-output dualities relative to goals, objectives and visions in the space of goal-objective dualities contained in the space imaginations and to actionize them onto the space of realities.

PROPOSITION 4.2.1.2: Thinking, Reasoning, Understanding, Execution (TRUE) and the Principle of Opposites

The central core of the educational enterprise must concentrate on teaching thinking and reasoning for knowing as the production of intellectual investment flows and not simply teaching the existing information-knowledge stocks which must be used judiciously as analytical instruments to illustrate problem-solution relational interdependencies in identifying problems and solutions, and the method and techniques of their relationality, avoiding unquestionable credulity of existing intellectual capital stocks, where such unquestionable credulity can produce zones of mental laziness, cognitive decay, cognitive imbecility to close the mind from the house of creative art of curiosity and critical inquiry through which knowing process and information-knowledge accumulation can progress over the social enterprise of the space of problem-solution dualities avoiding all mendacities, propaganda, disinformation and misinformation of all forms contained in the social philosophical consciencism at generation points.

The epistemic zones of absolute disagreements in the spaces of false-true dualities or acceptance-rejection dualities find their explanations in the spaces of analyses and the uses of classical paradigm of thought that is constrained by logical conditions of excluded middle within the general principle of opposites , information exactness and non-acceptance of contradictions as logical truth values in the space of absoluteness, where the non-acceptance of contradictions and excluded middles simply mean that opposites such as true and false cannot exist in the same space and time. The epistemic zones of relativity disagreements in the spaces of false-true dualities or acceptance-rejection dualities find their explanations in the spaces of analyses and the uses of the fuzzy paradigm of thought which allows logical conditions of continuum and unity, information inexactness and acceptances of contradictions as logical truth values in the space of relativity, where claims by neuro-decision-choice actions are constrained by fuzzy conditionality to define the applicable boundaries of thought and degrees of acceptance of claims. It is also the acceptance of the presence of contradictions that unites all areas of knowing and information-knowledge systems as self-correcting to engender the process of information-knowledge accumulation in all areas of knowing under the fuzzy-stochastic conditionality.

It is also through the presence of contradictions that problems are created, and solutions are found within the space of variety cost-benefit dualities, where variety problems are variety solutions and variety solutions are variety problems relative to goals, objectives, and visions as foundations for telescopic future which may be generated from dualistic-polar conditions of telescopic past and present varieties under the principle of opposites. Thinking and reasoning processes in knowing and information-knowledge production are negotiating activities between duals and between poles in continual balancing acts for continual interchanges of duals in dualities and interchanges of the poles in polarities in a system of negations of negations under categorial conversions in the space of destruction-replacement processes that contains the space of reversible-irreversible dualities of neuro-decision-choice actions, the understanding of which is essential for the understanding of environmental civilization in the epistemological space in terms of epistemological input-output dualities relative to the space of ontological input-output dualities, where certain outcomes of epistemological decisions may be irreversible in the ontological space. In this respect, the study and the development of the theories of irreversible decisions and the space of knowable-unknowable dualities are essential to the understanding of human civilization containing environmental civilization over the space of natural-social dualities contained in the space of ontological-epistemological dualities, where the environmental civilization draws on certain ancient practices of usage of resource of water, vegetation and elements in the global ecological system.

Every variety resides as a negative-positive duality and polarity, where the duality relates to identification conditions and polarity relates to transformation conditions. The dominance of negative dual reveals the identity of the negative duality, while the dominance of the positive dual reveals the identity of the positive duality. The negative duality resides in the negative pole while the positive duality resides in the positive pole. The dominance of the negative pole defines the negative polarity, while the dominance of positive pole defines the positive duality. The negative dual of the duality and the negative pole of polarity reside as a cost-benefit duality and cost-benefit polarity. The positive dual of the duality and the positive pole of polarity also reside as cost-benefit duality and cost-benefit polarity. The micro-thinking algorithms establish a toolbox of methods and techniques for reasoning between opposites of the duals of duality and poles of polarity optimizing benefits subject to costs in the process of claims and conclusions, in the spaces of identification problem-solution dualities and transformations problem-solution dualities. The thinking processes within an organic paradigm of thought provide a theoretical and algorithmic framework of toolboxes for reasoning to resolve the relational conflicts in the internal duals of dualities, internal poles of polarities and external dualities and polarities under the principle of opposites.

The algorithmic structures are driven by a thinking process to optimize real benefits subject to real costs or to optimize real costs subject to real benefits. The real variety costs and benefits are seen through the variety characteristic dispositions as abstracted from the ontological variety dispositions in the movements from knowledge by acquaintance to knowledge by description. It is through this organic dualistic-polar process that we have duality principle in the theory of decision-choice processes over the spaces of quantity-quality dualities, possibility-impossibility dualities and certainty-uncertainty dualities with mathematical, linguistic and non-mathematical optimization, involving maximization of the dual dispositions subject to the opposing dual dispositions to maintain the system of relational give-and take-sharing modes and the minimization of alternative dual dispositions subject to the opposing duals in the space of cost-benefit dualities as well as give rise to mini-max principles in thinking and reasoning in the spaces of intra-dualities inter-dualities, intra-polarities and inter-polarities. It is these dualistic-polar conditions that led Euler to state *that: Nothing happens in the universe that does not have a sense either certain maximum or minimum.* In the space of neuro-decision-choice actions of cognitive agents, the maximum and minimum are directly or indirectly operated over the space of cost-benefit dualities, where diversities are in relational continuum and unity over the space of relativity and in relational separation and disunity over the space of absoluteness. This mini-max principle has meaning in the space human actions under the principle of non-satiation.

The concepts of diversity and unity of knowing, and the diversity and unity of science accept the presence of contradictions that generate anticipations, expectations, and risk in all areas of knowing and neuro-decision-choice actions in relational give-and-take sharing modes. The anticipation, uncertainty, expectation, and risk find expressions and meanings in the space of variety cost-benefit dualities within the space of thinking-reasoning processes with inter-variety cost-benefit information to induce the ranking and execute decision-choice actions to create new forms of information-knowledge items. The problem-solution process as a condition of epistemic unity-linkage is thus connected to the *neuro-decision-choice space* under the real cost-benefit rationality over all areas of knowing from knowledge by acquaintance to knowledge by description without any exception. The neuro-decision-choice space covers all areas of neuro-decision-choice actions over the spaces of problem-solution dualities, question-answer dualities, and ignorance-knowledge dualities. All knowing and the practices of knowing are neuro-decision-choice determined and driven under human curiosity, motivation, survival, necessity, freedom and increasing comfort over the space of imagination-reality dualities, where thinking creates variety imaginations and reasoning create variety realities relative to the space of certainty-uncertainty dualities by applications through relational transports between the space of imaginations into the space or realities.

4.2.2 The Social Philosophical Consciencism and Thinking-Reasoning Processes in Knowing and Science

There are other important factors that may be considered as enhancements of the primary and derived principles of the unity of knowing, and the unity of science while we keep in mind that implied in the principles of unity are the principles diversity. These enhancing factors find residence in the social philosophical consciencism and are common to all areas of knowing whether one classifies the general space of knowing into categories of science, non-science, natural science, social science, engineering, non-engineering, explanatory science, or prescriptive science with further sub-categorial formations. The construct of the derived category of the principle of unity in this monograph also falls into the spaces of the problem-solution dualities and polarities. Similarly, the search for the enhancing factors for the primary and the derived categories of the principles of diversity and unity of knowing for our general understanding also falls into the space of problem-solution dualities under either transformation-substitution dynamics or destruction-replacement dynamics in the neuro-decision-choice space and its interactions with the information-knowledge space. All varieties in the neuro-decision-choice space are connected by thinking-reasoning processes in cognition, creation and innovation with the directions of social changes, the rise and fall of empires, the creation of independence and colonialism of nations, war-peace dualities, justice-injustice dualities, necessity-freedom dualities, continual expansion of information-knowledge accumulation and intellectual stock-flow disequilibrium dynamics that go to increase the width and depth of the social philosophical consciencism as in the intellectual internal and external covers of the social structure composed of the economic, political and legal sub-structures. The social philosophical consciencism contains all the information-knowledge elements of the social structure at each generational time point and over the space of static-dynamic dualities.

The point of essentiality finds an expression in the statement that every knowing and every neuro-decision-choice action are about cognitive activities through neuro-decision-choice actions over the space of problem-solution dualities in relation to the space of question-answer dualities constrained by cognitive capacity limitations in the epistemological space through the space of imagination-reality dualities contained in the space of actual-potential dualities. The spaces of problem-solution dualities and polarities find expressions and meanings in the space of varieties, where the elements are ranked by real cost-benefit information-knowledge structures under thinking-reasoning processes toward perfect understandings and perfect information-knowledge structures as the ultimate points to reach in the space of absoluteness from the space of relativity. The variety problem-solution dualities relate to the variety ignorance-knowledge dualities, in that, the variety ignorance is seen as a variety problem with undesirable real cost-benefit conditions in the space of the actuals contained in the space of actuals to be solved, and the variety solution is seen as a variety with desirable real cost-benefit conditions in the space of imaginations contained in the space of potentials to be found in the search space of problem-solution dualities of varieties.

The perfect understanding and the perfect knowledge imply the non-existence of ignorance, cognitive capacity limitation, uncertainty, risk and unintended consequences which define the state of complete information-knowledge system, where information is knowledge and knowledge is information and every neuro-decision-choice action is perfect and epistemological variety characteristic dispositions are the same as ontological variety characteristic dispositions and epistemological varieties are the same as ontological varieties without *epistemic distance* in the sense that variety realities are the same as variety actuals. In the perfect information-knowledge state, information is data, data is knowledge and knowledge is evidence and evidence is evidential things that may be interchangeably used in the thinking-reasoning processes in the decision-choice actions over the spaces of problem-solution dualities, question-answer dualities decidability-undecidability dualities and rejection-acceptance dualities with nonexistence of the space of imaginations such that the epistemic characteristic dispositions are the same as the ontic characteristic dispositions which implies the non-existence of the space of error-correction dualities which allow movements between the space of relativity and the space absoluteness. The information-knowledge perfect state is found in the actual-potential space contained in the ontological space and not in the space of imagination-reality dualities contained in the epistemological space in which neuro-decision-choice agents operate. Analytically, we must always keep in mind that the space of actual-potential dualities is under natural processes with information-knowledge equalities in states of input-output perfections and the space of imagination-reality dualities is under human processes with information-knowledge inequalities in the state of input-output imperfections, where every variety imagination has a supporting variety potential which may or may not be actualized and variety realities may deviate from variety actuals due to cognitive capacity limitations over the space of necessity-freedom dualities as reflected in the ruling social philosophical consciencism.

It is useful to keep in mind the difference between the concepts of actual and reality. Similarly, it is useful to also keep in mind the difference between the concepts of the space of actuals and the space of realities in relation to what there was (the telescopic past) and what there is (the telescopic present). The actual and the space of actuals are existence under natural perfection with undisputable states. The reality and the space of realities are existence under creation through human imperfections with cognitive capacity limitations and complicated by conditions of misinformation, disinformation and propaganda induced by preferences over the general space of cost-benefit dualities from the space of actuals. It is also useful to note the differences between the concepts of potential and the space of potentials on one hand and the concept of imagination and the space of imaginations on the other hand. The concepts of potential and space of potentials are related to natural processes that constitute the primary categories under perfect information. The concepts of imagination and the space of imaginations are related to human processes with cognitive capacity limitations constituting derived categories. The differences give rise to differences between the space of actual-potential dualities as the primary identities and the space of imagination-reality dualities as the derived identities under the spaces of principle of opposites with a relational continuum and unity, categorial intra-dependence and inter-dependence with differential give-and-take sharing modes. It is within the space of these differences that the space of ontic-epistemic dualities arise to give rise to the space of certainty-uncertainty dualities to produce the space of imagination that houses the spaces of possibility-impossibility dualities, probability-improbability dualities and possibility-probability dualities to give rise to the spaces of necessity-freedom dualities and anticipation-expectation dualities related to the space of doubt-surety dualities under the epistemic field, supported by relational and information fields.

Under imperfect information-knowledge systems, all institutions of learning, teaching, research, and policy are driven by the ideas generated over the space of problem-solution dualities and polarities in relation to cognitions of identifications and transformations of varieties that require conscious and unconscious searching activities over the universal space containing actual and potential varieties. The results of the problem-solution search over the variety space as intellectual investment flows are transmitted to update and increase the information-knowledge accumulation system as intellectual capital stocks in a continual process of universal information-knowledge stock-flow disequilibrium system that provides the history of the evolving socio-natural systems in disequilibrium dynamics. The general information-knowledge stock-flow disequilibrium process provides an increasing understanding of universal environment for socio-natural management and control by cognitive agents, where we learn more from solutions to problems as well as the rise of problems from solutions. The spaces of certainty-uncertainty dualities, doubt-surety dualities are testimonies of human cognitive capacity limitations at the level of acquaintance while the spaces of computability-uncomputability dualities, predictability-unpredictability dualities and acceptance-rejection dualities are evidential varieties in support of the presence of cognitive limitations at the level of description. The space of decidability-undecidability dualities involving cognitive limitations is due to the constraints from the contents of the social philosophical consciencism, where the social philosophical consciencism and the space of thinking-reasoning dualities reside in the give-and-take processes over the space of mutual constructionism-reductionism dualities. In this epistemic framework, the humanity has one way of knowing and acquiring information-knowledge about human-self and the surrounding environment which is basically neuro-decision-choice actions through thinking and reasoning, under the principle of opposites with relational continuum and unity, guided by social philosophical consciencism over the space of problem-solution dualities relative to the goal-objective space for establishing vision, aspirations, hope and a path of individual and social progresses guided by the contents of social philosophical consciencism which help to set the elements in the space of objectives, goals and visions contained in the space of imaginations in which are defined the spaces of possibility-impossibility dualities and probability-improbability dualities for neuro-decision-choice implementations to establish the cost-benefit elements in the space of realities relative to the space of acquaintance-description dualities.

Ironically, the increasing understanding through knowing and information-knowledge accumulation widens the space of ignorance since every variety problem-solution structure translates into the production of other variety solution-problem structures in the conflict zones of dualities and polarities with relational continua and unity in the space of relativity and not with excluded middles and relational separations. The whole processes of knowing and knowledge accumulations are continual dualistic-polar games of cognitive agents, where the games are between problem varieties and solution varieties in the infinite spaces of problem-solution dualities and ignorance-knowledge dualities, The areas of search are constrained by intellectual-time resources of the social setup. The more knowing we intellectually and successfully accomplish to widen the zone of knowledge, the more unknowing we intellectually and successfully acquire to widen the zone of ignorance over the space of ignorance-knowledge dualities. The increasing space of ignorance, resulting from the increasing knowing and information-knowledge accumulation in the relational connectivity among the neuro-decision-choice actions, problem-solution dualities and ignorance-knowledge dualities, is paradoxical over the space of absoluteness and finiteness under the classical paradigm of thought with the principle of excluded middle and non-acceptance of contradiction which generate anomalies in the spaces of thinking-reasoning dualities, fundamental-applied dualities and thought-practice dualities with continual restructuring of micro-paradigms in the sense of Kuhn [R3.53][R3.54]. This paradox is a paradigmatic paradox, where the problems of searches for completeness in the sense of Gödel [R26.2][R26.3] and problems of computability in the sense Turing [R26.6] [R26.7] are phantom problems in the sense of Max Planck [R8.44], where the phantom problem-solution dualities may be resolved with the fuzzy paradigm in the space of relativity under the principle of opposites with relational continua and unity, and acceptance of contradictions in socio-natural processes, where solutions to problems generate other problems in parent-offspring processes in the infinite space of primary-derived dualities contained in the space of imagination-reality dualities which is also contained in space of actual-potential dualities. It may also be pointed out that the cognitive anomalies are not restricted to quantitative space but also defined in qualitative space in the general space of quality-quantity dualities.

In the space of actual potential dualities, the concepts of completeness, termination and finality are undefined. Every logical foundation representing the conditions of absoluteness, completeness and finality will always run into some form of a paradox in thought and analyses since these concepts, in the last analysis, are reducible to conditions of equilibrium which define temporary conditions of stationary states in the dualistic-polar games in disequilibrium dynamics over the space of identification-transformation dualities. Every equilibrium is naturally temporary, except the matter-energy equilibrium which is taken to reside in permanent equilibrium flow-stock state, where all varieties revolve around it in disequilibrium states in such a way that the path of every variety transformation is connected temporary equilibrium points. The conditions of anomalies in the space of thinking-reasoning dualities are essentially the results of organic paradigm of thought and the type of information inputs into the neuro-decision-choice processes which are connected to the organic meaning and essence of life.

The perfect understanding and the perfect knowledge imply the non-existence of ignorance, cognitive capacity limitation, uncertainty, risk and unintended consequences which define the state of complete information-knowledge system, where information is knowledge and knowledge is information and every decision-choice action is perfect and the variety epistemological characteristic dispositions as abstracted from the ontological signal dispositions are the same as the variety ontological characteristic dispositions, epistemological varieties are the same as ontological varieties epistemic characteristic dispositions are the same as the ontic characteristic disposition without epistemic distances, ignorance, uncertainty and risk. In the perfect information-knowledge state, information is data, data is knowledge, and knowledge is evidence, and evidence is evidential things that may be interchangeably used in the thinking-reasoning processes in the neuro-decision-choice actions over the spaces of problem-solution dualities and decidability-undecidability dualities. This information-knowledge perfect state is found in the ontological space and not in the epistemological space.

Under imperfect information-knowledge systems, all institutions of learning, teaching, research, and policy are driven by the ideas which are generated over the space of problem-solution dualities and polarities in relation to cognition of identification and transformation of varieties that require conscious and unconscious search activities over the universal space containing actual and potential varieties. The results of the problem-solution search over the variety space are production of intellectual investment flows and transmitted to update and increase the information-knowledge accumulation system as the production of intellectual capital stocks in a continual process of the universal information-knowledge stock-flow disequilibrium processes. The general information-knowledge stock-flow disequilibrium processes provide an increasing understanding of universal environment for socio-natural management and control by cognitive agents. The spaces of certainty-uncertainty dualities and doubt-surety dualities are testimonies of cognitive capacity limitations at the level of acquaintance while the spaces of computability-uncomputability dualities and predictability-unpredictability dualities are cognitive limitations at the level of description. The space of decidability-undecidability dualities reveals cognitive limitations due to the constrains of social philosophical consciencism. In this epistemic framework, the humanity has one way of knowing and acquiring information-knowledge about human-self and the surrounding environment which is basically decision-choice actions over the space of problem-solution dualities relative to the goal-objective space for establishing visions, aspirations, hope and a path of social progress over the space of negative-positive dualities.

In fact, the existence of space of problem-solution games for cognitive agents provides a general *meaning of life,* while the way cognitive activities take place through neuro-decision-choice actions over the space of problem-solution dualities provides the general *essence of life* and its progress in the space of variety ignorance-knowledge dualities in the social set-up. Let us examine the relational connectivity between *problem-solution dualities* and *ignorance-knowledge dualities* in terms of when one may consider a successful variety problem-solution replacement and ignorance-knowledge replacement under the transformation-substitution dynamics between variety ignorance and variety knowledge given variety identifications conditions over the space of necessity-freedom dualities. It has been explained in the previous chapters that every variety is a problem-solution duality with a relative combination of *problem disposition* and *solution disposition,* where a variety is viewed as a problem or solution when the problem or solution disposition dominates the solution or problem disposition relative to a selected goal. Variety problem and variety solution are defined in terms of cost-benefit balances as seen by cognitive agents over the elements of the space of actual-potential dualities, where variety is a problem when the internal cost dual (characteristic disposition) dominates the internal benefit dual (characteristic disposition); and where the variety is a solution when the internal benefit dual (characteristic disposition) dominates internal cost dual (characteristic disposition) relative to goals, objective and vision [R8.16a].

Under the conditions of thought-practice dualities, every theory of epistemic significance must be epistemically a thought in presenting either explanatory or prescriptive conditions or both to fight to bring about some understanding of variety behaviors; it must be tender within the system of knowing to ascertain the comparative strengths and weaknesses of contributions of other theories; it must also be neuro-decision-choice action-based to accept conflicting outcomes with contradictions over the space of certainty-uncertainty dualities; it must contain substantial degrees of confidence to admit success-failure outcomes over the space of doubt-surety dualities; it should contain conditions of algorithmic humility to relate contents to the space of acceptance-rejection dualities in comparisons with other theories over the epistemic field; it should contain sufficient degrees of analytically paradigmatic strength to absolve criticisms and it should have logical flexibility to allow revisions and restructuring to withstand conditions over the space of necessity-freedom dualities in knowing and information-knowledge accumulation from the internal laws of thought and input-output process to satisfy the production-consumption chain dynamics.

Over the epistemic fields and characteristic-signal fields, every variety problem is a variety reality contained in the space of actual varieties, while every variety solution is a variety imagination also as an element contained in the space of potentials. The variety problems in the space of realities may be conceived as phantom or non-phantom problems, where the non-phantom problems may be ill-posed in such a way as to belong to the space of unsolved-unsolvable dualities. The extensive discussions on the nature and phenomena of phantom and non-phantom problems and other logical anomalies are undertaken in another monograph [R8.16b]. It is through the system of phantom problems that cognitive illusions and delusions are formed about existence, nonexistence, and morphology of varieties. The identification of a variety problem or a variety solution or both is a revelation of variety characteristic dispositions from a signal disposition and hence additions to variety information-knowledge accumulations as inputs into neuro-decision-choice actions. The combined knowing and knowledge production processes is an epistemic journey from the variety ignorance to the variety knowledge, where thinking-reasoning activities with information input-output structures are the cognitive vehicles for continual movements, where the characteristic disposition of the variety ignorance defines the initial conditions and necessity to know, and the characteristic disposition of variety knowledge defines the sufficiency and freedom of knowing within the spaces of problem-solution dualities and question-answer dualities as defined over both the epistemic field and characteristic-signal field for all areas of cognition.

Some important questions arise for cognition, and these questions encompass all areas of knowing whether the area is classified as science, non-science, social science, or non-social science. These questions involve the conditions of all epistemic journey from variety ignorance to variety knowledge and from variety knowledge to decision and action. When does one know that a variety has been transformed from either a variety problem to a variety solution or a variety solution to a variety problem? Similarly, when does one know that a *variety ignorance* has been transformed to a *variety knowledge*? In other words, when does the search journey crosses from the *zone of ignorance* to the *zone of knowledge* or *from a zone of problem characteristic disposition* to a *zone of solution characteristic disposition*? The set of conditions for neuro-decision-choice actions in relation to these questions is called the *epistemic* (or *cognitive*) *transversality conditions* to which we have discussed in another monograph [R8.16b].

These problems also involve the intra-dualistic-polar and inter-dualistic-polar journeys over the space of dualities and polarities. We must always keep in mind in knowing and information production that our awareness of problems is through variety signal dispositions which indicate the existence of the variety problems and a search for their characteristic dispositions to reveal the nature of the problem before entering the space of imaginations to search for variety solutions. At the space of imaginations, two neuro-decision-choice conditions are encountered. They are the identification neuro-decision-choice of variety solutions, goals, objective, and visions at the possibility space which houses necessities and anticipations for ranking and best selection; and the neuro-decision-choice implementations of substitutions of the best variety solutions for variety problems at the probability space that houses freedoms and expectations.

CHAPTER 5

FUZZY OPTIMIZATION OF IGNORANCE-KNOWLEDGE RELATIONAL DECOMPOSITION IN KNOWING THROUGH TRANSVERSALITY CONDITIONS

ABSTRACT

Chapter 5 is devoted to the fuzzy paradigm of thought as an organic (macro-) paradigm for thinking and reasoning in the development of the set of fuzzy meta-paradigms including the fuzzy optimization for the analysis of the conditions in the space of ignorance-knowledge dualities as an example for other dualistic-polar categorial varieties. From the cognitive framework of the fuzzy paradigm of thought, techniques and methods of fuzzy optimization are used to analyze the optimal conditions of neuro-decision-choice varieties in the space of ignorance-knowledge dualities, where variety ignorance and variety knowledge are mutually constraining for abstracting the optimality in defining the transversality conditions in the journey from the zone of ignorance to the zone of knowledge to specify the conditions for the variety identity in knowing. The conditions of optimality are used to decompose the zones of knowledge and ignorance for specifying the optimal knowledge through the fuzzy optimal decomposition. Discussions are made with examples in the use of fuzzy Numbers for optimizing ignorance-knowledge relation and the use of the conditions of the fuzzy optimality for the decomposition of the space of ignorance-knowledge dualities. The cumulative results of the chapter are the development of a theory of ignorance-knowledge dualities that allows the *specificationism* of the conditions of zones of ignorance and knowledge for claims of degrees of confidence of variety identities as known to be added to the accumulated information-knowledge stocks.

K. K. Dompere, *The Theory of Epistemic Fields*, Studies in Systems, Decision and Control 498, https://doi.org/10.1007/978-3-031-42470-0_5

5.1 AN INTRODUCTION TO A FUZZY COMPUTATIONAL APPROACH IN IGNORANCE-KNOWLEDGE DUALITIES AND POLARITIES

This chapter is devoted to the developmental analytics of the epistemic transversality conditions in journeys between duals of duality and poles of polarity in internal and external conditions of opposites. The approach is taken through the fuzzy paradigm of thought to abstract optimal paths for relational decompositions by examining and developing the epistemic transversality conditions for the points and lines of dualistic-polar separations of opposites, where then claims of knowledge or ignorance of varieties are examined against entropic conditions for information-knowledge quality controls. The approach is seen in terms of fuzzy information-processing which contains vagueness and incompleteness of experiential information as abstracted from the ontological signal dispositions at the space of acquaintance, where every variety find expressions in a system of dualities and polarities and every neuro-decision-choice action as defined in the space of cost-benefit dualities and rationality is seen as a process for an optimal dualistic-polar balance to generate the optimal points of separations.

5.2 TRANSVERSALITY CONDITIONS OF THE IGNORANCE- KNOWLEDGE GAMES UNDER THE CLASSICAL AND FUZZY PARADIGMS OF THOUGHT

The concepts and measurements of ignorance and knowledge about any variety in a dualistic-polar structures are fuzzy-stochastic variables generated by fuzzy-stochastic information systems, where a variety ignorance relates to a variety problem and variety knowledge relates to a variety solution in the space of problem-solution dualities in the sense that the spaces of problem-solution and question-answer dualities are related to the space of ignorance-knowledge (*onim-sua-*

ohu) dualities through analytical mappings, where *onim* means ignorance, *ohu* means knowledge and *sua* means knowing or learning through neuro-decision-choice journeys between variety ignorance and variety knowledge. In the space of learning, the *onim-sua-ohu* principle translates, in practice, into a simple idea that *he who knows not and learns will become he who knows*, and *he who knows not and learn not will always become he who knows not* as an element in the principle of opposites. In the space of imagination-reality dualities, the *onim-sua-ohu* principle translates into conditions that affect the transformation dynamics of the space of input-output dualities and the space of development-underdevelopment dualities in relation to the general principle of learning by doing over the space of production-consumption dualities in social set-ups and human existence. Increasing development is related to increasing learning by doing, while underdevelopment is related to unlearning by doing in the space of production of intellect investment flows to update intellectual capital stocks. The ignorance of any variety is a linguistic variable in terms of concepts and phenomena represented as a fuzzy set with specified degrees of set-belonging in terms of fuzzy membership functions. Similarly, the concept of knowledge about any variety is a linguistic variable in terms of concepts and phenomena also represented as a fuzzy set with defined degrees of set-belonging specified with fuzzy membership functions. These linguistic variables, also called fuzzy or fuzzy-stochastic variables, are expressed in degrees of set-belonging to account for vagueness and information-volume constraints that are expressed as variety qualitative and quantitative dispositions, where fuzzy information is imperfect with vagueness, inexactness and incompleteness, and classical information is imperfect with clarity, exactness, and incompleteness. The characteristics of any variety belong to the space of quantity-quality dualities to make up the characteristic disposition of the variety. The nature of qualitative and quantitative dispositions which allows neuro-decision-choice interpretations of identities of varieties is discussed in [R3.26] [R4] [R4.29] [R4.35] [R4.35] [R17.17] [R17.18].

In this structure, the epistemic transversality conditions in ignorance-knowledge dualities are fuzzy processes that are obtained as points of fuzzy optimality in relation to some elements in the space of goals and objectives. Let us keep in mind that each variety characteristic disposition is made up of negative characteristic disposition and the positive characteristic disposition. The negative characteristic disposition is the negative dual and the positive characteristic disposition is the positive dual, where each dual is defined in the space of cost-benefit dualities in the space of neuro-decision-choice actions over the space of imagination-reality dualities as abstracted from the space of actual-potential dualities. The space of cost-benefit dualities applies to the epistemological space, where the space of goals, objectives and vision is defined but not in the ontological space where nature operates.

In the neuro-decision-choice and variety transformation processes over the space of imagination-reality dualities, under the principle of opposites in relational continuum and unity, the behaviors of negative dispositions are constraints on the behaviors of the positive dispositions and vice versa to establish variety identities in all areas and times within the space of actual-potential dualities. In the space of imagination-reality dualities abstracted from the space of actual-potential duality, the variety cost dispositions are constraints on the variety benefit dispositions and vice versa in all neuro-decision-choice activities over the space of problem-solution dualities, and where variety constructions are constrained by variety destruction in the space of creation and the methodological constructionism is constrained by the methodological reductionism in the space of intellectual investment flows and intellectual capital accumulations. The relational structures of the constraints do not only provide a way of understanding the activities and outcomes in the space of construction-destruction dualities, war-peace dualities, love-hate dualities, and many dualities but they also provide us with a way for useful negotiations between opposites under relational continuum and unity that may bring about an extended peace under problem-solution disequilibrium dynamics and the corresponding understandings. The mutual dualistic-polar constraints on varieties also provide a way of understanding the foundations of all theories of optimization over the space of input-output dualities and in continual socio-natural productions of varieties in the general space of construction-destruction and production-consumption dualities. The epistemic path of knowing is that all cognitive processes are such that variety problems are viewed as *goods* in terms of benefits, they are also viewed as *bads* in terms of costs. The benefit of a variety problem is the opportunity cost of knowing what it presents, while the cost of a variety problem is the opportunity benefit of suffering that it presents. The cost-benefit conditions of variety problems are the reverse of cost-benefit conditions of variety solution as specified in the space of negative-positive dualities, where the net benefit of variety solution is the opportunity relief of net suffering while the cost of variety solution is the possibility of new problem or no further knowing.

In the space of absoluteness under the classical paradigm of thought, every representation must be logically and axiomatically complete to avoid the Gödel incompleteness conditions and conflicts that may arise in space of consistency-inconsistency dualities. With reference to any phenomenon of interest, within the space of decidability-undecidability dualities in the space of neuro-decision-choice actions without fuzzy-stochastic conditionality, the axiomatic model representation of any variety must meet completeness conditions in the space of absoluteness. These conditions are such that the system's propositions in the model, must reflect the ontological exactness of the variety characteristic dispositions. The rules of inference from the laws of thought must be made exactly as the objective natural laws governing the behavior of the varieties. Additionally, the defined initial conditions of the variety characteristic dispositions, as reveled by info-statics or conditions of categorial conversion, must correspond exactly to the ontological identities and behaviors of variety characteristic dispositions. In other words, the epistemological conditions of the varieties in the space of imagination-reality dualities must be made as exact as the ontological conditions of the varieties in the space of actual-potential dualities. The achievement of this ontological-epistemological parallelism in the space of knowing implies the non-existence of cognitive capacity limitations.

These completeness conditions of information are equivalent to Gödel-completeness and hence by the parent-offspring process, if the initial system is identifiable, decidable, computable, predictable, and transformable in the space of identification-transformation dualities then all subsequent points of identifiability, decidability, computability, predictability and transformability in the space of identification-transformation dualities must be derivable from the initial truth of the characteristic dispositions of identifiability, decidability, computability predictability, and transformability of the epistemic representations of the epistemic system.

To achieve this completeness, the knowledge by acquaintance must be exact, true, and empirically perfect with perfect experiential information with information-knowledge equality. The laws of thought at the epistemological space must be the same as the objective laws governing the behaviors of varieties at the ontological space. These completeness conditions are the demands of the epistemic efficiency of the use of the classical paradigm of thought. They require 1) a *perfect information* at the space of acquaintance, 2) a *perfect language* at the space of recording and communication, and 3) a perfect *paradigm of thought* at the space of thinking and reasoning, and with no deficiencies of vagueness, subjectivity, and volume-limitations. The completeness conditions for the use of the classical paradigm of thought are unattainable due to the *cognitive capacity limitations* of cognitive agents and the varieties under the principle of the excluded middle. The completeness conditions are the demands for working within the space of absoluteness in the process of knowing, where the paradigm of thought represents the general principles in thinking and reasoning in the space of decidability-undecidability dualities relative to the spaces of problem-solution and question-answer dualities, where logical perfection is demanded in knowing and science.

It is, here that there are un-attainment of conditions of information perfection to work within the space of absolutes, that the space of relativity acquires an important role in all areas of knowing and information-knowledge development under information imperfection, where a need arises to create either ignorance-knowledge or problem-solution epistemic transversality conditions in gradual processes to help the establishments of the zones of effectiveness of information-knowledge claims, where the transversality conditions are derived from the space of possibility-probability dualities contained in the space of imaginations within the space of relativity to allow comparability of degrees of claims and acceptance. Here, the use of the fuzzy paradigm of thought, composed of its logic and mathematics with a relational continuum and unity offers a powerful analytical advantage over the classical paradigm of thought with its logic and mathematics under the conditions of excluded middle and non-acceptance of contradictions, where the truth value of every proposition is a listed set of two elements of the form $\{0,1\}$ to indicate a membership value of complete knowledge (1) or complete ignorance with a value (0) under the principle of opposites, where there is always a need to claim one or the other and not both of the opposites in all neuro-decision-choice actions. Over the space of absoluteness under the classical paradigm of thought, a variety is either a cost or a benefit and not both, it is either a problem or a solution and not both, where a statement is either true or false and not both and many such dualistic-polar structures that find expressions over the spaces of relativity and neuro-decision-choice actions. At this point, it is useful to suggest that there are epistemic connection between $\{0,1\}$ and the {nothingness, somethingness}, where (0) relates to nothingness and (1) relates to somethingness in the space of absoluteness. Part of this relational structure has been examined in [R8.16] and the examination will continue at any epistemic relevance. The membership characteristic function of the set belonging in the classical paradigm of thought may be specified as $\mu_{(\cdot)} \in \{0,1\}$ for all areas of knowing.

All decision-choice activities including knowing as an intellectual production of investment flows, and information-knowledge accumulation as an production of intellectual stock-flow dynamics must balance real cost-benefit and negative-positive dispositions or both in the teste-preference system of varieties under the cognitive *principle of non-satiation,* where every variety is simultaneously a cost and a benefit, and simultaneously a negative and a positive in dualistic existence and not either simple as one, the other and not both. In the classical exact representation of either knowledge or ignorance over the space of absoluteness, there is no logical relationality of degrees of belonging for either ignorance-knowledge or problem-solution interdependencies. Similarly, there is either a variety knowledge disposition or a variety ignorance disposition and not ignorance-knowledge simultaneity about a variety disposition. This is a system of classical paradigmatic paradoxes for which the classical paradigm of thought is one that also leads to the Gödel incompleteness theorems when one seeks generalizations that include extensions to varieties in the space of relativity, where the generalization cannot include varieties in the space of excluded middle. The incompleteness theorem and the completeness conditions in knowing and science must be related to two fundamental structures of *experiential information* and the *paradigm of thought*, where the paradigm of thought is constructed to avoid degrees of vagueness and subjective judgements over the space of decidability-undecidability dualities through a gradual process over the spaces of thinking-reasoning dualities and mistake-correction dualities.

The resolution of the paradox presented in the incompleteness theorem, for example, requires that in the logic of excluded middle, all the opposites are separated such that every proposition exists as a logical point which translates to either a knowledge or an ignorance and not both, where the system sits on a relational no-give-and-take sharing mode in intra-dualities and inter-dualities of varieties in the space of knowing. This classical paradigm applies to the extremes in a digital form and has a limited application in the understanding of continual neuro-decision-choice process in the space of knowing with a self-correcting information-knowledge system, where there are ignorance-knowledge variety transformations in the space of problem-solution dualities to account for changing combinations of qualitative-quantitative dispositions and the general dynamics of stock-flow conditions of the continual information-knowledge production, especially in the varieties of social events and obscurities of natural events.

Every complete generalization within the classical paradigm of thought is constrained by the varieties in the set of excluded middle which holds the system of contradictions. Any serious theory of unity of knowledge whether science or non-science cannot be meaningfully constructed without contradictions since the classical paradigm of thought has a contradiction with itself as has being pointed out for example by Russel [R3.861 [R4.7] [R18.58]. Contradictions are important for changes and changes generate contradictions to promote conditions of information-knowledge system's self-corrections of elements in the space of relativity containing the space of cognitive travelling to reach any element in the space of absoluteness. Generally, contradictions generate epistemic conflicts and tensions which form the foundations of intellectual stock-flow disequilibrium dynamics that maintain information-knowledge system as a self-correcting entity generating an increasing knowledge stocks through variety solutions and an increasing ignorance through variety problems, where problems are solutions and solutions are problems within the space of goals, objectives and visions contained in the space of imagination-reality dualities.

The classical paradigm of thought with the principle of excluded middle has an important epistemic difficulty, where considered in its true-false claims cannot fulfil its main proposition of thinking and reasoning that every statement is either true or false but not both. The classical paradigm which is consistent with the principle of opposite with relational separation and disunity, therefore, does not provide a path of ignorance-knowledge dynamics, and a continual journey over the space of problem-solution dualities. Knowledge search and discovery are processes of neuro-decision-choice actions over the space of problem-solution dualities with error-correction processes that propel the knowing actions through the space of ignorance-knowledge dualities, where ignorance dispositions dominate the knowledge disposition. Knowledge search in all areas of knowing is a gradual production process of intellectual investment flows passing through many degrees of doubt before claiming a partial certainty, where surety is relatively valued over doubt with fuzzy-stochastic conditionality.

With the classical paradigm of thought, one may claim an ignorance or a knowledge about a variety and not ignorance-knowledge relative characteristic proportions about a variety due to the use of the principle of excluded middle to constrain the areas of search. In this respect, what is the role of statistical reasoning in the acceptance process? In the information-knowledge search process to solve the variety identification problem, there are journeys on many paths of knowing that connect a variety ignorance to a variety knowledge on which cognitive agents may travel. The establishment of the journey over the continuous problem-solution knowing paths to cross either the *epistemic identification transversality conditions* or the *epistemic transformation transversality conditions* requires the acceptance of the principle of opposites with a relational continuum, and unity, and a compactible logical approach with an epistemic tracking of points of degrees of successes and degrees of failure on the lines of connectedness. The multiplicity of journeys over the knowing paths between ignorance and knowledge of varieties is also a common element of all areas of knowing even though the connecting paths may vary from area to area in the knowing process. The epistemic paths of knowing may vary from different information-knowledge areas and varieties due to differences in quality-quantity characteristic combinations in the dispositions. The appropriate logical approach for complete search is found in the fuzzy paradigm of thought which is consistent with the principle of opposites with relational continua and unity under the conditions of dualistic-polar conflicts, where claims of knowledge are fuzzy-stochastic conditional, and the conditionality takes care of an acceptable degree of fuzzy-stochastic uncertainty and information-knowledge entropy over the space of subjective-objective dualities relative to decidability-undecidability dualities. The epistemic conditionality tells us the conditions about the degree of doubt or surety associated with a claim of knowledge, ignorance, solution, or problem in the zones of the space of decidability-undecidability dualities.

In the fuzzy paradigm of thought, all propositions of ignorance and knowledge over the epistemological space are represented as sets of degrees of belonging to the opposites, where every knowledge disposition is proportional to ignorance disposition , truth disposition is proportional to falsity disposition , and sorrow disposition is proportional to happiness disposition, war disposition is proportional to peace disposition, joy disposition is proportional to sorrow proposition, nothingness disposition is proportional to somethingness disposition and many other such dispositions under the principle of opposites with relational continua and unity without which the meaning of these concepts are vacuous over the space of relativity. In this epistemic respect, a claim of existence of a problem, solution, truth, false, ignorance, knowledge, peace, war, and others are all linguistic variables in fuzzy set under dualistic-polar conditions over the neuro-decision-choice space in cognitive actions under a system of preferences expressed over the space of cost-benefit dualities.

The solutions to the claims of their meanings are formulated as fuzzy decision problems with the applications of methods and techniques of fuzzy optimization of the membership characteristic functions of the knowledge dual subset, subject to the membership characteristic function of the ignorance dual subset to abstract the optimal point of demarcation in proportionality of belonging. The *optimal fuzzy point* is the *epistemic transversality point* and the vertical line through this point is the *epistemic transversality line* between variety ignorance and variety knowledge generating connected zones of ignorance and knowledge. The similar transversality points may be obtained for other linguistic dualities such as problem-solution duality, war-peace duality, and many others. It is these dualistic-polar conditions of simultaneity of existence of duals in the same space and time that define the conditions of continuity of information-knowledge processes in variety identifications and transformations through the neuro-decision-choice actions over the spaces of problem-solution dualities and ignorance-knowledge dualities. In the knowing process, the variety identification processes create initial conditions and necessity for neuro-decision-choice actions relative to preferences and desirability, while the transformation processes create the sufficient conditions and freedom for neuro-decision-choice actions under the principle of non-satiation over the space of cost-benefit dualities over the space of unchanged-change dualities. The transversality-condition points of the dualistic-polar zones help to facilitate neuro-decision-choice actions to claim the dominance of a dual and a pole.

5.3 FUZZY REPRESENTATIONAL ANALYTICS OF IGNORANCE-KNOWLEDGE DUALITIES

As we have discussed in previous chapters the principle of opposites finds meaning over the space of relativity including the extremes which find meanings and expressions over the space of absoluteness. The principle of opposites is composed of the subprinciples of duality and polarity. The principle of opposites with relational separation, excluded middle, unacceptance of contradictions with no-give-and-take sharing modes produces diversities and disunity with *exact classical information* without variety connectivity and with the use of the classical paradigm of thought over the space of thinking-reasoning dualities. The principle of opposites with relational continuum and unity with give-and-take sharing modes, acceptance of contradiction, non-excluded middle, with variety connectivity produces diversities and relational unity with *inexact fuzzy information* and the use of the fuzzy paradigm of thought that allows the development of epistemic analysis with subjective judgments which are essential in all neuro-decision-choice processes including epistemic transversality conditions.

The point and the line indicating the *epistemic transversality conditions* in knowing through the space of problem-solution dualities is $\mathbf{x}^* \in (\mathbf{A} \cap \mathbf{B}) = \mathbf{D}$ where $(\mathbf{A})$ is a fuzzy set of either solution or knowing with a membership characteristic function $\mu_{\mathbf{A}}(\mathbf{x}) \in [0,1]$, and where every degree of solution or knowledge has a corresponding degree of problem or ignorance with $(\mathbf{B})$ as a fuzzy set of problem or ignorance with a membership characteristic function

$\mu_{\mathbf{B}}(\mathbf{x}) \in [0,1]$, where every degree of problem or ignorance has a corresponding degree of solution or knowledge with membership characteristic condition of the form $\mu_{\mathbf{A}}(\mathbf{x}) + \mu_{\mathbf{B}}(\mathbf{x}) = 1$ such that we seek an optimal separation point $\alpha^* \in [0,1]$ of the form $\mu_{\mathbf{D}}(\mathbf{x}^*) = \mu_{\mathbf{A} \cap \mathbf{B}}(\mathbf{x}^*) = \mu_{\mathbf{A}}(\mathbf{x}^*) = \mu_{\mathbf{B}}(\mathbf{x}^*) = \alpha^* \in [0,1]$. This α^*–point of demarcation is the transversality point which is then used to induce a fuzzy decomposition into dualistic-polar dominance in the game of variety ignorance-knowledge relation within the polarity into actual and potential poles as well as in the duality into either problem-solution or ignorance-knowing duals that relate to info-static and info-dynamic structures of the space of variety problem-solution or ignorance-knowledge dualities, where the extreme points of 0 and 1 sit in relational continuum and unity with other infinite points of dual such that all the elements in the fuzzy paradigm have analytical abilities to examine elements in the space of absoluteness. The fuzzy paradigm of thought is applicable to the information in the spaces of relativity and absoluteness with either exact, inexact, complete, and incomplete information structures. The classical paradigm of thought is applicable to information in the space of absoluteness only with exact, clear, complete, and incomplete information structures.

The great advantage of the fuzzy mathematical component of the fuzzy paradigm is the availability of a set of mathematical functions and the flexible processes of their modifications to incorporate subjective judgments for creativity, innovation and curiosity that may be used in dualistic-polar representations of fuzzy-stochastic information which allows one to speak of fuzzy-random variable and random-fuzzy variable in fuzzy-stochastic spaces with possibility-probability hybrid quantities in relation to elements in the space of varieties and relate to the concepts and phenomena of certainty and uncertainty. The complex variables in the space of certainty-uncertainty dualities may be split into fuzzy, random and hybrid variables to represent real phenomena in socio-natural processes depending on the assumed information structure as input into the elements in the space of neuro-decision-choice actions. There is nothing in the space of ontological-epistemological duality that is not governed by the conditions of the principle of opposites in terms of dualistic-polar structures such that our number system has negative and positive structures, and our languages have structures of antonyms and synonyms which are defined in with fuzziness in the space of relativity. The processes of knowing and information-knowledge accumulations are fuzzy processes, where the dualistic-polar conditions of the games of variety transformation, change or no change of the battle of the duals and poles as mutually constraining are made explicit in the zones of opportunity in the space of cost-benefit dualities in the destruction of the actual and the search for variety replacement form the space of the variety potentials through the space of imagination-reality dualities, where the space of ignorance-knowledge dualities with the corresponding space of uncertainty-certainty dualities is generated by cognitive capacity limitations of cognitive agents from acquaintances to descriptions. Generally, every neuro-decision-choice activity in the epistemological space is real cost-benefit-driven under cost-benefit conditions and rationality based on the fuzzy paradigm of thought that shapes thinking-reasoning processes in approximations over the spaces of problem-solution dualities and relativity, where knowing moves gradually with error-correction process from the zone of ignorance to the zone of knowledge under the conditions of ignorance-knowledge-relativity dispositions.

In general, therefore, the use of the basic analytical constructionism-reductionism duality within any paradigm of thought in understanding the process of the continual change and problem-solution dynamics is to simultaneously work from the principle of opposites composed of duality and polarity with relational continuum and unity from the first principle by rejecting the principle of excluded middle, and accepting contradictions as valid in thought and existence in combination with the fuzzy paradigm of thought over inexact information representations and analytical constructs of information-knowledge systems with epistemic conditionality to state the conditions in which the cognitive capacity limitations constrain varieties in the space of acceptance-rejection dualities. The presence of cognitive capacity limitations is understood in terms of conflicts between ontic variety characteristic dispositions and epistemic variety characteristic dispositions that give rise to the space of variety ignorance distribution which give rise to uncertainty distribution. The process of variety transformation is understood in terms of conflicts and games of opposites over the space of internal-external dualities, where knowledge is the payoff through which existing variety ignorance-problems are destroyed and transformed into new knowledge solutions about varieties from the space of imagination as a sub-space of the potential space, where every variety actual and variety potential are simultaneously problems and solutions in the destruction-replacement dynamics. Let us keep in mind that in the destruction-replacement dynamics, the destruction of a variety problem and the replacement of a variety solution generate new problem or problems which then present new variety information-knowledge structures to the stock-flow disequilibrium dynamics to provide a continual opportunities to expand the information-knowledge system through neuro-decision-choice actions over the space of problem-solution dualities, where knowing is a production of intellectual investment flows and an information-knowledge accumulation is a production of intellectual capital stocks over the space of input-output duality of real production under the cost-benefit system of static-dynamic processes.

5.3.1 Fuzzy Numbers for Optimizing Ignorance-Knowledge Relational Decomposition.

In the analysis of polarity and duality with relational continuum and unity through the principle of opposites with the fuzzy paradigm of thought, several membership functions dealing with simultaneous representations of the duals and poles are provided in [R4.10] [R5.13] [R5.35]. The analytical important point that must always be remembered is that all cognitive activities take place as variety reality in the space of actuals or variety imagination in the space of potentials with problem-solving actions, where such actions are information-decision-choice driven on cost-benefit conditions in replacing variety problems with variety solutions, where variety benefits outweigh the variety costs. The problem-solution activities are dynamic under the principle of opposites to continually generate new information-knowledge structures, not in one area of knowing but in all areas of cognition, not in the areas of diversities and disunity but in the areas of diversities and unity, not in relational excluded middles but in relational continua, and not in vertical relations of importance but in horizontal relations of importance. From the space of information-decision-choice integrative processes, every information-knowledge item is reduceable to either real cost disposition or real benefit disposition in intra-variety or inter-variety conditions of existence over the space of neuro-decision-choice actions.

It may be noted that the developments of mathematical and non-mathematical theories of approximation in the pace of variety problem-solution dualities and the space of computable-incomputable dualities find expressions and meanings within the principle of opposites with relational continuum and unity which allows the justification of the information-decision-choice practices in addition to communications over the space of source-destination dualities of approximations, relative values and comparative analytics with qualitative-quantitative dispositions. It is the presence of the relational connectivity of areas of variety knowing that provides justifications for continual creations of different areas of knowing and justifications for collaborative activities over the space of problem-solution dualities for continual developments of the space of cognitive ignorance-knowledge relationality within the space of imagination-reality dualities contained in the space of actual-potential dualities and polarities.

The principle of opposites composed of dualities and polarities may be viewed in terms of color-rainbow spectrum or sound spectrum, where each color type or sound type migrates or fades into one another by qualitative graduation. Similarly, the numbers of our number system gradually fade to one another irrespective of the base. It is this fading and gradual migration that allow the construct of simple fractions providing infinite numbers between any two digits such as (0) and (1) or between(m) and (n) to also allow us to speak of proportionality and fractals over the space of thinking-reasoning dualities. The migration or the fading phenomenon also illustrates the knowing process and information-knowledge accumulation in that the conditions of theory construction do not show the information-knowledge personality of the theorist as much as they reveal the thinking-reasoning personality of the theorist under cognitive capacity limitations in the space of problem-solution dualities for variety identifications and transformations.

The fuzzy mathematical representations of the principle of opposites include different info-static conditions to deal with the elements in the space of identification problem-solution dualities and info-dynamic conditions to deal with the elements in the space of transformation problem-solution dualities in terms of representation of the duals in duality, the poles in polarity and the respective dualities within respective poles of the polarities of varieties, and in relations to all transformations for the analytical constructs of the theory of ignorance-knowledge processes in the production of knowing as *intellectual investment flows* to update the information-knowledge stocks in the universal system of knowledge as an *intellectual capital*, and its dynamic behavior in disequilibrium processes through stock-flow conditions. Let us keep in mind that in the space of opposites, the theory of a dual or pole is also a theory of the other dual or pole such that the theory of knowledge is a theory ignorance and similarly the theory human problem solving is also the theory of human-problem generation in the space of varieties that reveals the analytical power of the fuzzy paradigm of thought over the space of representation-analysis dualities. For example, one can use the triangular fuzzy number (TFN) for the degree of ignorance-knowledge duality which is specified graphically as in Figure 5.3.1.1. In figure 5.3.1.1, the area between the horizontal line passing through the point G and the horizontal line defining the value of one (1) is an *irreducible zone of ignorance* (basic ignorance) due to cognitive capacity limitations which constrain variety information-knowledge accumulation to reach completeness. Similarly, the area between the horizontal line passing through the point F and the line defining zero (0) is an *irreducible zone of knowledge* (basic knowledge) due to unintended awareness of initial conditions for information-knowledge accumulation in the space of human existence encompassing all different kinds of acquaintance. In the knowing and information-knowledge processes, we have the basic knowledge and the basic ignorance as defining the intellectual stock-flow dynamics within the space of variety ignorance-knowledge dualities, where the basic ignorance and the basic knowledge reside in dualistic-polar conditions in continuum and unity.

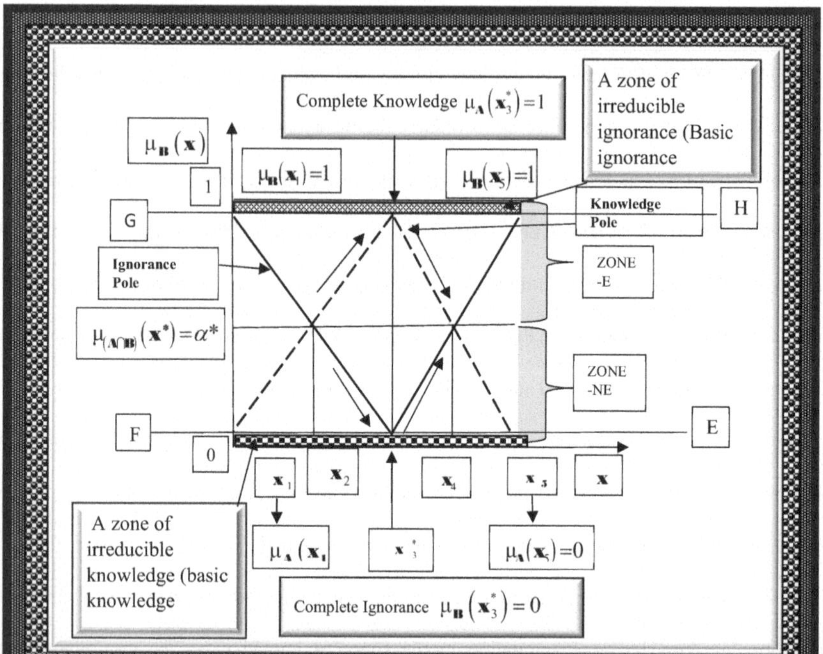

Figure5.3.1.1 Triangular Fuzzy Number (TFN) Representation of Ignorance-Knowledge Polarity

The TFN (triangular fuzzy number) is such that the increasing left upper arrow in the ZONE-E indicates an increasing knowledge. Translated in transformation conditions, an increasing knowledge is being gained about the variety as more of the variety characteristics are revealed, while the ignorance is on the decline to reveal the identity of the variety. The decreasing right lower arrow in the ZONE-E indicates a decreasing ignorance and translated in transformation conditions such that the variety is revealing its identity to solve the variety identification problem as increasing research is undertaken. In terms of polarity, any pole in the ZONE-E holds dominance over any pole in the ZONE-NE which is dominated in the problem-solution games of the ignorance-knowledge transformations. Between $\mathbf{x}_1$ and $\mathbf{x}_2$ as well as between $\mathbf{x}_4$ and $\mathbf{x}_5$ the ignorance (actual) pole dominates the knowledge (potential) pole of the polarity in knowing and in the ignorance-knowledge transformation process, where the actual pole may be seen in terms of reality pole and the potential pole may also be seen in terms of imagination pole. The points $\mathbf{x}_2$ and $\mathbf{x}_4$ are the breakeven points that define the transversality points or the tipping points in the ignorance-knowledge transformation games over the space of problem-solution dualities, where either the ignorance (actual) pole is fading away and the knowledge (potential) pole is emerging or the actualized ignorance pole is fading away to make room for a new actualized knowledge potential. Between $\mathbf{x}_2$ and $\mathbf{x}_4$ the knowledge (potential) pole of the actual-potential polarity dominates the ignorance (actual) pole in the variety-problem-solution transformation processes, where the variety potential has been actualized with the destruction of the variety problem with the substitution of the variety solution through the point F and the line defining zero is an *irreducible zone of ignorance* due to the information-knowledge accumulation. Epistemically, let always keep in mind of the complexities of spaces, interactive spaces, union spaces, intersection spaces that produce relational containments, conflicts, unity, cooperation, and give-and-take-sharing systems.

Analytically, we may specify the mathematical representation of the TFN for ignorance-knowledge polarity with internal dualities. There are two ways of representation of membership characteristic functions under the principle of opposites with relational continuum and unity to maintain the dualistic-polar give-and-take sharing processes among the opposites. One way is *open-ended* with no endpoints while the other way is a *close-ended* with the endpoints. The functional specification with endpoints means that the transformation system is such that there is a possibility to attain the conditions where completeness and incompleteness reside in each other so that [0, 1] and [1,0] are dualistically legitimate logical pairs which implies the attainment of a complete perfection or ignorance-knowledge categorial-conversion process in a continuous characteristic membership function, where the attainment of such perfection is constrained by cognitive capacity limitations. The case where completeness and incompleteness reside in each other is represented by exponential and z-fuzzy characteristic functions or similar ones different from the triangular fuzzy number or trapezoidal fuzzy number or similar ones and may be specified as in eqn. (5.3.1.1).

$$\left\{\begin{array}{l} \lim\limits_{\mathbf{x}\to\infty}\mu_{\mathbf{A}}(\mathbf{x})=\begin{cases}1 \Rightarrow \text{Zone of Complete knowledge}\\ \qquad\text{OR}\\ 0 \Rightarrow \text{Zone of Complete ignorance}\end{cases}\\ \lim\limits_{\mathbf{x}\to 0}\mu_{\mathbf{B}}(\mathbf{x})=\begin{cases}1 \Rightarrow \text{Zone of Complete ignorance}\\ \qquad\text{OR}\\ 0 \Rightarrow \text{Zone of Complete knowledge}\end{cases}\end{array}\right\}, \text{Closed End-points, } \mu_{\mathbf{A}}(\mathbf{x})\in[0,1] \quad (5.3.1.1)$$

The functional specification with no endpoints $(0,1)\subset[0,1]$ means that the transformation system or the categorial-conversion system never attains complete perfection in the limiting process of knowing due to cognitive capacity limitations of cognitive agents or attain complete imperfection due to the presence of initial natural awareness. The imperfection conditions point to several important revelations in the space of neuro-decision-choice system. Every claim is not absolute but under epistemic conditionality of degree of doubt whether communicated or not with error-correction processes. In this respect, claims of either true or false is epistemically conditional whether such conditionality is communicated or not to allow for errors and corrections. The general neuro-decision-choice system always works toward the state of perfection in the space of variety problem-solution dualities relative to the space of destruction-substitution processes without attaining the state of complete perfection which is consistent with the principle of non-satiation and socio-economic development over the space of relativity. In this situation, we have:

$$\left.\begin{cases} \lim\limits_{\mathbf{x}\to\infty}\mu_{\mathbf{A}}(\mathbf{x})=\begin{cases} <1 \Rightarrow \text{Less than and Toward Complete knowledge} \\ \qquad \text{OR} \\ >0 \Rightarrow \text{More than Complete ignorance} \end{cases} \\ \lim\limits_{\mathbf{x}\to 0}\mu_{\mathbf{B}}(\mathbf{x})=\begin{cases} <1 \Rightarrow \text{Less than and Towaerd Complete ignorance} \\ \qquad \text{OR} \\ >0 \Rightarrow \text{More than Complete lack of knowledge} \end{cases} \end{cases}\right\} \text{Open Endpoints}, \mu_{\mathbf{A}}(\mathbf{x})\in(0,1) \quad (5.3.1.2)$$

In general, we can specify any membership characteristic function to represent eqns. (5.3.1.1) and (5.3.1.2). The nature of mathematical representation will depend on the nature of the question-answer and the problem-solution dualities. In my research agenda, my focus is on establishing the epistemic relations among *fuzzy information-knowledge structures* and *neuro-decision-choice systems* over the spaces of *problem-solution dualities* and *question-answer dualities* under *the principle of opposites* with diversity in relational continua and unity and the generalized application of the fuzzy paradigm of thought to unity of knowing and science. At this point of discussion on the problem of epistemic transversality conditions on the journey from ignorance to knowledge in all areas of knowing, eqns. (5.3.1.1) and (5.3.1..2) combine to establish general conditions between the fuzzy paradigm of thought and the principle of opposites with *relational continua and unity* for information theory, data analytics, theory of evidence, theory of facts, theories of misinformation and disinformation, and the theory of knowledge under fuzzy information structures to allow for the analysis of qualitative and quantitative variety characteristic dispositions in soft and hard data over information-knowledge disequilibrium dynamics, as well as to incorporate subjectivity from neuro-decision-choice actions (See [R17.17] for definitions of information, data, fact, evidence and knowledge; see also [R17.18] for laws of info-dynamics for continual variety creation and [R4.7], for discussions on fuzzy and classical paradigms of thought and see [R21.5] for the test of validity and acceptance through analytical reductionism).

The conceptual frame and the epistemic foundations for the development of a general theory of information, theories of abstract and ordinary languages, theories of small-big data analytics, theory of evidence, complexity theory, predictive analytics, explanatory analytics, prescriptive analytics, planning analytics, engineering analytics and the general theory of knowledge, irrespective of the area of knowing are provided in the theory of knowledge square [R4.10], fuzzy rationality [R4.7], the theory of exact and inexact sciences [R3.26], the theory of info-statics [R17.17], the theory of info-dynamics [R17.18] and the theory of fuzzy-stochastic entropy [R21.5]. This monograph as an addition to these previous works presents a theory of decomposition of the problem-solution and question-answer spaces for the development of categories, institutions, and departments in the information-knowledge processes over the spaces of static-dynamic dualities and theory-application dualities. The central claim linking all these works is that information is a property of every variety under knowing which finds meaning over the space of problem-solution dualities as well as finds utility as an input of neuro-decision-choice processes in relation to variety identifications and variety transformations as seen in the space of ignorance-knowledge dualities under the principle of opposites in relational continua and unity in intra-categories and inter-categories, where knowing is the production of intellectual investment flows and information-knowledge accumulation is the production of intellectual capital stocks over the space of input-output dualities for neuro-decision-choice actions.

The unity principle of knowing, the diversity principle of problem-solution process, other forms of knowing, other principle of knowing and the problem-solution processes are to recognize the interactive audacity of neuro-decision-choice dynamics over the spaces of variety ignorance-knowledge dualities, variety problem-solution dualities and imagination-reality dualities. It is through the general principle of unity of knowing that complexity theory arises. It is through the *general diversity principle* of problem-solution processes under different variety characteristic-signal dispositions that the theory of synergy emerges. It is through the existence of categories with differential categorial characteristic-signal dispositions that different theories arise concerning the information-knowledge characteristics of identities and behaviors of categorial members and their degrees of membership. It is the set of conditions of unity principle of knowing, also as the set of conditions of general problem-solution dualities that provides the defining conditions of the unity principle of science and unity principle of the human knowledge heritage over the space of production-consumption dualities contained in the space of input-output dualities.

Alternatively viewed, the diversity and unity principle of general knowing is the *primary category of diversity and unity* while the diversity and unity principle of science and non-science is its *derived category.* It is the set of conditions of diversity principle of problem-solution dualities, also as the set of conditions of ignorance-knowledge dualities that provides the defining conditions of diversities in problem-solution dualities in the form of categorial interdependencies for all knowing including science and non-science under a system of give-and-take sharing relations. In this respect, there is the general theory of knowing as production of intellectual investment flows and information-knowledge accumulation as the production of intellectual capital stocks over the epistemic field with relational and information fields. Given the general theory of knowledge as the primary category, there are specific theories from different categories of the epistemic field with specific relations and information as sub-theories as derived categories contained in the general theory without which their validities are not justifiable. This statement is equivalent to the previous statement on organic paradigm with macro-algorithm and derived categories of micro-paradigms with micro-algorithms.

Viewed in another direction, the unity principle of knowing is the *primary category of unity* while the diversity principle of problem-solution dualities is its *derived category* for interdependencies within the space of problem-solution dualities. The information-knowledge process in relation to the unity and disunity of knowing and diversities in problem-solution dualities, is such that information is an input into neuro-decision-choice actions to obtain solutions to identification and transformation problems, the outcomes of which are the solutions to identifications and transformations of variety ignorance also as information for the next round of neuro-decision-choice actions over the space of problem-solution dualities. The relational conditions of problem-solution structures are also information over the space of ignorance-knowledge dualities. In a sense, the world of knowing is determined by neuro-decision-choice actions and information-knowledge system over the space of input-output processes, where there are always information-knowledge inequalities raising the question as to whether information or knowledge is the input into neuro-decision-choice processes. It is also these variety information-knowledge inequalities that contribute to neuro-decision-choice outcomes over the space of success-failure dualities and influence the nature of epistemic self-correction processes of neuro-decision-choice systems.

In ignorance-knowledge investigative analytics, eqn. (5.3.1.1) combines with the general conditions between the classical paradigm of thought and the principle of opposite with the *excluded middle*, the developments of information theory, data analytics, theory of evidence and the theory of knowledge under the classical information structure have very little emphasis on the role of human decision-choice judgment in ignorance-knowledge claims through neuro-decision-choice processes as gradual and limiting processes over the spaces of ignorance-knowledge and uncertainty-certainty dualities to claim outcomes. Any limiting epistemic process is a gradual process of approximation that must indicate a law of motion for connecting epistemic states. In the classical paradigmatic case, $\left[\mu_{\mathbf{A}}(\mathbf{x}) \in \{0,1\} \text{ if } \mathbf{x} \in \{\mathbf{x}_2, \mathbf{x}_4\}\right]$ as the classical excluded middle leads to a reasoning process in such a way that either $\left(\mu_{\mathbf{A}}(\mathbf{x}) = 1 \text{ if } \mathbf{x} = \mathbf{x}_3^*\right)$ or $\left(\mu_{\mathbf{A}}(\mathbf{x}) = 0 \text{ if } \mathbf{x} \leq \mathbf{x}_1 \text{ or } \mathbf{x} \geq \mathbf{x}_4\right)$ and not both for knowledge claims. Similarly, $\left(\mu_{\mathbf{B}}(\mathbf{x}) = 1 \text{ if } \mathbf{x} \leq \mathbf{x}_1 \text{ or } \mathbf{x} \geq \mathbf{x}_4\right)$ or $\left(\mu_{\mathbf{B}}(\mathbf{x}) = 0 \text{ if } \mathbf{x} = \mathbf{x}_3^*\right)$ and not both for ignorance claims. There is no partial knowledge and partial ignorance. There is either a complete knowledge or a complete ignorance as established in the space of absoluteness where gradual approximation with a law of approximation is nonexistent.

The condition $\mu_{\mathbf{B}}(\mathbf{x}) \in (0,1) \text{ if } \mathbf{x} \in (\mathbf{x}_2, \mathbf{x}_4)$ as the excluded middle under the principle of opposites is not available in the classical false-truth value system. Our general neuro-decision-choice operations work with approximations in arriving at conclusions and claims within the space of true-false dualities where error-correction processes provide epistemic motion to connect neuro-decision-choice points. The whole of our knowing as an intellectual investment process works through gradual steps and approximations from the zone of ignorance acquiring increasing variety characteristics as an equation of motion over the space of quantity-quality dualities to the zone of knowledge over the space of relativity, where the theory of knowledge is simply not about the study of criteria of claims but also about the study of process of moving from the criterion space to the spaces of decidability-undecidability dualities and doubt-surety dualities in increasing degrees of cognitive confidence, which also include the instruments for travelling from the zones of ignorance to the zones of knowledge. Let us note that our judicial neuro-decision-choice process in the space of innocence-guilty dualities moves through evidential processes where information-knowledge accumulations are abstracted from different witnesses or evidential things or both.

Generally, the system of the membership functions for the dualistic-polar conditions in the ignorance-knowledge transformation in relation to Figure 5.3.1.1 is given in eqn. (5.3.1.3).

$$\left.\begin{array}{l}
\mu_{\mathbf{A}}(\mathbf{x})=\left\{\begin{array}{l}1 \text{ if } \mathbf{x}=\mathbf{x}_3^* \\ \in(0,1) \text{ if } \mathbf{x}\in(\mathbf{x}_2,\mathbf{x}_4) \\ 0 \text{ if } \mathbf{x}\le\mathbf{x}_1 \text{ or } \mathbf{x}\ge\mathbf{x}_4\end{array}\right\} \text{ Potential(Knowledge) Pole} \\
\mu_{\mathbf{B}}(\mathbf{x})=\left\{\begin{array}{l}0 \text{ if } \mathbf{x}=\mathbf{x}_3^* \\ \in(0,1) \text{ if } \mathbf{x}\in(\mathbf{x}_2,\mathbf{x}_4) \\ 1 \text{ if } \mathbf{x}\le\mathbf{x}_1 \text{ or } \mathbf{x}\ge\mathbf{x}_4\end{array}\right\} \text{ Actual (Ignorance) Pole} \\
\mu_{\mathbf{A}}(\mathbf{x}_3^*)=\mu_{\mathbf{B}}(\mathbf{x}_3^*)=\mu_{\mathbf{B}}(\mathbf{x}^*)\wedge\mu_{\mathbf{B}}(\mathbf{x}^*) \text{ transveral points}
\end{array}\right\}, \text{ Ignorance-Knowledge Polarity} \quad (5.3.1.3)$$

Similarly, the analysis of actual-potential polarity may be represented in a dualistic-polar form of ignorance (actual or reality) pole and knowledge (potential, or imagination) pole of variety transformation in terms of dynamics of problem-solution dualities that may be mapped onto question-answer dualities. The laws of motion are such that as the membership function of the knowledge characteristic set increases, the membership function of the ignorance characteristic set falls under the dualistic-game conflicts, where the variety is said to be knowledge when the knowledge characteristic disposition dominate the ignorance characteristic disposition and may specified as:

$$\text{Ignorance-Knowledge motion}\begin{cases}\mu_{\mathbf{A}}(\mathbf{x}),\dot{\mu}_{\mathbf{A}}(\mathbf{x})\geq 0, & \text{motion on knowledge} \quad (a)\\ \mu_{\mathbf{B}}(\mathbf{x}),\dot{\mu}_{\mathbf{B}}(\mathbf{x})\leq 0, & \text{motion on ignorance} \quad (b)\\ \dot{\mu}_{\mathbf{A}}(\mathbf{x})=\dot{\mu}_{\mathbf{B}}(\mathbf{x})=0, & \text{transversality point} \quad (c)\end{cases} \tag{5.3.1.4}$$

The conditions of eqn. (5.3.1.4) over the space of cost-benefit dualities, are such that as marginal benefit is increasing in (a) marginal cost is decreasing in (b) such that the transversality condition is the point of equality of cost and benefit at the margin of motions (c) where the transversality conditions are translated as stopping rule in the space of neuro-decision-choice actions.

Every pole of a polarity has an ignorance sub-set and a knowledge characteristic sub-set in such a framework that in ignorance resides knowledge, and in knowledge resides ignorance in relationally dominating structures of dualistic-polar dynamic games of transformation and change. It is these negative-positive dualistic-polar conditions of varieties that give rise to the conditions of disappearance of the old and the emergence of the new of the same essence to establish the life-death duality, where the new resides in the old and vice versa and life resides in the death and vice versa under the matter-energy principles for continual transformations and information-knowledge stock-flow disequilibrium dynamics as input-output processes into neuro-decision-choice actions under the conditions in the space of cost-benefit dualities, where the structures of the input-output processes form the general foundations of the space of production-consumption dualities to generate the space of variety supply-demand dualities over the space of imagination-reality dualities.

5.3.2 An Example of Continuous Fuzzy Membership Functions for the Analytics of the Transversality Conditions over the Space of Ignorance-Knowledge Dualities

The analytical simultaneity of either ignorance and knowledge or problem and solution in each pole is a powerful methodological integration of the fuzzy paradigm of thought and the principle of opposites with relational continuum and unity that offers an approach to deal with qualitative-quantitative processes of transformation under any subjective-objective phenomenon over the epistemological space through the space of imagination-reality dualities. Let us keep in mind that knowing is a cognitive journey from ignorance to knowledge regarding variety identification and variety transformation. Similarly, transformations are cognitive journeys between problems and solutions based on thinking and reasoning actions.

5.3.2.1 A Geometric Representation with a Continuous Fuzzy Membership Function for the Analytics of the Transversality Conditions over the Space of Ignorance-Knowledge Dualities

Like the use of triangular fuzzy number (TFN) in dualistic-polar analytics, one may choose a continuous functional representation of the membership set. These functions include exponential, s-shape, ramp, and many others depending on the nature of the transformation decision-choice activities over the space of the problem-solution dualities [R5.35]. An example of the use of continuous characteristic membership function in dualistic-polar problem-solution analytics is show as an epistemic geometry in Figure 5.3.2.1, where the set $\mathbf{B}$ represents the set of degrees of belonging to a zone of ignorance and corresponding to it is a membership characteristic function of the form $\mu_{\mathbf{B}}(\mathbf{x})$ with ignorance zonal representation $\left(\mathbf{B},\mu_{\mathbf{B}}(\mathbf{x})\right)$. There is also a set $\mathbf{A}$ representing the set of degrees of belonging to the zone of knowledge with a membership characteristic function of the form $\mu_{\mathbf{A}}(\mathbf{x})$ with knowledge zonal representation $(\mathbf{A}, \mu_{\mathbf{A}}(\mathbf{x}))$. From these zonal structures within the space of either ignorance-knowledge dualities or problem-solution dualities, we may provide a neuro-decision-choice representational structure of the interactively dualistic-polar analytics with continuous characteristic membership functions and resolution of the dualistic conflicts, where the zones of ignorance-knowledge conflicts $\mathbb{Z}^{\mathbf{C}}$ may be specified as $\mathbb{Z}^{\mathbf{C}} = \{(\mu_{\mathbf{A}}(\mathbf{x})) - (\mu_{\mathbf{B}}(\mathbf{x})) \gtreqless 0\}$. In this representation, $\mathbb{Z}^{\mathbf{C}} < 0 \Rightarrow \text{ignorance dominance}$, $\mathbb{Z}^{\mathbf{C}} > 0 \Rightarrow \text{knowledge dominance}$ and $\mathbb{Z}^{\mathbf{C}} = 0 \Rightarrow \text{ignorance-knowledge indifference}$. These conditions are established by the neuro-decision-choice processes over the space of imagination-reality dualities contained in the space of actual-potential dualities, where the space of realities is a sub-space of the space of actuals and the space of imaginations is a sub-space of the space of potentials, all of which are constrained by dualistic-polar conditions. The dualistic-polar conditions illustrate that each dual or pole has the characteristics of its opposite dual or pole to maintain the relational continuum, unity, and give-and-take-sharing modes over the space of destruction-replacement dualities within the space of input-output processes. The spaces of actuals and potentials are ontologically defined with ontic variety characteristic dispositions under information-knowledge perfection and equalities. The spaces of realities and imaginations are epistemologically defined with epistemic variety characteristic dispositions under information-knowledge imperfection and inequalities.

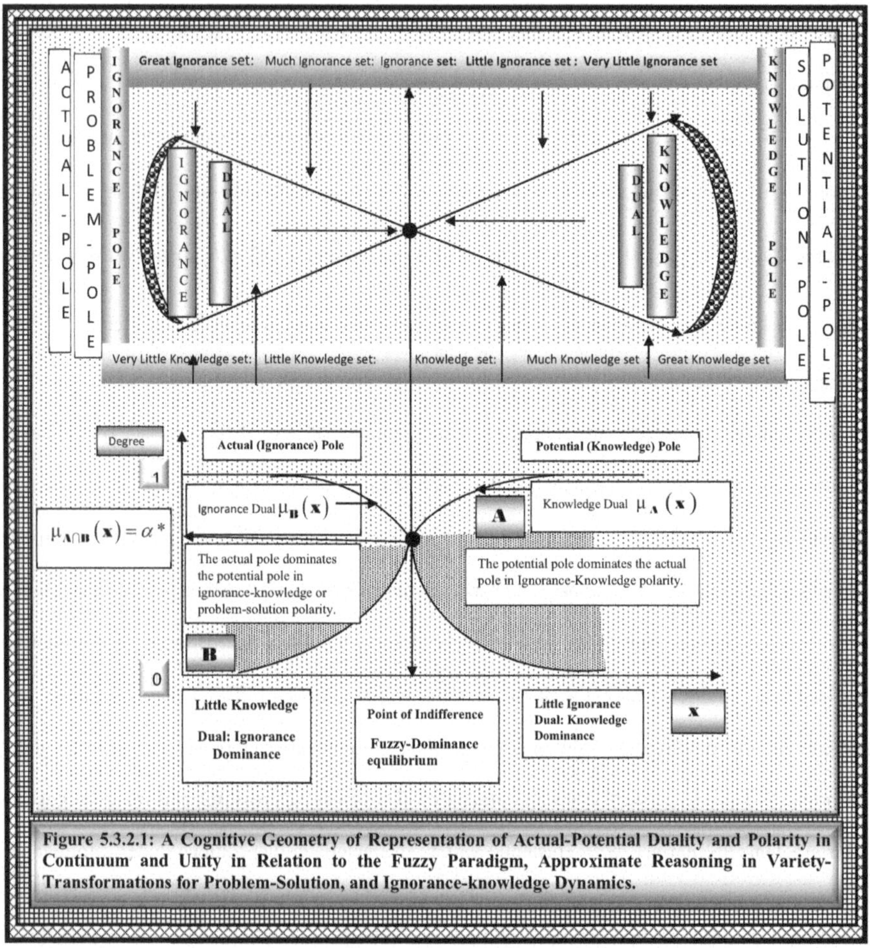

Figure 5.3.2.1: A Cognitive Geometry of Representation of Actual-Potential Duality and Polarity in Continuum and Unity in Relation to the Fuzzy Paradigm, Approximate Reasoning in Variety-Transformations for Problem-Solution, and Ignorance-knowledge Dynamics.

The neuro-decision-choice process is such that the maximization of the knowledge (benefit) fuzzy membership characteristic function is constrained by the ignorance (cost) fuzzy characteristic membership function while a minimization of the ignorance (cost) fuzzy characteristic membership function is constrained by the knowledge (benefit)fuzzy characteristic membership function under the *asantrofi-anoma rationality*. The conditions of this process of neuro-decision-choice action are the foundation on the development of theory of optimization and other forms of operations research and analytics where the neuro-decision-choice actions are to resolve the constraint-objective conflicts between goals and constraints. It is here that one must understand that the goal of every analytic, involved in either variety maintenance or change over the space of varieties as seen from the space of goal-objective dualities is either benefit-maximization with cost-constraints or cost minimization with benefit constraints by providing relevant socio-natural information-knowledge conditions as inputs, or by providing efficient algorithms as thinking procedures for reasoning to execute the programs of neuro-decision-choice actions, where the relevant algorithms become external thinking to be used by any neuro-decision-choice agent to guide the user's reasoning.

It is also here that *complexity analytics* reduce cost and increase benefit by providing simplicity to neuro-decision-choice actions over all areas of knowing and neuro-decision-choice actions in relation to the elements of the space of problem-solution dualities under both the fundamental and applied conditions of information-knowledge systems. Any theory provides a thinking algorithmic framework to guide reasoning over the space of problem-solution dualities in the production and use of knowledge, where knowledge is both an input and an output. In this respect, a theory is simply teaching (or learning) thinking algorithms of decision-choice actions as important skills over an area of the space of problem-solution dualities. As such the most important element in teaching or learning is not in relation to the existing intellectual capital stocks but in relation to the development of thinking and reasoning skills over the space of problem-solution dualities that generate knowing as the production of intellectual investment flows and information-knowledge accumulation as the production of intellectual capital stocks. Let us remember that the thinking algorithms are meta-paradigms within an organic or mega-paradigm the definitions of which have been offered. For example, the classical theories of optimization are meta-paradigms in the classical paradigm of thought with classical information structures, while the theories of fuzzy optimization are meta-paradigms in the fuzzy paradigm or thought with fuzzy information structures.

5.3.2.2 Fuzzy Continuous Mathematical Representation and Analytics of the Transversality Conditions

Mathematically, we may specify the dualistic-polar representation of either the ignorance-knowledge or problem-solution polarity of a variety transformation dynamics in Figure (5.3.2.1) in terms of respective poles as:

$$\left.\begin{array}{l} \mathbf{A}=\left\{\left(\mathbf{x},\mu_{\mathbf{A}}(\mathbf{x})\right) \mid x\in\mathbf{E},\ \mu_{\mathbf{A}}(\mathbf{x})\in[0,1] \text{ and } \left({}^{d\mu}/_{d\mathbf{x}}\right)\geq 0\right\} \text{ Potential Pole} \\ \\ \mathbf{B}=\left\{\left(\mathbf{x},\mu_{\mathbf{B}}(\mathbf{x})\right) \mid x\in\mathbf{E},\ \mu_{\mathbf{B}}(\mathbf{x})\in[0,1] \text{ and } \left({}^{d\mu}/_{d\mathbf{x}}\right)\leq 0\right\} \text{ Actual Pole} \\ \text{where}\left[\mathbf{E}=(\mathfrak{P}\otimes\mathfrak{S})\right], \mathfrak{P}=\text{problem space and } \mathfrak{S}=\text{the solution space} \end{array}\right\} \text{Ignorance-knowledge Polarity} \quad (5.3.2.1)$$

The potential pole with a residing duality represents any variety that is not known to be in existence in a coordination or place of relevance and hence covers all variety solutions, all new information-knowledge structures in theory and practice, and all non-existing or potential varieties. The conditions of the potential pole may also be replaced by the conditions of imagination pole with residing imagination duality. The actual pole with a residing duality represents any variety that is either known or unknown to be in existence and hence covers all variety problems, all, or some accumulated existing information-knowledge structures and all varieties in the space of actuals.

Similarly, the conditions of actual pole may be replaced by the conditions of reality pole with the residing duality, where the space of imagination-reality dualities resides in the space of actual-potential dualities and the space of imagination-reality polarities resides in the space of actual-potential polarities with relational continuum, unity and give-and-take sharing processes with dualistic-polar wanting modes. Given the fuzzy-set representation of dualistic-polar processes, we may then define a dualistic-polar fuzzy game as a fuzzy-transformation decision problem in the form $\Delta = \{\mathbf{A} \cap \mathbf{B} \mid \mathbf{A}, \mathbf{B} \subset \mathbf{E}\}$, whose membership characteristic function may be defined from the combination of the membership characteristic functions of ignorance and knowledge as in eqn. (5.3.2.1), where $\left[\mathbf{E} = (\mathfrak{P} \otimes \mathfrak{S})\right]$ is the space of variety problem-solution or ignorance-knowledge dualities and existence as:

$$\Delta = \left\{\left(\mathbf{x}, \mu_{\Delta}(\mathbf{x})\right) \mid \mu_{\Delta}(\mathbf{x}) = \left[\mu_{\mathbf{A}}(\mathbf{x}) \wedge \mu_{\mathbf{B}}(\mathbf{x})\right] \text{ and } \mathbf{x} \in \mathbf{E}\right\} \tag{5.3.2.2}$$

The variety-transformation decision problem is to find the transversality point or the line that may be used for an optimal fuzzy decomposition into an ignorance (problem), zone and a knowledge (solution) zone $\left(\mathcal{Z}^{\mathbf{c}} \gtreqless 0\right)$ for the search of an understanding and use by cognitive agents. The solution is obtained from the fuzzy optimization of the membership decision function of eqn. (5.3.2.2) by converting it into a fuzzy mathematical programming problem as in eqn. (5.3.2.3, A and B) such that the membership characteristic function of knowledge is maximized subject to the membership characteristic function of ignorance, where the alternative specification is to minimize the characteristic function of ignorance subject to the characteristic function of knowledge.

$$\operatorname*{opt}_{\mathbf{x}\in\mathbf{E}} \mu_{\Delta}(\mathbf{x}) = \begin{cases} \max_{x\in\mathbf{Q}} \mu_{\mathbf{A}}(\mathbf{x}) \\ \quad \text{s.t } \mathbf{Q} = \{\mathbf{x}\in\mathbf{E} \mid [\mu_{\mathbf{B}}(\mathbf{x}) - \mu_{\mathbf{A}}(\mathbf{x})] \geq 0\} \end{cases} \tag{5.3.2.3A}$$

Alternatively:

$$\operatorname*{opt}_{\mathbf{x}\in\mathbf{E}} \mu_{\Delta}(\mathbf{x}) = \begin{cases} \min_{x\in\mathbf{Q}} \mu_{\mathbf{B}}(\mathbf{x}) \\ \quad \text{s.t } \mathbf{Q} = \{\mathbf{x}\in\mathbf{E} \mid [\mu_{\mathbf{B}}(\mathbf{x}) - \mu_{\mathbf{A}}(\mathbf{x})] \geq 0\} \end{cases} \tag{5.3.2.3B}$$

The method for abstracting the algorithmic solution may be found in [R5.13] [R5.37] [R5.41] [R6] [R6.3] [R6.6] [R6.8]. The solution to the fuzzy optimization to the variety ignorance-knowledge epistemic transversality problem may be characterized as:

$$\text{Opt } \mu_{\Delta}(\mathbf{x}^*) = \{(\mathbf{x}^*, \mu_{\Delta}(\mathbf{x}^*)) \mid \mu_{\mathbf{A}}(\mathbf{x}^*) = \mu_{\mathbf{B}}(x^*) = \mu_{\Delta}(\mathbf{x}^*) = \alpha \in [0,1], \mathbf{x}^* \in \mathbf{E}\} \tag{5.3.2.4}$$

The value $\alpha^* \in [0,1]$ of eqn. (5.2.2.4) is the epistemic transversality point and the vertical line through the point $\alpha^* \in [0,1]$ is the epistemic tranversality line between the zone of variety's existence and the zone of variety's nonexistence, between the zone of ignorance and the zone of knowledge , and in general between the zone of negative dualistic-polar dominance and the zone of positive dualistic-polar dominance in their characteristic dispositions. This value is used for establishing the needed fuzzy decomposition of space of knowing into the zone of ignorance and the zone of knowledge in the form which is used in the acceptance of neuro-decision-choice action over the space of decidability-undecidability dualities as well as over the space of acceptance-rejection dualities as specified in eqn. (5.3.2. 5) in terms of a stopping, a halting or a tipping-point rule in the algorithm, where every fuzzy paradigmatic neuro-decision-choice rule has a fuzzy epistemic conditionality of relativity of form $\alpha \in (0,1)$ and the classical paradigmatic decision-choice rule has a classical epistemic conditionality of absoluteness of the form $\{0,1\}$. We may define the knowledge and ignorance zones as:

$$\left.\begin{array}{l}
\mathbf{A}=\{(\mathbf{x},\mu_{\mathbf{A}})\,|\,\mathbf{x}>\mathbf{x}^* \text{ and } \mu_{\mathbf{A}}(\mathbf{x})\geq\mu_{\mathbf{A}}(\mathbf{x}_{\mathbf{L}})=\alpha_{\mathbf{M}}>\alpha^*,\ \forall\mathbf{x}\in\mathbf{E}\}\ \text{Knowledge Zone}\\
\mathbf{B}=\{(\mathbf{x},\mu_{\mathbf{B}})\,|\,\mathbf{x}\leq\mathbf{x}^* \text{ and } \mu_{\mathbf{B}}(\mathbf{x})\leq\mu_{\mathbf{B}}(\mathbf{x}_{\mathbf{M}})=\alpha_{\mathbf{M}}<\alpha^*,\ \forall\mathbf{x}\in\mathbf{E}\}\ \text{Ignorance Zone}\\
\mathbf{A}\cap\mathbf{B}=\mathbf{C}=(\alpha_{\mathbf{M}},\alpha_{\mathbf{L}})=\text{The Zone of ignorance-knowledge conflicting claims}\\
\mu_{\mathbb{A}}(\mathbf{x}^*)=\alpha^* \text{ is the transversality point with a vertical line as}\\
\qquad\text{a transversality line}\\
\text{where } \mu_{\mathbf{A}}(\mathbf{x}_{\mathbf{M}})=\mu_{\mathbf{B}}(\mathbf{x}_{\mathbf{L}})=\alpha_{\mathbf{M}} \text{ and } \mu_{\mathbf{A}}(\mathbf{x}_{\mathbf{L}})=\mu_{\mathbf{B}}(\mathbf{x}_{\mathbf{M}})=\alpha_{\mathbf{L}}
\end{array}\right\}\ \text{Actual-Potential Polarity} \qquad (5.3.2.5)$$

In eqn.(5.3.2.5), $\alpha_{\mathbf{L}}=\mu_{\mathbf{A}}(\mathbf{x}_{\mathbf{L}})$ is the minimum point of zone of *safety claim* of knowledge acceptance in the sense that the neuro-decision-choice evaluation concludes that the knowledge content dominates the ignorance content of the variety of interest and hence the nature of the variety has been identified, The point $\alpha_{\mathbf{M}}=\mu_{\mathbf{B}}(\mathbf{x}_{\mathbf{M}})$ is the maximum point of the *zone of safety* for the claim of ignorance acceptance in the sense that the neuro-decision-choice evaluation accepts the ignorance content to completely dominate the knowledge content of the relevant variety and hence there is not enough knowledge to claim the identity of the variety in question and hence the identity of the variety is unknown. The rectangle $\mu_{\mathbf{B}}(\mathbf{x}_{\mathbf{M}})-\mu_{\mathbf{A}}(\mathbf{x}_{\mathbf{M}})-\mu_{\mathbf{B}}(\mathbf{x}_{\mathbf{L}})-\mu_{\mathbf{A}}(\mathbf{x}_{\mathbf{L}})$ equals to (EFGH) of Figure 5.3.1.1 defines the ignorance-knowledge zone of conflicts in identity claims of the information-knowledge about the relevant variety. The rectangle $\mu_{\mathbf{B}}(\mathbf{x}_{\mathbf{M}})$-$\mu_{\mathbf{A}}(\mathbf{x}_{\mathbf{M}})$-$\mu_{\mathbf{B}}(\mathbf{x}_{\mathbf{L}})$-$\mu_{\mathbf{A}}(\mathbf{x}_{\mathbf{L}})$=(EFGH) is the *zone of critical dispute* between ignorance-knowledge claims of cognitive agents. The point

$\alpha^* = \mu_\Delta(\mathbf{x}^*) = \frac{1}{2}$ in dualistic-polar games over the space of uncertainty-certainty dualities provides the fuzzy-stochastic dualistic-polar justification for *the principle of insufficient reasoning* or *the principle of indifference* in the space of neuro-decision-choice actions. The nature of the zones will depend on the dualistic-polar characteristic membership functions, the nature of the dualities, polarities, and their relationships. the principles of insufficient reasoning or indifference are explainable from the conditions of the space of epistemic distances as generate from the space of ontic-epistemic dualities contained in the space of ontological-epistemological dualities relative to the space of imagination containing the space of possibility-probability dualities. in this framework, it is not enough to human ignorance that generates uncertainty but to provide the explanatory path of its existence. The fuzzy paradigm of thought allow the analyses of all elements in the neuro-decision-choice space including systems of criteria, objectives and constraints over the space of individual-collective dualities, where the methods of fuzzy aggregations and fuzzy arithmetic become analytically useful.

From Figure 5.3.2.1, let GH be a linear approximation of the membership function of ignorance and OP the linear approximation of membership function of knowledge around the neighborhood of the transversality point, then we can represent the ignorance-knowledge zone of conflicting claims of variety identity as in Figure 5.3.2.2, where variety identity is knowledge. The experiential information-knowledge structure is such that we may specify a degree of epistemic surety as (ε) with a knowledge-acceptance confidence as $\beta_{\mathbf{L}} = (\alpha^* + \varepsilon) \geq \alpha_{\mathbf{L}}$ and an ignorance acceptance confidence as $\beta_{\mathbf{M}} = (\alpha^* - \varepsilon) \leq \alpha_{\mathbf{M}}$. The understanding of this dualistic-polar approach to decision dynamics and approximation with the fuzzy paradigm of thought, one may consult the general judicial decisions over the space of justice-injustice duality, guilty-nonguilty dualities and similarly one may consult the structure of medical decisions with or without medications over the space of sickness-wellness dualities, life-death dualities or cure-no-cure dualities, where the opposites are viewed as fuzzy variables, and all these belong to the space of problem-solution dualities as well the space of ignorance-knowledge dualities to give rise to the meanings of research, knowing, finding, learning and the system of experts. The conditions of $(\varepsilon\text{-surety})$ may be related to the classical definition of knowledge as justified true belief (JTB), where the $(\varepsilon\text{-surety})$ relates to subjective assessments of the claims and variety entropies. The $(\varepsilon\text{-surety})$ may vary over different areas of knowing and information-knowledge productions.

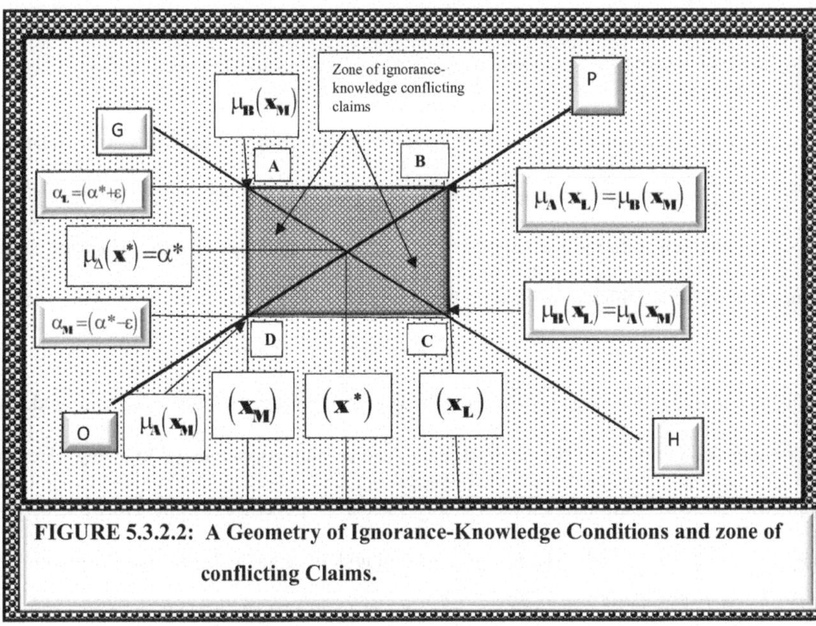

FIGURE 5.3.2.2: A Geometry of Ignorance-Knowledge Conditions and zone of conflicting Claims.

The membership functions in the dualistic-polar representations of varieties under the principle of opposites in relational continua and unity are many to capture vagueness, subjectivity, inexactness and exactness in qualitative dispositions, approximations, inaccuracies, errors and inexactness in the space of quantitative and qualitative dualistic-polar dispositions over all categories of problem-solution dualities as well as all areas of knowing through the dynamic space of parameters to capture differences in specific categorial problem-solution conditions. The possible variations of the parameters of the membership characteristic functions have the epistemic flexibility to allow for continual modifications to fit different areas of knowing, problem-solution dualities as well as to allow neuro-decision-choice actions to operate in soft, hard, non-fuzzy and fuzzy computing areas for thinking-reasoning processes in the spaces of problem-solution and ignorance-knowledge dualities in relation to information-knowledge stock-flow disequilibrium process.

The epistemic structure is such that, given the principle of opposites with relational continuum and unity, there is a continual relational connectivity over the space of relativity among the spaces of problem-solution, question-answer, ignorance-knowledge dualities which are further connected to the mind, thinking-reasoning processes, the space of neuro-decision-choice actions and the information-knowledge stock-flow disequilibrium dynamics in approximations that establish the meaning and essence of life and the beliefs of purpose of human existence under the matter-energy conditions and the neuro-decision-choice approximations of knowing of events within self-correction information-knowledge systems as governed by the principle of non-satiation over the space of relativity toward a state of undefined and unknown perfection, towards which the whole acts of thinking and reasoning are to be directed to balance the forces between opposites to achieve desired ends under the principle of non-satiation. It is through the balancing acts of thinking-reasoning processes between duals of dualities and between poles of polarities that errors are intentionally and unintentionally made for either self-correction or misinformation and disinformation to affect the distributions of the elements in the space of cost-benefit dualities.

The flexibility and the large class of the functional forms of the fuzzy membership construction in the processes of knowing with the neuro-decision-choice actions over the space of problem-solution dualities, allows one to specify various set-indicator functions that are neuro-decision-choice-based in creating the inter-dependent sets from fuzzy categories in the use of the classical paradigmatic analysis through fuzzy decompositions, where the extreme true-values of one or zero in the limits are accepted as approximations over the space of absoluteness in dealing with the development of the analytical framework for understanding the internal transformation of problem-solution dualities and ignorance-knowledge dualities as neuro-decision-choice determined in the space of relativity which may be subject to change, depending on the conditions of problem-solution needs. In this regard every claim of knowledge or ignorance, true or false and good or bad is an absolute approximation in the space of relativity in terms of relative dualistic-polar dominance in the sense that variety knowledge disposition dominates the variety ignorance disposition and vice versa. The added advantage of the flexibility of fuzzy membership functional forms over the probabilistic functional forms is that they incorporate natural linguistic vagueness and inexact thinking and reasoning as natural processes of thoughts and actions, where the elements in the family of abstract languages (FAL) used to approximate thoughts in the family of ordinary languages (FOL) provides frameworks in the development of the theory of problem solving from the fuzzy domain structures as well as constructions of dictionaries of meanings of words in the space of synonym-antonym dualities which are organized in the degrees of meaning and differences. As it has been discussed in [R3.26] [R4.7] [R4.8] [R4.10], the developments of definitions, explications and dictionaries are impossible without the conditions of vagueness in real and normal varieties in order to present conditions of the development of theories of definition in terms of synonyms and antonyms under the principles of opposites with relational continuum and unity that are governed by the fuzzy paradigm of thought over the space of relativity [R3.82][R10.11], where synonyms and antonyms are ranked in the degrees of closeness over the space of relativity and substitutions are made in terms of either decreasing or increasing degrees of belonging.

The development of the general theory of problem solving, the epistemic journey from ignorance to knowledge and the test of validity of claims of either knowledge or ignorance, true or false, yes, or no, pain or pleasure, new or old, solid, or liquid, life or death and many others of dualistic-polar structures are not limited to some areas of knowing. In fact, they are parts of the unity principle of opposites and the conditions of approximate reasoning in all areas of knowing in the sense that the positive or negative results from the study of elemental behaviors of the spaces of problem-solution, question-answer and ignorance-knowledge dualities in one area of knowledge enhance the guidelines for other areas of knowing as seen from the framework of *ontological information* as the third dimension of existence, and where time is seen as the fourth dimension of epistemological existence in addition to the three ontological elements in understanding the epistemic conditions over the space of variety *certainty-uncertainty* dualities as a dynamic phenomenon of problem-solution, question-answer and ignorance-knowledge dualities in the space of the general human neuro-decision-choice activities, and a continual search for a perfect state of being under changing generational preference schemes and the principle of non-satiation as knowing expands to increase the production of intellectual investment flows and information-knowledge accumulates to update the intellectual capital stocks as inputs for further neuro-decision-choice actions over the space of problem-solution dualities under the conditions of the spaces of production-consumption, supply-demand and input-output dualities.

Information, as the third dimension of existence, leads us to examine some important concepts in the space of thinking-reasoning processes in all cognitive activities. Here, the fuzzy-stochastic uncertainties induced by information-process within the space of imagination-reality dualities contained in the space of actual-potential duality may be intentionally or unintentionally amplified using neuro-decision-choice modules to alter the essential variety signal dispositions through disinformation and misinformation and to corrupt the decision-choice inputs to affect the senders' desired output outcomes of the receiver's decision-choice actions. Among some of the essential amplifications of fuzzy-stochastic uncertainties are *phantom problems*, *ill-posed problems*, *fake news, misinformation, disinformation, lies, deceptions, fact-denies and propaganda* within the principles of unity and diversity in knowing and science. Disinformation and misinformation are not limited to one area of knowing.

The effects of the nature of information are, however, amplified in the neuro-decision-choice actions in the basic structure composed of economic, political, and legal structures where scientific knowledge and non-scientific knowledge may be manipulated for cost-benefit advantages in the space of dualistic-polar games. The conditions of the elements of amplification and their effects in the principles of diversity and unity of knowing and science must be examined. Misinformation and disinformation in scientific and non-scientific spaces of knowing may be weaponized as political, economic, and legal instruments for oppression and injustice over the space of cost-benefit dualities where these weapons and the manners of the process of weaponization become integrated into the social philosophical consciencism to create structural variety good or variety bad to increase the space of systemic risk and racism. The conditions of some of these concepts will be examined in the next chapter. (For more discussions see [R7.17] [R7.18A]). Finally, it may be pointed out that the fuzzy characteristic membership structure of $\mu_{(\cdot)} \in [0,1]$, just like the classical functional membership belonging structure of $\mu_{(\cdot)} \in \{0,1\}$, relates to $\{\text{nothingness, somthingness}\}$ where nothingness and somethingness are the extremes belonging to the space of absoluteness and $(0,1)$ relates to $(\text{nothingness, somthingness})$ in the space of relativity with relational continuum and unity without the endpoints and $[0,1]$ relates to the space of $(0,1) \cup \{0,1\} = [0,1]$ that contains the excluded and the non-excluded middles [R4.7][R8.16]. Part of this conceptual relation has been discussed and will be further discussed as needs arise, under the principle of opposites somethingness resides in nothingness and nothingness is resides in somethingness.

CHAPTER 6

THE CONCEPTS OF THE SPACE OF IMAGINATION-REALITY DUALITIES, POSSIBILITYAND PROBABILITY IN THE PRINCIPLES OF DIVERSITY AND UNITY OF KNOWING AND SCIENCE IN THE FUZZY-STOCHASTIC SPACE

ABSTRACT: CHAPTER 6

In chapter 6, we explore certain conditions of conceptual and epistemic anomalies such as paradox, phantom, possible world, impossible world, imagination, fantasy, curiosity, and creativity in relation to reality, actuality, potentiality, and other such analytical structures that emerge in knowing as a production of intellectual investment flows and information-knowledge accumulation as a production of intellectual stock-flow activities. These concepts are placed in a system of neuro-decision-choice actions in relation to elements in the spaces of decidability-undecidability dualities and input-output dualities. The epistemic concern here, is how these anomalies relate to the space of possibility-probability dualities that houses the spaces of necessity-freedom dualities and risk-benefit dualities as one journeys between variety ignorance and variety knowledge in the space of ignorance-knowledge dualities relative to the space of intellectual investments and information-capital accumulations over the space of individual-collective dualities. The discussions on the anomalies are then linked to the discussions of categorial conversions in [R17.15] and to the discussions of social philosophical consciencism in [R17.16], where the contents in philosophical consciencism include the anomalies, conceptual weapons that generate structural conditions in the spaces of justice-injustice, war-peace and other dualistic-polar structures. The space of imagination-reality dualities within the epistemological space is introduced as an analytical space and shown to be a sub-space of the space of actual-potential

K. K. Dompere, *The Theory of Epistemic Fields*, Studies in Systems, Decision and Control 498, https://doi.org/10.1007/978-3-031-42470-0_6

dualities within the ontological space. The space of imaginations is shown as a derived space from the space of the potential as the primary space and then shown to contain the possibility and probability spaces, while the space of realities is shown to be contained by the space of the actuals and then shown to contain disinformation, misinformation, lies, phantomness, illusions and other similar varieties. The concepts and phenomena of possibility and possibility are defined and explicated with comparative differences and similarities differently from the existing ones. The space of imagination-reality dualities is argued to generate all the intellectual anomalies, the structures of which are discussed relative to the space of actual-potential dualities which contains perfect variety characteristic dispositions.

6.1: SOME REMARKS ON DISINFORMATION AND MISINFORMATION

In the previous chapter, we discussed the definition, meaning, and measure of epistemic transversality conditions in the ignorance-knowledge transitions with the fuzzy paradigm of thought, The epistemic transversality conditions are then connected to the theory of epistemic field with relational and information fields. The concepts of both ignorance and knowledge were considered as linguistic variables with vagueness which accept comparative degrees in shades of meaning and with inexact structures which accept comparative degrees in approximations, in understanding and composition of their relational values which are always determined by fuzzy neuro-decision-choice actions with fuzzy-stochastic conditionality. Similarly, both the concepts of variety problem and variety solution were considered as linguistic variables that accept comparative degrees in shades of meaning, understanding and composition of their relational values which are always determined by fuzzy neuro-decision-choice actions with a fuzzy-stochastic conditionality to the degree to which they are affected by cognitive capacity limitations at the levels of acquaintance and paradigms of thought. Every variety is seen as composed of a problem-solution duality and as an element in the space of problem-solution dualities contained in the space of imagination-reality dualities, where every variety problem is a variety solution in the space of goals, objective, and visions. The problem-solution dualities with the corresponding question-answer dualities are translated into variety ignorance-knowledge dualities in the space of ignorance-knowledge dualities. The problem-solution dualities and the ignorance-knowledge dualities are then translated into real cost-benefit dualities and then into imagination-reality dualities contained in the space of actual-potential dualities in the dualistic-polar system of variety identifications and transformations. The variety problem and the variety solution are defined in terms of relative characteristic dispositions, where a variety problem is defined with the conditions, where the internal cost characteristic disposition outweighs the internal benefit characteristic disposition; and similarly, a variety solution is defined with the condition, where the internal benefit characteristic disposition outweighs the internal cost characteristic disposition, where the total variety characteristic disposition is the sum of internal benefit and cost dispositions relative to some elements in the space of goals, objectives and visions.

Every variety, whether considered as negative, positive, cost, benefit, problem, solution goal, objective or vision has a corresponding characteristic disposition that establishes its identity which is revealed through a signal disposition as the message with contents as the characteristic disposition. The concept of information of any variety is thus defined as a characteristic-signal disposition in the space of source-destination dualities for the establishment of similarity and differences under relational continua, unity, and give-and-take-sharing modes. The characteristic disposition defines the identity while the distributions of the characteristic dispositions communicate the nature and distribution of varieties over the space of varieties to form categories of belonging. The signal disposition is always the message that carries the characteristic disposition as the content of the message over the space of source-destination dualities among varieties in the space of varieties which contains the space of actual-potential dualities containing the space of imagination-reality dualities which contains the space of problem-solution dualities which then contains the space of cost-benefit dualities.

The signal disposition is sent from the variety source to the variety destination, where the sender controls the variety signal dispositions of the variety characteristic dispositions, and the receiver must decipher the signal disposition to abstract the characteristic disposition to reveal the identity of the transmitted variety either in the real or the nominal space. The characteristic-signal messaging processes are such that the variety signal dispositions may be tampered with by the sender from the source in disguise to misrepresent the variety characteristic disposition for the purpose of affecting the nature of decisions of the receivers at the destination and change the cost-benefit relational terms over the space of problem-solution dualities to abstract advantage. The misrepresentations take the form of misinformation, disinformation, propaganda, fake news, and many deceptive forms of characteristic distortions through the signal dispositions.

6.1.1: The Effectiveness of Information-Knowledge Misrepresentation

The effectiveness of the misrepresentation of the variety characteristic dispositions lies in the input-output dynamics, where information-knowledge items are the inputs into neuro-decision-choice modules of individuals and the collective, and where the decision-choice outputs are also information-knowledge items as inputs into new neuro-decision-choice actions. The information as input-output carries with it cost-benefit implication relative to the varieties in the in the space of problem-solution dualities The degree of the effectiveness of tampering with the variety characteristic dispositions through the signal dispositions will depend on whether the input into neuro-decision-choice action is taken as information or knowledge, where knowledge as a variety characteristic disposition is abstracted and refined from information as signal disposition at the destination for an input acceptance before it is used as an input into the neuro-decision-choice action, or the signal disposition is taken as knowledge disposition and used as raw input into the neuro-decision-choice process without any intellectual massaging. This tampering process to gain advantage over the space of cost-benefit dualities in the social set-up has an important corrupting influence on the epistemic transversality conditions, where the neuro-decision-choice agent may be restricted to remain in an unknown zone of irreducible ignorance or an unknown zone of irreducible problems as happens in neocolonial states with an increasing real cost over the dualistic-polar game space such as agent-principal games, incumbent-fringe games, war-peace games, imperial-neocolonial games, labor-capital games and many others that are too many to mention.

The ignorance-knowledge epistemic transversality conditions through the space of problem-solution dualities over the dualistic-polar space of varieties are intimately and inseparably connected to cognitive activities in search of an optimal variety solution among many variety solutions to a defined variety problem relative to a goal over the space of subjective-objective dualities, where the defined variety problem and the variety goal are also neuro-decision-choice determined with an information-knowledge structure. Every variety problem and every variety solution are defined in relation to a goal-objective element in the goal-objective space, where neuro-decision-choice actions are undertaken. The neuro-decision-choice processes to set variety solutions against variety problems are motions over the space of qualitative-quantitative dualities where the inputs are fuels to bring about outputs as fuels for another round of the processes. The results of the neuro-decision-choice actions become information-knowledge additions to intellectual capital accumulation. Let us keep in mind that every variety exists as a problem-solution duality in characteristic dispositions, and a variety is said to be a problem (solution) when the problem (solution) disposition dominates the solution (problem) disposition as viewed in the space of cost-benefit dualities by cognitive agents toward a goal-objective element or elements, where variety problems and variety solutions are defined in cost-benefit relativity space. A variety is said to be a problem when the internal cost dual dominates the internal benefit dual; it is said to be a solution when the internal benefit dual dominates the internal cost dual relative to goals, objectives, or visions. All problems exist in the space of realities contained in the space of actuals, and all solutions, in addition to goals, objectives and visions exist in the space of imaginations which is contained in the space of potentials. Let us keep in mind that cost-benefit differences between variety reality and variety imagination relative to an element in the space of goals, objectives and visions create conditions for cognitive motions for neuro-decision-choice actions under the principle of non-satiation, where to every variety reality there is a better variety imagination with increasing demand for information-knowledge items. Let us keep in mind that there is no space of problem-solution dualities if there is no space of goals, objectives, and visions relative to the space of varieties under the principle of non-satiation.

The concepts of the imagination-reality, actual-potential, problem-solution, ignorance-knowledge, quality-quantity, subjective-objective, precise-imprecise, vagueness-clarity and exact-inexact dualities are in relational continua and unity and connected to all areas of knowing without exception into the general unity of information-knowledge accumulation as intellectual production in the space of relativity. The concept of relativity implies the existence of dualistic-polar conditions with simultaneity of opposites in the same space and time. The concept of unity implies diversity. The concept of diversity may imply unity with relational continuum and give-and-take sharing processes for connectivity among varieties, or disunity with relational separation and no-give-and-take sharing processes for dis-connectivity among varieties. The essential question is: what are the similarities and differences among these concepts of dualities? The answer to this question will provide a pathway to define and understand the concepts of *phantom problem, unresolvable problem, ill-posed problem, fake news,* and *propaganda, lies, disinformation and misinformation* in the principles of diversity and unity of knowing and science under neuro-decision-choice processes over the space of ignorance-knowledge dualities. The concepts of *disinformation, misinformation and propaganda* that are more pronounced in social neuro-decision-choice activities and policy developments have been discussed in other monographs [R17.17] [R17.18]. The same definitional and analytical approaches are not only useful but common and unifying across all areas of knowing and will be used here to establish the differences, similarities and identities of the concepts of phantom problem, ill-posed problem, unresolved problem, and unsolvable problem over the epistemic field with the corresponding relational and information fields in the space of neuro-decision-choice actions, where fake news are not mistakes but reside in the space of intentionality of neuro-decision-choice actions to abstract rents.

In the theory of info-statics, the universal existence is said to compose of matter, energy and information, where energy is an inseparable property of matter, and information is an inseparable property of all nominal and real existence and necessary for real and nominal definitions and explications in the field of communication, knowing and universal existence. Different matter-energy forms exist as varieties, where every variety is identified by *information.* The information is defined by characteristic disposition at the ontological space, and by characteristic-signal disposition at the epistemological space, where the characteristic disposition is the *content of the information,* and the signal disposition is the *message* in the source-destination messaging systems. In this framework, all objects, processes, concepts, ideas, vocabularies and thought exist as real and nominal varieties with content dispositions and signal dispositions to facilitate distinctions, similarities, communications, meaning, thinking, reasoning and understanding of varieties and categorial varieties in relation to diversity, continua and universal unity. The characteristic-signal disposition is the general definition of information, where this characteristic-signal disposition helps to define categories, groups, sets, rings and belonging. Based on this characteristic-signal disposition, things and varieties are distinguished with identities and naming, where a name is a characteristic disposition which is a unique set of characteristics and any variety with this unique set of characteristics belongs to a distinct category.

These dualistic-polar concepts and phenomena may also be seen in relation to the past-present-future time (information) trinity, the useful understanding of them is the *asantofi-anoma problem* composed of the three information-time dualities and polarities of past-present dualities and polarities, the present-future dualities and polarities and past-future dualities and polarities. Corresponding to them, we have the concepts of the spaces of potential-possibility dualities, imagination-reality dualities, possibility-probability dualities, probability-actual dualities, and actual-potential dualities, where many structurally dualistic-polar combinations may be formed from them. The space of possibility-probability dualities relates to the initial and sufficient conditions of change and transformation of varieties in the space of actual-potential dualities, where necessity resides in possibility, freedom resides in probability, conformity resides in the principle of knowledge credibility and progress resides in the principle of knowledge doubt in the space imagination-reality dualities. These dualistic-polar structural approaches and spaces of dualistic-polar structural approaches seem to be the best appropriate to the analysis of problem-solution dualities, ignorance-knowledge dualities and necessity-freedom dualities in variety identification and transformation processes, where necessity residing in categorial conversion indicates the initial conditions in the space of internal-external dualities.

The concept of possibility-probability dualities is an analytical path due to cognitive capacity limitations to help the process of understanding of the cognitive agents' knowing and information-knowledge processes of epistemic journeys from the zones of ignorance to the zones of knowledge over the space of actual-potential dualities and polarities through the space of imagination-reality dualities and polarities, where human actions operate. The possibility indicates necessity for change while the probability indicates freedom to change based on knowing processes and information-knowledge of intellectual capital stocks and the conditions of cost-benefit dualities relative to the elements in the space of goals, objectives, and visions under the principle of non-satiation. In terms of knowing and information-knowledge accumulation, what are the differences and similarities between the concepts of possibility and possible worlds, and between possibility space and space of possible worlds, and how will they help in conceptualizing phantom problem, ill-posed problem, fake news, misinformation, disinformation, alternative fact, unsolved problem, unsolvable problem, lies, alternative facts, propaganda, the space of propaganda and many such varieties , especially where information is seen as inputs into neuro-decision-choice actions?. The phenomena of disinformation, misinformation, propaganda have been discussed in previous monographs [R4.8] [R4.9] [R4.10], where information is the characteristic-signal disposition, misinformation is a distorted signal disposition and disinformation is the destruction of the existing variety characteristic-signal disposition.

The understanding of these phenomena of intentional and unintentional misrepresentations including others in our cognitive system regarding knowing as the production of intellectual investment flows and information-knowledge accumulation as the production of intellectual capital stocks may start by recognizing the epistemic construct of the space of imagination-reality dualities contained in actual-potential dualities with continuum and unity, where the space of the actuals contains actual varieties in terms of *what there was*, that gave rise to *what there is* which may give rise to *what there would be* through variety transformations by socio-natural processes into the potential space; and where the potential space contains potential varieties that may be transformed by socio-natural processes into the space of the actuals containing the space of realities.

The development of knowing and information-knowledge accumulation centers around variety identification and variety transformation within the space of ontological-epistemological dualities as essential analytical space for all neuro-decision-choice actions over the space of imagination-reality dualities. The space of actual-potential dualities is the essential and indispensable phenomenon to all the natural processes, while the space of imagination-reality dualities is the essential and indispensable phenomenon for social processes. The ontological space contains all varieties in the space of actual-potential dualities while the epistemological space presents a space of knowing and development of information-knowledge accumulation about the ontological varieties from the spaces of acquaintance and paradigms of thought at the directions of neuro-decision-choice actions in the space of continual variety identification, transformation and production, where the variety identification initializes the conditions for variety transformation in the possibility space and variety transformations produce changes in relation and property in the probability space. The system of productions actualizes dualistic-polar conditions of variety transformation in the space of imagination-reality dualities in which all neuro-decision-choice agents operate.

6.1.2: Epistemic Connectivity Analytics of Relevant Spaces

The ontological space constitutes the primary category of the identity space while the epistemological space constitutes the derived category of its identity space in knowing and neuro-decision-choice actions. Similarly, the space of the actual-potential dualities constitutes the primary identity space while the space of imagination-reality dualities constitutes its derived identity space. Every variety in the ontological space is an identity with perfect variety characteristic disposition, while every variety in the epistemological space is a derived identity with imperfect characteristic disposition. Any ontological variety may serve as identity to more than one derived variety over the epistemological space. All these analytics point to the idea that the connectivity of the spaces of the actuals and potentials on one hand, and spaces of imaginations and realities on the other hand is identification-transformation dynamics induced by *socio-natural processes* as the *actual-potential connectivity* and the *imagination-reality connectivity* in relational the continuum and unity. The connectivity between the ontological and epistemological spaces on one hand and between the spaces of imaginations and realities on the other is *neuro-decision-choice processes* as an *ontological-epistemological connectivity* and an *imagination-reality connectivity* also in the relational continuum and unity, all of which are connected to knowing as the production of intellectual investment flows and information-knowledge accumulation as the production of intellectual capital stocks to maintain the intellectual stock-flow disequilibrium dynamics in support of neuro-decision-choice disequilibrium dynamics.

The quality and the usefulness of the production of intellectual investment flows and the production of intellectual capital stocks in the epistemological space at any time point depend on the available epistemic instruments of production and reproduction of real life from the space of acquaintance to the space of descriptions. Most of the transformation activities over the ontological space are natural processes from the space of primary variety identities, where such variety identities propel the ontological items into the space of imagination-reality dualities by cognitive actions. Most of the transformation activities over the epistemological space are social processes from the space of derived identities, where such variety identities propel the epistemological items into the space ignorance-knowledge dualities. The set of all types of relational connectivity may be seen in terms of *fuzzy conformal membership mapping* such that the structure and path of parent-offspring replacement processes are maintained over the space of methodological constructionism-reductionism dualities within any given paradigm of thought.

All the activities at the epistemological space are social processes in pure and applied systems under the control of neuro-decision-choice actions to ascertain information-knowledge conditions about ontological varieties as inputs into the system of neuro-decision-choice actions for further variety and information-knowledge productions. To assist the development of analytical and conceptual systems in the knowing and knowledge production and the rise of anomalies, we introduce the notion of two general subspaces in the epistemological space. They are the space of *real phenomena* that has correspondence with ontological phenomena. We also have the space of *nominal phenomena*, the elements of which may or may not have correspondence with any element in the ontological space such that the epistemological characteristic dispositions have no ontological signal dispositions and hence no correspondence with the characteristic dispositions in the ontological space which contains the space of the actual-potential dualities and polarities in relational continua and unity under the principle of opposites. It is within the space of ontological-epistemological non-correspondences of characteristic dispositions that dreams and conditions of paremiology may be explained.

Every variety in the potential space has a pathway to actualization but not every element in the space of possible-world-impossible-world dualities has a pathway to actualization in the sense that the characteristic disposition of the variety has no internal or external process or technology of actualization and hence belongs to the space of impossibility-improbability dualities. Every variety that has either internal or external process and technology of actualization belongs to the space of possibility-probability dualities. The conditions of the space of actualization-non-actualization dualities will help with the understanding of the rise of different anomalies in knowing and information-knowledge processes from variety production to variety quality controls over the spaces of true-false dualities, doubt-surety dualities, acceptance-rejection dualities, and uncertainty-certainty dualities in the space of relativity. Let us keep in mind that the selections of varieties into spaces are done with cognitive actions through thinking-reasoning processes with dualistic-polar transversality conditions over the epistemic field supported by relational and information fields.

6.2 CHARACTERISTIC-SIGNAL ANALYTICS IN THE PRINCIPLES OF DIVERSITY AND UNITY IN KNOWING AND SCIENCE

The applications of the toolbox of characteristic-signal disposition as defining the meaning and contents of information within the principles of diversity and unity of knowing require several epistemic steps which are general to all areas of knowing. Every area of knowing has a category of varieties with common characteristic disposition to define the common identities. All areas of knowing in social set-ups and enterprises of information-knowledge production require a language for recording of the experiential information from the acquaintance space and analysis toward the elements of problem-solution dualities in a paradigmatic space as well as for communication over the space of source-destination dualities. The language may be a member of either the family of ordinary languages [FOL] or the family of abstract languages [FAL] or both [R17.17] [R17,18]. Any language may be either verbal, written or both. The verbal alone leads to transmitting of knowing and information-knowledge stocks by oral tradition with all the risks of system of distortions, misleading, forgetfulness and poor storage. The verbal and written languages seem to be the most efficient form for dealing with knowing and information-knowledge production as stock-flow disequilibrium dynamics with self-correcting capabilities over the spaces of problem-solution dualities and ignorance-knowledge dualities.

In general, the *neuro-decision-choice processes* as the *ontological-epistemological connectivity* in knowing and information-knowledge development are such that there is the ontological information $(\mathbb{Z}_{\Omega})$ defined by an ontological characteristic disposition $(\mathbb{X}_{\Omega})$ and a corresponding signal disposition $(\mathbb{S}_{\Omega})$ such that the ontological information is $(\mathbb{Z}_{\Omega} = \mathbb{X}_{\Omega} \otimes \mathbb{S}_{\Omega})$ where (Ω) is the ontological space. There is also the epistemological information, $\mathbb{Z}_{\mathfrak{E}} = (\mathbb{X}_{\mathfrak{E}} \otimes \mathbb{S}_{\mathfrak{E}})$, where $(\mathbb{X}_{\mathfrak{E}})$ is epistemological characteristic disposition which is abstracted from the ontological signal disposition through an *interpretive(decipher) function* $\mathfrak{I}(\bullet)$ such that $\{\mathbb{X}_{\mathfrak{E}} = \mathfrak{I}(\mathbb{S}_{\Omega})\}$, where $\{\mathbb{X}_{\mathfrak{E}} = \mathfrak{I}(\mathbb{S}_{\Omega}) \subset \mathbb{X}_{\Omega}\}$ is the epistemological characteristic disposition contained in the ontological characteristic disposition and hence $\mathbb{Z}_{\mathfrak{E}} = (\mathfrak{I}(\mathbb{S}_{\Omega}) \otimes \mathbb{S}_{\mathfrak{E}})$ and $(\mathfrak{E})$ is the epistemological space. Similarly, the information of the ontological variety is specified as $\mathbb{Z}_{\Omega_v} = (\mathbb{X}_{\Omega_v} \otimes \mathbb{S}_{\Omega_v})$ while the corresponding information of the epistemological variety is specified as $\mathbb{Z}_{\mathfrak{E}_v} = (\mathbb{X}_{\mathfrak{E}_v} \otimes \mathbb{S}_{\mathfrak{E}_v})$, with $\mathbb{X}_{\mathfrak{E}_v} = \mathfrak{I}(\mathbb{S}_{\Omega_v})$ and $\mathbb{S}_{\mathfrak{E}_v}$ is the corresponding variety epistemological signal disposition and hence $\mathbb{Z}_{\mathfrak{E}_v} = (\mathfrak{I}(\mathbb{S}_{\Omega_v}) \otimes \mathbb{S}_{\mathfrak{E}_v})$ with the conditions that $\mathbb{X}_{\mathfrak{E}_v} = \mathfrak{I}(\mathbb{S}_{\Omega_v}) \subseteq \mathbb{X}_{\Omega_v}$ and hence

$\#\left(\mathbb{X}_{\mathfrak{C}_\nu}\right)=\#\left(\mathfrak{I}\left(\mathbb{S}_{\Omega_\nu}\right)\right)\le\#\left(\mathbb{X}_{\Omega_\nu}\right)$ under the assumption that the characteristics in the disposition are countable. The efficiency of the decipher (decoding) function, $\left(\mathfrak{I}(\bullet)\right)$ always depends on the cognitive capacity limitations which are influenced by the evolving social philosophical consciencism given the cognitive abilities and the social set-up and that the coding of the signal dispositions corresponds to the ontological characteristic dispositions without camouflage from deceptions.

The ontological information constitutes the intrinsic characteristics in the basic form as the primary category of information representing the true and natural identities of the varieties. The variety ontological characteristic dispositions are the variety ontic characteristic dispositions in perfect states which constitute the variety identities. The epistemological information is the same as the knowledge by acquaintance or experiential information which is obtained by cognitive action on the ontological signal dispositions through acquaintances as a derived category of information representing the derived characteristic identities of varieties from the ontological signal dispositions over the epistemological space and corresponding to them are created variety signal dispositions which may be constrained by ontological variety signal dispositions such that $\left(\mathbb{S}_{\mathfrak{C}_\nu}\subseteq\mathbb{S}_{\Omega_\nu}\right)$and $\left(\#\left(\mathbb{S}_{\mathfrak{C}_\nu}\right)\le\#\left(\mathbb{S}_{\Omega_\nu}\right)\right)$. The space of acquaintance is broadly defined to include physical and non-physical spectra where the instruments of acquaintance are many and may be inter-supportive, conflicting, or non-inter-supportive in ascertaining the variety characteristic dispositions to reveal their identities.

The epistemological information, $\mathbb{Z}_{\mathfrak{e}_\nu} = \left(\mathbb{X}_{\mathfrak{e}_\nu} \otimes \mathbb{S}_{\mathfrak{e}_\nu}\right)$ is *epistemic* and may differ from the ontological information $\mathbb{Z}_{\Omega_\nu} = \left(\mathbb{X}_{\Omega_\nu} \otimes \mathbb{S}_{\Omega_\nu}\right)$ which is *ontic* for any given variety $(\nu \in \mathbb{V})$ due to cognitive capacity limitations of cognitive agents over the space of acquaintance, hence $\mathbb{Z}_{\mathfrak{e}_\nu} \subseteq \mathbb{Z}_{\Omega_\nu}$ due to the outcome of the interpretive (decipher) function $\mathfrak{I}(\bullet)$. The differences in the ontic and epistemic variety characteristic dispositions are the variety *epistemic distances* (φ) between the ontological and epistemological variety information and the rise of possible errors of identities of the varieties and neuro-decision-choice errors, where varieties are wrongly identified. The epistemic variety distance of $(\nu \in \mathbb{V})$ may be abstractly specified as $\varphi_\nu = \left(\#\mathbb{X}_{\Omega_\nu} - \#\mathbb{X}_{\mathfrak{e}_\nu}\right) = \left(\#\mathbb{X}_{\Omega_\nu} - \#\mathfrak{I}\left(\mathbb{S}_{\Omega_\nu}\right)\right)$. The epistemic distance is the variety ignorance and $\varpi_\nu = \left(\frac{\varphi_\nu}{\#\mathbb{X}_{\Omega_\nu}}\right) = \left(\frac{\left(\#\mathbb{X}_{\Omega_\nu} - \#\mathfrak{I}\left(\mathbb{S}_{\Omega_\nu}\right)\right)}{\#\mathbb{X}_{\Omega_\nu}}\right)$ is the measure of the *degree of ignorance* in the variable ignorance-knowledge duality which may relate to the error of knowing and dealt with by fuzzy-statistical estimate and analysis, where the fuzzy-statistical estimate is a hybrid number constructed from the space of possibility-probability dualities which is related to the spaces of *anticipation-expectation dualities* and *necessity-freedom dualities*. Corresponding to a degree of ignorance around any variety is always a degree of knowledge of the form $(1-\varpi_\nu)$. The fuzzy-statistical theory involves the use of hybrid numbers in the epistemological information system to develop relevant measures to quantify doubt and surety to assist neuro-decision-choice actions over the space of decidability-undecidability dualities for actions over the spaces of acceptance-rejection dualities and certainty-uncertainty dualities with fuzzy-stochastic conditionality, where anomalies may arise in the use of a particular paradigm of thought.

Questions arise as to whether the degree of ignorance is the same as the degree of uncertainty, the degree of knowledge is the same as the degree of certainty, where ignorance may be taken to produce uncertainty, knowledge may be taken to produce certainty, and how are these concepts related to other concepts that affect human actions? The answers to these questions will be discussed in [R8.16b] and taken up in a separate monograph on the theory of certainty-uncertainty duality. The definitions and discussions of the anomalies in this monograph will be structured to be relevant to all areas of neuro-decision-choice actions over the spaces of problem-solution dualities and ignorance-knowledge dualities keeping in mind that all neuro-decision-choice actions are human-behavioral based and all human behaviors are neuro-decision-choice action-based relative to the space of information-knowledge dualities, where anomalies, mistakes, variety epistemic distances and error-correction processes arise in the space of ontic-epistemic dualities. Let us examine the phenomenon of the *cost-benefit duality* as it relates to the variety knowing, and the information-knowledge production over the space of ontological-epistemological duality through the space of imagination-reality dualities contained in the space of actual-potential dualities.

6.2.1: Cost-Benefit Analytics over the Problem-Solution Dualities and the Rise of General Anomalies in Knowing, Information-Knowledge System and Paradigms of Thought

The Cost-benefit conceptual system has been central in the development of the ideas in this monograph, especially for the definitional system regarding neuro-decision-choice actions over all the spaces of problem-solution dualities and the conceptual extensions in terms of the principles of diversity and unity of knowing and science. The elements in the variety space are not only identified as existing in the space of positive-negative dualities and polarities but that each dual and each pole exist in the space of real cost-benefit dualities and polarities over the space of relativity, where the real cost-benefit duals provide the internal values of the varieties and the real cost-benefit dualities provide the comparative analytics of the varieties in ranking as seen by neuro-decision-choice agents. In this respect, all variety characteristic-signal dispositions as information over the epistemological space are reducible to cost-benefit characteristic dispositions as information input-output structures into the space of neuro-decision-choice actions under a paradigm of thought over the space of imagination-reality dualities, where costs are seen as undesirables and benefits are seen as desirables by neuro-decision-choice agents. Benefit is used to rank increasing variety preferences holding cost constant and cost is used to rank decreasing variety preferences holding benefit constant while relative cost-benefit reality is used to rank variety preferences over the general decision-choice space, where the cost and the benefit are defined over the space of real-nominal dualities [R5.13] [R5.14].

In a simplified structure, all neuro-decision-choice actions are undertaken in the space of desirable-undesirable dualities, where desirability relates to conditions of benefits and undesirability relates to conditions of costs. All definitions revealing the real and nominal identities and transformations of varieties find expressions and meanings over the space of real cost-benefit dualities and polarities that resides in the space of *asantrofi-anoma* conditions as derivatives from *nyansapo principle* presenting beginning-end dualities. The asantrofi-anoma conditions present to us a philosophical bird (variety) that simultaneously contain (*musuo*) cost and (*serade*) benefit for every variety in the space of variety selections by neuro-decision-choice agents, where one cannot simply select the benefit (*serade*) without simultaneously selecting the corresponding cost (*musuo*) of the same variety. The asantrofi-anoma conditions illustrate the unified nature of the variety existence over the decision-choice space as the space of cost (*musuo*) characteristics, benefit (*serade*) characteristics, conditions of relativity and rationality of choice based on cost-benefit (*musuo-serade*) balances in continua and unity in never-ending processes of transformations and identifications, where the variety preferences are presented in costs relative to benefits in the domain of the cognitive existence demanding the use of the cost-benefit analytics for variety ranking before an optimal variety selection is made among a set of decision-choice varieties.

The cost-benefit analytics, in whatever paradigm of thought that one selects, is always in relation to parent-offspring (predecessor-successor) dynamics in the adinkra conceptual tradition of *nyansampo principle* projecting the problem-solution disequilibrium dynamics over the space of input-output processes that defines the space of opportunities forgone and opportunities gained, where the opportunities forgone are the costs and the opportunities gained are the benefits. This is the science of neuro-decision-choice actions as projected by the *asantrofi-anoma* conceptual system. All analysis, statements, decision-choice actions, conclusions in information-knowledge space are paradigm-dependent over the space of cost-benefit dualities and hence are reducible to real cost-benefit dispositions with an epistemic conditionality, which may be conformally mapped onto the space of nominal cost-benefit dispositions with a corresponding epistemic conditionality. The epistemic conditionality defines the applicable areas of effective claims relative to transversality conditions, keeping in mind the epsilon-surety condition introduced to relate to the zones of knowledge-ignorance acceptances and related to the justified true belief.

In this respect, the phenomena of variety *epistemic distance* as a measure of distance between variety ignorance-knowledge duality, phantom problem, ill-posed problem, well-posed problem, solvable problem, unsolvable problem, and unsolved problems, fake news, misinformation, disinformation and propaganda, credulity, doubt, conformity, and thinking are all cost-benefit information based in a paradigm of thought, the developments of which depend on the experiential information from the space of acquaintances. In this respect, the principle of credulity in knowing generates conformity in thought and action in the space of ignorance-knowledge dualities while the principle of doubt generates thinking in thought and action in the space of ignorance-knowledge dualities $\mathfrak{Q} = (\mathfrak{U} \otimes \mathfrak{K}) = (\mathfrak{U} \cup \mathfrak{K})$, where $(\mathfrak{k} \in \mathfrak{K}) =$ knowledge space and $(\mathrm{u} \in \mathfrak{U}) =$ ignorance space with a fuzzy conditionality $(\mathfrak{U} \cap \mathfrak{K}) \neq \varnothing$. The definitions and analysis of the concepts of possibility, possible world and their opposites must derive their properties from cost-benefit characteristic dispositions, where neuro-decision-choice actions are made to set real variety benefits against real variety costs over the space of real and nominal cost-benefit dualities. In this respect, all varieties and corresponding definitions have their identities and misidentifies to reside in the space of the characteristic dispositions, where neuro-decision-choice actions are activities to balance the forces between duals, between poles and between dualities and polarities, and where the outcomes in the space of imagination-reality dualities are reducible to the elements in the space of cost-benefit dualities without remainders.

It is useful to make a note between ontic characteristic disposition and epistemic characteristic disposition if we are to understand human actions over the spaces of success-failure dualities and error-correction processes. The differences are due to the presence of cognitive capacity limitations of multiplicity of forms that restrain the correctness of the cognitive actions over all spaces of knowing and the space of individual-collective dualities. The ontic characteristics are the primary identities and associated with the variety conditions of the ontological space. The epistemic characteristics are derived by cognitive actions from the ontological characteristics as derived identities, through decoding by thinking-reasoning processes that are cognitively defective to bring about errors in outcomes generating the variety epistemic distances. The epistemic distance is composed of two forms. The knowledge-by-acquaintance epistemic distance and the knowledge-by-description epistemic distance, where the objectives of the journeys from the knowledge by acquaintance to knowledge by description are to reduce the knowledge-by-acquaintance epistemic distance. The existence of variety epistemic distances must be related to knowing and information-knowledge outcomes in the space identification doubt-surety dualities while the services of information-knowledge stocks as inputs in to the space of neuro-decision-choice actions must be related to the space of transformation doubt-surety dualities. In other words, there are two important elements that we must pay attention to in relation to the space of success-failure dualities relative to neuro-decision-choice processes. They are the ontic-epistemic variety conditions and identification-transformation variety conditions within the space of ontological-epistemological dualities. The variety epistemic distance is the difference between the variety ontic characteristic dispositions and the variety epistemic characteristic dispositions. Let us give a formal proposition for the variety epistemic distance.

PROPOSITION 6.2.1.1: Epistemic Distance

Given the ontological information, $\mathbb{Z}_{\Omega_v} = \left(\mathbb{X}_{\Omega_v} \otimes \mathbb{S}_{\Omega_v}\right)$ as an ontic representation and the epistemological information, $\mathbb{Z}_{\mathfrak{E}_v} = \left(\mathbb{X}_{\mathfrak{E}_v} \otimes \mathbb{S}_{\mathfrak{E}_v}\right)$, where $\mathbb{X}_{\mathfrak{E}_v} = \mathfrak{J}\left(\mathbb{S}_{\Omega_v}\right)$ is an epistemic representation for any variety $\left(v \in \mathbb{V}\right)$, then a *variety epistemic distance* $\left(\varphi_v\right)$ and degree of ignorance $\left(\varpi_v\right)$ are said to exist relative to a neuro-decision-choice module if:

a) $\left\{\mathbb{X}_{\mathfrak{E}_v} = \mathfrak{J}\left(\mathbb{S}_{\Omega_v}\right) \subset \mathbb{X}_{\Omega_v}\right\}$,

b) $\left(\mathbb{X}_{\Omega_v} \cap \mathbb{X}_{\mathfrak{C}_v}\right) = \mathbb{X}_{\mathfrak{C}_v} = \mathfrak{I}\left(\mathbb{S}_{\Omega_v}\right)$ and

c) $\left(\mathbb{X}_{\Omega_v} \cup \mathbb{X}_{\mathfrak{C}_v}\right) = \mathbb{X}_{\Omega_v}$ such that

$$\text{d) } \varphi_v = \left(\#\mathbb{X}_{\Omega_v} - \#\mathbb{X}_{\mathfrak{C}_v}\right)\begin{cases} = 0 \Rightarrow \text{perfect variety identification (complete knowledge)} \Rightarrow \left(\mathbb{X}_{\mathfrak{C}_v} = \mathfrak{I}\left(\mathbb{S}_{\Omega_v}\right) \subseteqq \mathbb{X}_{\Omega_v}\right) \\ > 0 \Rightarrow \text{possible error in variety identification (ignorance-knowledge structure)} \end{cases}$$

$$\text{e) } \varpi_v = \left(\frac{\varphi_v}{\#\mathbb{X}_{\Omega_v}}\right) = \frac{\left(\#\mathbb{X}_{\Omega_v} - \#\mathbb{X}_{\mathfrak{C}_v}\right)}{\left(\#\mathbb{X}_{\Omega_v}\right)} = \text{Relativity degree of ignorance}$$

$$\text{f) } \begin{cases} \left(\lim\limits_{\#\mathbb{X}_{\mathfrak{C}_v} \to \#\mathbb{X}_{\Omega_v}} \varpi_v\right) = \left(\lim\limits_{\#\mathbb{X}_{\mathfrak{C}_v} \to \#\mathbb{X}_{\Omega_v}} \left(\frac{\varphi_v}{\#\mathbb{X}_{\Omega_v}}\right)\right) = \left(\lim\limits_{\#\mathbb{X}_{\mathfrak{C}_v} \to \#\mathbb{X}_{\Omega_v}} \frac{\left(\#\mathbb{X}_{\Omega_v} - \#\mathbb{X}_{\mathfrak{C}_v}\right)}{\left(\#\mathbb{X}_{\Omega_v}\right)}\right) = 0 \Rightarrow \text{complete knowledge} \\ \left(\lim\limits_{\#\mathbb{X}_{\mathfrak{C}_v} \to 0} \varpi_v\right) = \left(\lim\limits_{\#\mathbb{X}_{\mathfrak{C}_v} \to 0} \left(\frac{\varphi_v}{\#\mathbb{X}_{\Omega_v}}\right)\right) = \left(\lim\limits_{\#\mathbb{X}_{\mathfrak{C}_v} \to 0} \frac{\left(\#\mathbb{X}_{\Omega_v} - \#\mathbb{X}_{\mathfrak{C}_v}\right)}{\left(\#\mathbb{X}_{\Omega_v}\right)}\right) = 1 \quad \Rightarrow \text{complete ignorance} \end{cases}$$

The defined epistemic distance, measure of degrees of ignorance and proposition 6.2.1.1 as dualistic-polar relations between the spaces of actual-potential dualities and ontological-epistemological polarities will help in defining the concepts of *phantom problem, ill-posed problem, unsolvable problem, misinformation, disinformation, alternative fact* and *faked news* and other relevant ones in general as they are seen in relations to the space of problem-solution dualities, variety identifications and transformations in the space of neuro-decision-choice actions over the space of cost-benefit dualities. The epistemic distance is defined in terms of the differences between ontic characteristic dispositions and the corresponding epistemic characteristic dispositions. In the case of knowledge by acquaintance, the conditions $(\varpi_v = 0)$ implies a perfect acquaintance with a correct variety identification in the sense that all the ontic characteristics of $(v \in \mathbb{V})$ have been identified and hence the relativity degree of ignorance is practically zero and the relativity degree of knowledge is one. The correct identifications of varieties also imply correct variety problem identifications $(\mathfrak{p}_\nu \in \mathfrak{P})$ as well as variety solution identifications $(\mathfrak{s}_\nu \in \mathfrak{S})$. To find correct variety solutions requires correct identifications of variety problems, where the variety problems reside in the space of reality $(\mathfrak{H})$ as a sub-space of the space of actuals $(\mathfrak{A})$ in that $(\mathfrak{p}_\nu \in \mathfrak{P} \subset \mathfrak{H} \subset \mathfrak{A} \mid \nu \in \mathbb{V})$ and the variety solutions reside in the space of imaginations $(\mathfrak{G})$ as a subspace of the space of potential $(\mathfrak{U})$ in that $(\mathfrak{s}_\nu \in \mathfrak{S} \subset \mathfrak{G} \subset \mathfrak{U} \mid \nu \in \mathbb{V})$. The concept of imperfection finds expressions in the relationship between ontological and epistemological characteristic dispositions in both nominal and real spaces which is the case, where $\varphi_\nu = \left(\#\mathbb{X}_{\Omega_\nu} - \#\mathbb{X}_{\mathfrak{C}_\nu}\right) > 0 \Rightarrow \mathbb{X}_{\mathfrak{C}_\nu} = \mathfrak{I}\left(\mathbb{S}_{\Omega_\nu}\right) \subset \mathbb{X}_{\Omega_\nu}$ and $\mathbb{X}_{\mathfrak{C}_\nu} \cap \mathbb{X}_{\Omega_\nu} \neq \varnothing$ from the proposition 6.2.1.1 with $\mathfrak{I}(\bullet)$ representing the decipher of the decoding function. The collection of all the epistemic distances is the space ignorance, $\Phi = \sum_{\nu \in \mathbb{V}} \varphi_\nu = \sum_{\nu \in \mathbb{V}} \left(\#\mathbb{X}_{\Omega_\nu} - \#\mathbb{X}_{\mathfrak{C}_\nu}\right)$. The case where $\varphi_\nu = \left(\#\mathbb{X}_{\Omega_\nu} - \#\mathbb{X}_{\mathfrak{C}_\nu}\right) = 0 \Rightarrow \left(\mathbb{X}_{\Omega_\nu} \cap \mathbb{X}_{\mathfrak{C}_\nu}\right) = \mathfrak{I}\left(\mathbb{S}_{\Omega_\nu}\right) = \mathbb{X}_{\Omega_\nu}$ and non-variety ignorance.

The objective in this epistemic approach is to seek conditions of generality in information-knowledge-decision interactive processes irrespective of the area of knowing and to find the logical foundation for the concepts and phenomena of doubt, imagination, hesitation, anticipation, expectation, and others. In other words, to examine the set of conditions that is common to all areas of knowing and thus unites knowing as the production of intellectual investment flows and information-knowledge accumulation as the production of intellectual capital stocks in line with the economic theory of production-consumption dualities seen in terms of the dynamics of input-output dualities irrespective of whether an area of knowing is science, non-science, mathematics, technology, economics or engineering with their subcategories constituting departments. The condition (f) of the proposition (6.2.1.1) points to the idea that knowing is a limiting process to gradually increase the size of epistemic characteristic dispositions to be in equality with the ontic characteristic dispositions of varieties. The objective also allows for the specification of a framework to develop a theory of uncertainty that is connected to the theories of possibility and probability.

6.2.2: An Extended Representation of the Knowledge Square with Spaces of Dualities

In the theory of knowledge square, a simplified notion of knowledge square was presented with a corresponding cognitive geometry [R4.10], an example of which is presented in Fig 6.2.2.1. where the knowledge of the universe is obtained through a sequential know-how of *what there was*, *what there is* and *what there would be* about the varieties in the universe. In this respect, we specify the potential space of knowing as $\left(\mathfrak{U}, \mathbb{I}_{\mathfrak{U}}^{\infty}\right)$, where $(\mathfrak{U})$ is the space of potential varieties with an index set $\left(\mathbb{I}_{\mathfrak{U}}^{\infty}\right)$, the possibility space of acquaintance as $\left(\mathfrak{P}, \mathbb{I}_{\mathfrak{P}}\right)$, where $(\mathfrak{P})$ is the space of possible varieties from acquaintance and $\left(\mathbb{I}_{\mathfrak{P}}\right)$ is the index set of the possible varieties for identifications and transformations from the elements in the acquaintance space, the probability space of certainty and doubt as $\left(\mathfrak{B}, \mathbb{I}_{\mathfrak{B}}\right)$, where $(\mathfrak{B})$ is the space of knowledge with doubt and $\left(\mathbb{I}_{\mathfrak{B}}\right)$ is the index set of probable varieties for identifications and transformations, and the space of the actual of what there is as $\left(\mathfrak{A}, \mathbb{I}_{\mathfrak{A}}\right)$, where $(\mathfrak{A})$ is the space of actual varieties and $\left(\mathbb{I}_{\mathfrak{A}}\right)$ is its index set with the sequential paths of knowing and knowledge development as shown in Figure 6.2.2.1 diagram of the knowledge square, where every variety of knowing goes through potential space, possibility space, the probability space and then actual space of knowing. In this respect, it is useful to define and explicate the concepts of potential, actual, possibility and probability and show how they affect knowing, and information-knowledge production.

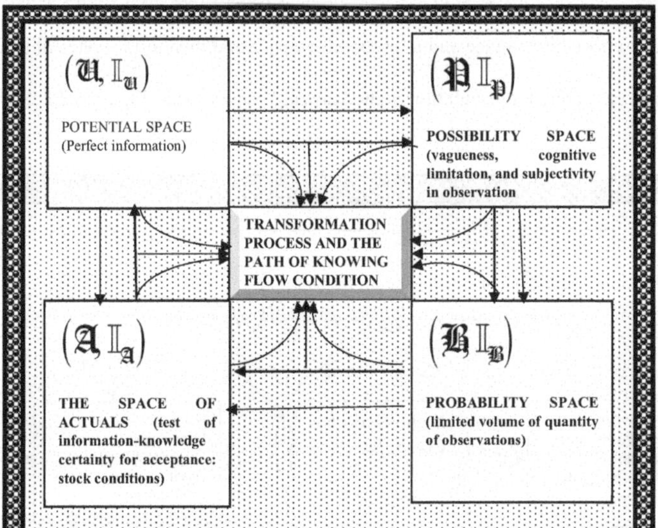

FIGURE 6.2.2.1: The Epistemic Geometry of the Variety knowing from Potential as the point of all knowable Varieties to the Possible as the Point of Information Acquaintance to Probable as the point of Information Doubt to the Actual as the Point of Degree of Information-Knowledge Doubt and back to the Potential in a never-ending Process in the Information-Knowledge Search about what *there was, is and would be* of Universal Objective Set.

The epistemic structure as presented in figure 6.2.2.1 is not complete does not provide us sufficient story of the process of knowing as the production of intellectual investment flows. The incompleteness comes from the manner in which the possibility and probability are linked to the structure that is connected to neuro-decision-choice actions over the spaces of knowing as the production of intellectual investment flows and information-knowledge accumulation as the production of intellectual capital stocks, where the variety potentials generate the space of potentials, the variety actuals generate the space of actuals, the variety possibilities generate the possibility space and the variety probabilities generate the probability space, where the knowing process and the information-knowledge process are linked by neuro-decision-choice actions from the potential space to possibility space and to probability space and then to the space of actuals to present an extended epistemic representation in the knowing process as in Figure 6.2.2.2.

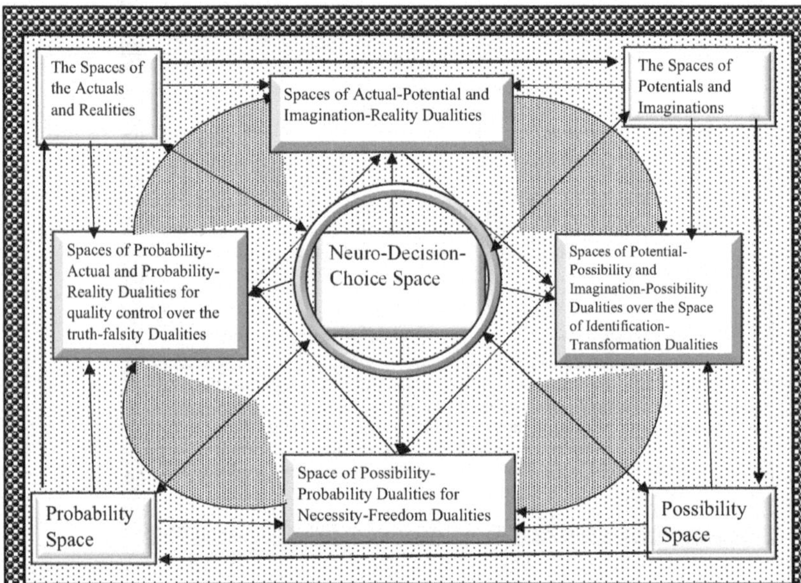

Figure 6.2.2.2: A Cognitive Geometry of a Space of Dualities for Neuro-Decision-Choice actions in the Space of Imagination-Reality Dualities as a Sub-Space of Actual-Potential Dualities

The system of cognitive capacity limitations gives rise to the volume incompleteness and vagueness in the observations and representations in many forms in the space of acquaintance to give rise to variety degree ignorance. The volume incompleteness and vagueness in observations and representations of variety characteristic dispositions then give rise to the phenomena of imaginations and the space of imaginations derived from the space of potentials, where the phenomenon of imagination is driven by the phenomenon of degree of ignorance or the degree of knowledge in terms of variety characteristic dispositions. The imaginations and the space of imaginations give rise to possibilities and the possibility space and connected to probabilities and probability space within the space of neuro-decision-choice actions with doubts and surety over the space of quality-quantity dualities in knowing, vocabulary and multiple meanings in languages over the communicational space of source-destination dualities and actions over the success-failure dualities. The linkage between the space of potentials and the space of actuals on one hand, and the possibility space and the probability space on the other hand in the enterprise of knowing is the space of imaginations and the space of realities as represented in Figure 6.2.2.3. which combines figure 6.2.2.1 with figure 6.2.2.2 through decompositions and re-compositions.

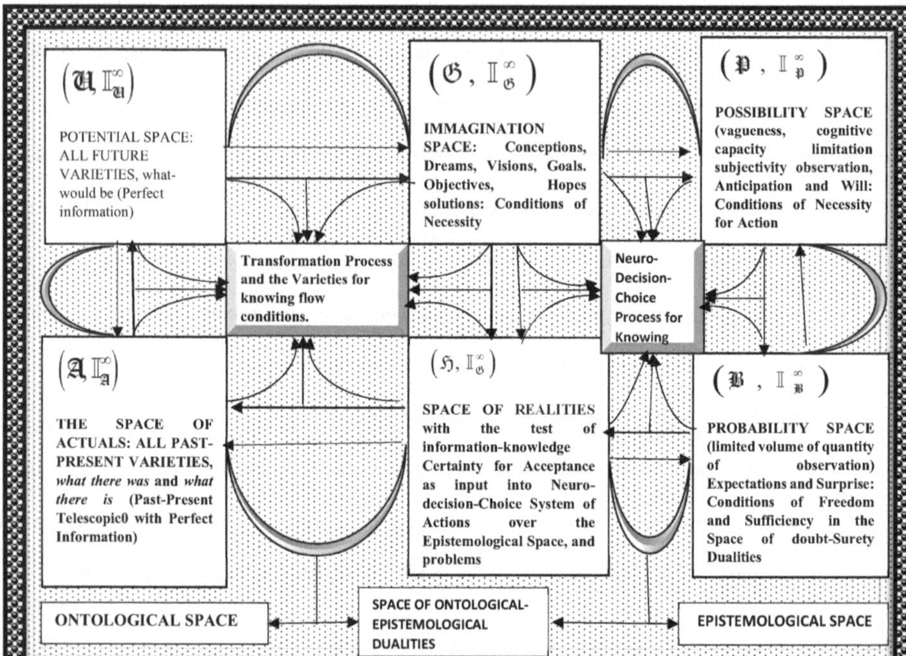

FIGURE 6.2.2.3: The Epistemic Geometry of the Variety knowing from Potential as the Space of all knowable Varieties to the Possible as the Space of variety Acquaintance to Probable as the Space of Information Doubt to the Actual as the Space of Degree of Information-Knowledge Uncertainty and back to the Potential in a never-ending Process in the Information-Knowledge Search about *what there was, what there is and what would be* of the Universal Objective Set.

The epistemic geometry of figure 6.2.2.3 may also be viewed as paremiologically cognitive geometry requiring some logical connectivity for interpretational completeness in understanding, why $(\mathfrak{G}, \mathbb{I}^{\infty}_{\mathfrak{G}})$ is the space of imaginations $(\mathfrak{G})$ with an index set $(\mathbb{I}^{\infty}_{\mathfrak{G}})$ and $(\mathfrak{H})$ is the space of realities with an index set $(\mathbb{I}^{\infty}_{\mathfrak{H}})$ and $\{(\mathfrak{G} \otimes \mathfrak{H}) = (\mathfrak{G} \cup \mathfrak{H}), \mathbb{I}^{\infty}\}$ is the space of imagination-reality dualities with an index set $(\mathbb{I}^{\infty}_{\mathfrak{G}})$ under dualistic-polar conditions and fuzzy-stochastic conditionality with a generic $\{(\mathfrak{g},\mathfrak{h}) \in (\mathfrak{G} \otimes \mathfrak{H}), \mathbb{I}^{\infty}\}$, where every variety imagination has a corresponding variety reality and vice versa. The epistemically structural geometry provides a foundation for analyzing the relationships among the variety ignorance, the phenomena of uncertainty and the space of imaginations that contains possibility and probability, from which success-failure dualities emerge, where the space of imaginations is a sub-space of the space of potentials.

It is useful to keep in mind that all neuro-decision-choice actions regarding variety identifications and variety transformations take place over the space of imagination-reality dualities $(\mathfrak{H} \otimes \mathfrak{G})$ with a generic element $(\mathfrak{h},\mathfrak{g}) \in (\mathfrak{H} \otimes \mathfrak{G})$, where $(\mathfrak{h} \in \mathfrak{H})$ is the space of realities and with the dualistic-polar condition of the form $(\mathfrak{H} \otimes \mathfrak{G}) = (\mathfrak{H} \cup \mathfrak{G})$, with $(\mathfrak{G} \cap \mathfrak{H}) \neq \varnothing$ defining a relational continuum and unity. The structure $(\mathfrak{g} \in \mathfrak{G}) = \text{the space of imaginations}$ which also houses the space of possibility-impossibility dualities $(\mathfrak{P} \otimes \mathfrak{M})$ with a generic element $(\mathfrak{p},\mathfrak{m}) \in (\mathfrak{P} \otimes \mathfrak{M})$, where $(\mathfrak{p} \in \mathfrak{P})$ is the possibility space and with a fuzzy-stochastic dualistic-polar restriction $(\mathfrak{P} \otimes \mathfrak{M}) = (\mathfrak{P} \cup \mathfrak{M}), (\mathfrak{P} \cap \mathfrak{M}) \neq \varnothing$ where $(\mathfrak{m} \in \mathfrak{M}) = \text{impossibility space}$. The space of possibility-impossibility dualities contains *necessities* and *anticipations* of transformations. Similarly, the space of imagination-reality dualities houses the space of probability-improbability dualities $(\mathfrak{B} \otimes \mathfrak{H})$ with a generic element $(\mathfrak{b} \otimes \mathfrak{h}) \in (\mathfrak{B} \otimes \mathfrak{H})$ where $(\mathfrak{b} \in \mathfrak{B}) = \text{probability space}$ and $(\mathfrak{h} \in \mathfrak{H}) = \text{improbability space}$ with a fuzzy dualistic-polar restriction $(\mathfrak{B} \otimes \mathfrak{H}) = (\mathfrak{B} \cup \mathfrak{H}), (\mathfrak{B} \cap \mathfrak{H}) \neq \varnothing$. The space of probability contains *freedoms* and *expectations,* and the space of possibility-probability dualities contains the spaces of necessity-freedom dualities and anticipation-expectation dualities. The extension of the knowledge square to include the spaces of imaginations, realities and imagination-reality is to show the logical places of neuro-decision-choice actions in the knowing processes and places in relation to possibility, probability doubt, among other elements that are facilitators or inhibitors of outcomes over the space of success-failure dualities. The extension also allows critical examinations of qualitative elements such as curiosity, courage, creativity, and other relevant elements over the space of problem-solution dualities.

All spaces reside in dualistic-polar conditions in intra-space and inter-space structures with various types of fuzzy dualistic-polar conditionality which implies that if $\mathbf{A}, \mathbf{B}, \mathbf{C}$ and $\mathbf{D}$ are either sets, categories or spaces then $\mathbf{D} = \mathbf{A} \cup \mathbf{B}$ and $\mathbf{C} = \mathbf{A} \otimes \mathbf{B}$ implies that $\mathbf{A} \cap \mathbf{B} \neq \varnothing$ with $(\mathbf{a}, \mathbf{b}) \in \mathbf{C}$ and $(\mathbf{a}, \mathbf{b}) \in \mathbf{D}$. All the spaces and the types of fuzzy dualistic-polar conditionality can be translated into information conditions structured in characteristic-signal dispositions. The fuzzy-dualistic-polar conditionality shows that $\mathbf{A} \cap \mathbf{B} \neq \varnothing \Rightarrow$ relational continuum $\mathbf{D} = \mathbf{A} \cup \mathbf{B} \Rightarrow$ realational unity while the condition $(\mathbf{a}, \mathbf{b}) \in \mathbf{C}$ and $(\mathbf{a}, \mathbf{b}) \in \mathbf{D}$ implies the general principle of opposites. The concepts and the phenomena of the spaces of imaginations, realities, possibility, probability and the corresponding dualities and spaces are extremely important in the processes of knowing and information-knowledge production as economic activities of neuro-decision-choice agents. The possibility and probability and corresponding dualities and spaces find residence in the space of variety imaginations. The variety solutions, visions, objectives, goals, and ideas also reside in the space of variety imaginations We shall however, deal with the conditions of possibility and probability in this chapter and then deal with the conditions of imagination, reality and the corresponding dualities and spaces in the next monograph. The construct of the individual spaces are taken from the spaces of dualities based on internal characteristic dominance of the varieties with the conditions of fuzzy indicator functions such that if $\mathbf{A} \cup \mathbf{B} = \mathbf{A} \otimes \mathbf{B} \Rightarrow$ set $\mathbf{A}$ is defined by the conditions $(\mathbb{X}_{\mathbf{A}} \ggg \mathbb{X}_{\mathbf{B}})$ and similarly, a set $\mathbf{B}$ is defined by the conditions $(\mathbb{X}_{\mathbf{B}} \ggg \mathbb{X}_{\mathbf{A}})$ where the symbol $(\ggg)$ represents dualistic-polar dominance and $\mathbf{A}$ and $\mathbf{B}$ are the internal duals of duality $\mathbf{C}$ or $\mathbf{D}$ such that $\mathbb{X}_{\mathbf{C}} = (\mathbb{X}_{\mathbf{A}} \cup \mathbb{X}_{\mathbf{B}})$ with $(\mathbb{X}_{\mathbf{A}} \cap \mathbb{X}_{\mathbf{B}}) \neq \varnothing$ or $\mathbb{X}_{\mathbf{D}} = (\mathbb{X}_{\mathbf{A}} \cup \mathbb{X}_{\mathbf{B}})$ with $(\mathbb{X}_{\mathbf{A}} \cap \mathbb{X}_{\mathbf{B}}) \neq \varnothing$.

6.2.2.1: The Phenomena of Possibility and Possibility Space in the Extended Knowledge Square for Knowing and Information-knowledge production.

Let us now examine the concepts and phenomena of possibility and impossibility with possibility space, impossibility space and the space of possibility-impossibility Dualities and how they help the understanding of the relationship between telescopic present and telescopic future relative to *what there is* and *what there would be* under the knowing processes of variety outcomes of socio-natural processes. The concepts and phenomena of possibility, impossibility, possibility space, impossibility space and the space of possibility-impossibility dualities are conceptual varieties within the space of cognitive capacity limitations, where the telescopic future can only be conceived in terms cognitive projections of variety characteristics of socio-natural processes from the telescopic present over the spaces of identification problem-solution dualities and transformation problem-solution dualities under the conditions of knowing and information-knowledge production also over the spaces of fundamentals, applied and fundamental-applied dualities as one travels over the spaces of ignorance-knowledge dualities and static-dynamic dualities. Let us provide some working definitions that will help in the understanding of knowing as production decisions of intellectual investment flows, and information-knowledge accumulation as the production decisions of intellectual capital stocks as well as the understanding of information-knowledge quality-control decisions and input-output decisions with cognitive capacity-limitations over the spaces of acceptance-rejection dualities and credulity-doubt dualities. The system of definitions will also help the understanding of the classical theory of knowledge over the space of correctness-incorrectness dualities. The definitions will place variety distinctions of concepts and phenomena and their implications in general communications over the source-destination dualities. These definitions will be presented as verbal ones and algebraic ones in complementation. The analytical process is simply to present the conceptual character of the variety phenomenon as qualitative representation and then the symbolic and computational character as quantitative representation over the space of quality-quantity dualities.

DEFINITION 6.2.2.1A: Possibility, Impossibility, Possibility Space, and Impossibility Space (verbal)

A possibility is an epistemic phenomenon of a variety that has capacity conditions for *what would be* with a defined characteristic disposition and a presumption in the Sankofa-anoma traditions from the space of potentials as distinct from *what there was* (past) and *what there is* (present) from the space of actuals to provide distinguishing characteristics equipped with a categorial convertor (functor) as actualizer of telescopic future varieties in the space of imaginations, relative to a variety conception by cognitive agents, based on defined aspects of epistemic characteristics of telescopic past varieties, telescopic present varieties and combinations of them in an infinite information-time horizon within the system of information-time dualities in the space of realities from the infinite conditions of ontic characteristics of potential existence in relation to the infinite conditions of actual existence in conceptualizing the cognitive space of the future unknown varieties which contains variety anticipations, necessities and initial conditions from the variety characteristic dispositions of *what there was* and *what there is* as restricted by cognitive capacity limitations. The conditions of incapability, non-anticipation, lack of necessity, lack of categorial convertor (functor) and lack of initial conditions define the epistemic phenomenon of *impossibility* in the cognitive space of knowing and information-knowledge production in relation to *what there would not be* relative to the space of actuals. The collection of all the varieties of *what there would be* is the *possibility space* which is connected to the subspace of the epistemological space, and the collection of all of *what there would not be* is the *impossibility space* which is also connected to the epistemological space through the *space of imaginations* $(\mathfrak{G})$. The union and the Cartesian product of possibility and impossibility spaces under dualistic-polar structures define the space of possibility-impossibility dualities with fuzzy-conditionality restrictions, presumptions, and non-evidential things.

DEFINITION 6.2.2.1B: Possibility Space and Impossibility Space (Algebraic)

The possibility space $(\mathfrak{p} \in \mathfrak{P})$ is a collection of all the telescopic future varieties considered as *what would be* by cognitive agents with a characteristic disposition $\left(\mathbb{X}_{(\mathfrak{p}_v)} \in \mathfrak{G}\right)$ with a corresponding functor $\mathfrak{f}(\bullet)$ belonging to the space of functors $(\mathfrak{f} \in \mathbf{F})$ from the space of imaginations $(\mathfrak{G} \subset \mathfrak{U})$ as a conformal mapping into the space of realities $(\mathfrak{H})$ such that $\left\{\mathfrak{f}\left(\mathbb{X}_{(\mathfrak{p}_v)}\right) = \mathfrak{h} \in (\mathfrak{H})\right\}$, where the possibility space is a union and Cartesian product of the spaces of information-decision activities $(\mathfrak{d} \in \mathfrak{D})$ and spaces of imagination-reality dualities $(\mathfrak{H} \otimes \mathfrak{G}) = (\mathfrak{H} \cup \mathfrak{G})$, with a characteristic disposition $\left(\mathbb{X}_{((\mathfrak{H} \otimes \mathfrak{G}))}\right)$, a space of problem-solution dualities$(\rho \in \Re)$ with a characteristic disposition $(\mathbb{X}_{(\Re)})$, where the space of neuro-decision-choice activities involves sequential interactions of acquaintance space $(\mathbb{A})$, information-knowledge processing space(Π) as the space of paradigms for *ignorance reduction* and knowledge enhancements passing through the space of variety ignorance-knowledge dualities $\mathfrak{Q} = (\mathfrak{U} \otimes \mathfrak{K}) = (\mathfrak{U} \cup \mathfrak{K})$ and the information-knowledge-certainty space$(\mathfrak{K})$ for *ignorance-doubt reduction* through the actions of $(\mathfrak{f} \in \mathbf{F})$, $\mathbf{F} = (\mathcal{P} \otimes \mathfrak{T}) = (\mathcal{P} \cup \mathfrak{T})$ $(\mathcal{P}) =$ spaced of processes , $(\mathfrak{T}) =$ space of technologies, $\mathfrak{f} = (p, t) \in \mathbf{F}$ where $(\mathfrak{U}) =$ space of ignorance . Similarly, the impossibility space $(\mathfrak{m} \in \mathfrak{M})$ is the collection of all the telescopic future varieties considered by cognitive agents as incapable of *there would be* with a characteristic disposition$(\mathbb{X}_{(\mathfrak{M})})$ with a generic element $(\mathfrak{m} \in \mathfrak{M})$ such that $\nexists\ \mathfrak{f} \in \mathbf{F} \ni \mathfrak{f}(\mathbb{X}_{(\mathfrak{p}_v)}) = \mathfrak{h} \in \mathfrak{H}$.

Algebraically, let$(\mathbb{A})$ represent the acquaintance space with a generic element $(\alpha \in \mathbb{A})$ with a characteristic disposition $(\mathbb{X}_{\mathbb{A}})$, $(\mathbb{V})$ is the space of varieties with a generic element$(v \in \mathbb{V})$ and with characteristic disposition $(\mathbb{X}_{\mathbb{V}})$, $(\mathfrak{B})$ is the probability space with a generic element $(\mathfrak{b} \in \mathfrak{B})$ and characteristic disposition $(\mathbb{X}_{(\mathfrak{B})})$, (Ω) is the ontological space where $(\mathbb{Z}_{\Omega} = \mathbb{X}_{\Omega} \otimes \mathbb{S}_{\Omega})$ is the ontological information with a generic element$(z_{\omega} \in \mathbb{Z}_{\Omega})$, $(\mathfrak{E})$ is the epistemological space, where $(\mathbb{Z}_{\mathfrak{e}})$ is the epistemological information with a generic element$\{(z_{\mathfrak{e}} \in \mathbb{Z}_{\mathfrak{e}} = (\mathbb{X}_{\Omega} \otimes \mathbb{S}_{\Omega})) \mid \mathfrak{e} \in \mathfrak{E}\}$, $(\mathfrak{K})$ is the information-knowledge space with a generic element $(\mathfrak{k} \in \mathfrak{K})$, and information

$(\mathbb{Z}_{\mathfrak{K}} = (\mathbb{X}_{\mathfrak{K}} \otimes \mathbb{S}_{\mathfrak{K}})), (\mathbb{U})$ is the space of certainty-uncertainty dualities with a generic element $(\upsilon \in \mathbb{U})$ and information $\{(z_\upsilon \in \mathbb{Z}_{\mathbb{U}} = (\mathbb{X}_{\mathbb{U}} \otimes \mathbb{S}_{\mathbb{U}})) \mid \upsilon \in \mathbb{U}\}$. The uncertainty space is contained in the space of certainty-uncertainty dualities which is a conformal mapping from the space of ontic-epistemic dualities $\mathfrak{O} = (\mathbb{X}_{\Omega_\nu} \otimes \mathbb{X}_{\mathfrak{E}_\nu}) = (\mathbb{X}_{\Omega_\nu} \otimes \mathfrak{J}(\mathbb{S}_{\Omega_\nu}))$ which may similarly be defined as $\mathfrak{O} = (\mathbb{X}_{\Omega_\nu} \cup \mathbb{X}_{\mathfrak{E}_\nu}) = (\mathbb{X}_{\Omega_\nu} \cup \mathfrak{J}(\mathbb{S}_{\Omega_\nu}))$ with dualistic-polar conditions of the form $(\mathbb{X}_{\Omega_\nu} \cap \mathbb{X}_{\mathfrak{E}_\nu}) = (\mathbb{X}_{\Omega_\nu} \cap \mathfrak{J}(\mathbb{S}_{\Omega_\nu})) \neq \varnothing$, where the certainty-certainty duality is of the form $\mathbb{U} = \mathfrak{f}(\mathbb{X}_{\Omega_\nu} \otimes \mathbb{X}_{\mathfrak{E}_\nu}) = \mathfrak{f}(\mathbb{X}_{\Omega_\nu} \otimes \mathfrak{J}(\mathbb{S}_{\Omega_\nu}))$, where explicitly, the space of uncertainties is a result of a mapping from the space of epistemic distances $(\mathfrak{U})$ = space of ignorance , and (Π) is the space of paradigms of thought for neuro-decision-choice processing actions with a generic element $(\eta \in \Pi)$ on information. The space of functors is $(\mathbf{F})$ which is specified as $\mathbf{F} = (\mathcal{P} \otimes \mathfrak{T}) = (\mathcal{P} \cup \mathfrak{T})$ or categorial converters with a generic element $(\mathfrak{f} \in \mathbf{F}, \mathbf{c} \in \mathbf{C})$ and a corresponding characteristic disposition $\{(\mathbb{X}_{\mathbf{F}}), (\mathbb{X}_{\mathbf{C}})\}$, where imaginations and space of imaginations are conformal mappings from the space of uncertainties $(\upsilon \in \mathbb{U})$ as a conformal maps from the space of ontic-epistemological dualities. Given these symbolic representations of spaces and sub-spaces, the algebraic structure of the possibility space is abstracted from the space of imaginations as a sub-space of epistemological space abstracted from the space of potentials, and represented as a Cartesian product with a complex system of unions, intersections and containments of all relevant spaces under dualistic-polar structures in the production-consumption processes of intellectual investment flows and information-capital accumulation as intellectual capital stocks, where all information $(z \in \mathbb{Z})$ is defined as characteristic-signal disposition $(\mathbb{Z}=\mathbb{X}\otimes\mathbb{S})$, where $(\mathbb{X})$ is the contents of the information and $(\mathbb{S})$ is the messaging signals in representations and communications and all variety identities find expressions in the epistemic field $(\mathscr{E})$ with an information field $(\mathscr{I})$ and relational field $(\mathscr{R})$. The possibility space is in the form:

$$\left.\begin{array}{l}
\mathfrak{P}=\left(\Omega\otimes\mathbb{V}\otimes\mathbb{U}\otimes\mathfrak{E}\otimes\mathfrak{G}\otimes\mathfrak{D}\otimes\mathbb{Z}_{\mathfrak{e}}\otimes(\mathbf{F})\right)=\left\{\mathfrak{p}=\left(\omega,\nu,\mu,\mathfrak{d},z_{\mathfrak{t}},\mathfrak{g},\mathfrak{f}\right)\mid\omega\in\Omega,\mathfrak{e}\in\mathfrak{E},\nu\in\mathbb{V},\mathfrak{g}\in\mathfrak{G},\mathfrak{d}\in\mathfrak{D},z_{\mathfrak{t}}\in\mathbb{Z}_{\mathfrak{e}}\left(\mathfrak{f}\in\mathbf{F}\right)\right\} \\
\text{where the space of certainty-uncertainty dualities is specified as} \\
\mathbb{U}=\left(\mathbb{A}\otimes\mathbb{Z}_{\Omega}\otimes\mathfrak{B}\otimes\mathfrak{D}\right)=\left\{\upsilon=\left(\alpha,z_{\omega},\mathfrak{b},\mathfrak{d}\right)\mid\alpha\in\mathbb{A},z_{\omega}\in\mathbb{Z}_{\Omega},,\mathfrak{b}\in\mathfrak{B},\mathfrak{d}\in\mathfrak{D}\right\}=\mathfrak{I}\left(\mathbb{X}_{\Omega_{\nu}}\otimes\mathbb{X}_{\mathfrak{e}_{\nu}}\right)=\mathfrak{I}\left(\mathbb{X}_{\Omega_{\nu}}\otimes\mathfrak{I}\left(\mathbb{S}_{\Omega_{\nu}}\right)\right) \\
\mathbb{A}=\left(\mathbb{V}\otimes\mathbb{Z}_{\Omega}\otimes\Pi\otimes\mathfrak{D}\right)=\left\{\alpha=\left(\nu,z_{\omega},\eta,\mathfrak{d}\right)\mid\nu\in\mathbb{V},z_{\omega}\in\mathbb{Z}_{\Omega},\eta\in\Pi,\mathfrak{d}\in\mathfrak{D}\right\} \\
\mathfrak{E}=\left(\mathbb{U}\cup\mathfrak{R}\right)=\left\{\mathfrak{e}=\left(\upsilon,\rho\right)\mid\upsilon\in\mathbb{U},\rho\in\mathfrak{R}\right\} \\
\mathfrak{B}=\left(\mathbb{Z}_{\mathfrak{e}}\otimes\mathfrak{G}\otimes\mathfrak{D}\otimes\mathbf{F}\right)=\left\{\mathfrak{b}=\left(z_{\mathfrak{t}},\mathfrak{g},\mathfrak{d},\mathfrak{f}\right)\mid z_{\mathfrak{t}}\in\mathbb{Z}_{\mathfrak{e}},\mathfrak{g}\in\mathfrak{G},\mathfrak{d}\in\mathfrak{D},\mathfrak{f}\in\mathbf{F}\right\} \\
\mathfrak{D}=\left(\mathbb{V}\otimes\Pi\otimes\mathbb{Z}_{\mathfrak{e}}\otimes\mathfrak{D}\otimes\mathfrak{K},\mathbf{F}\right)=\left\{\mathfrak{d}=\left(\nu,\eta,z_{\mathfrak{t}},\mathfrak{d},\mathfrak{k},\mathfrak{f}\right)\mid\nu\in\mathbb{V},\eta\in\Pi\otimes\mathbb{Z}_{\mathfrak{e}},\mathfrak{d}\in\mathfrak{D},\mathfrak{k}\in\mathfrak{K},\mathfrak{f}\in\mathbf{F}\right\} \\
\Omega=\mathfrak{A}\otimes\mathfrak{U}=\mathfrak{A}\cup\mathfrak{U},\ \text{with}\ \mathfrak{A}\cap\mathfrak{U}\neq\varnothing,\mathfrak{u}\in\mathfrak{U},\mathfrak{a}\in\mathfrak{A} \\
\mathfrak{G}=\left(\mathfrak{B}\cup\mathfrak{P}\cup\mathbf{F}\right)\subset\mathfrak{U}\ \text{with}\ \mathfrak{B}\cap\mathfrak{P}\cap\mathbf{F}\neq\varnothing,\mathfrak{b}\in\mathfrak{B},\mathfrak{p}\in\mathfrak{P},\mathfrak{f}\in\mathbf{F},\left(\mathfrak{b},\mathfrak{p},\mathfrak{f}\right)\in\left(\mathfrak{B}\otimes\mathfrak{P}\otimes\mathbf{F}\right)
\end{array}\right\}$$

(6.2.2.1.1)

Similarly, we may define the epistemological space as:

$$\left.\begin{array}{l}
\mathfrak{E}=\left(\mathbb{U}\cup\mathfrak{R}\right)=\left\{\mathfrak{e}=\left(\upsilon,\rho\right)\mid\upsilon\in\mathbb{U},\rho\in\mathfrak{R}\right\}, \\
\text{where the space of problem-solution dualities is specified as:} \\
\mathfrak{R}=\left(\mathbb{Z}_{\mathfrak{e}}\otimes\mathbb{V}\otimes\Pi\otimes\mathfrak{D}\otimes\mathfrak{K}\right)=\left\{\rho=\left(z_{\mathfrak{t}},\nu,\eta,\mathfrak{d},\mathfrak{k}\right)\mid z_{\mathfrak{t}}\in\mathbb{Z}_{\mathfrak{e}},\nu\in\mathbb{V},\eta\in\Pi,\mathfrak{d}\in\mathfrak{D},\mathfrak{k}\in\mathfrak{K}\right\} \\
\mathbb{Z}_{\mathfrak{e}}=\left(\mathbb{X}_{\mathfrak{e}}\otimes\mathbb{S}_{\mathfrak{e}}\right)=\left\{z_{\mathfrak{t}}=\left(x_{\mathfrak{t}},s_{\mathfrak{t}}\right)\mid x_{\mathfrak{t}}\in\mathbb{X}_{\mathfrak{e}},s_{\mathfrak{t}}\in\mathbb{S}_{\mathfrak{e}}\right\} \\
\mathbb{Z}_{\Omega}=\left(\mathbb{X}_{\Omega}\otimes\mathbb{S}_{\Omega}\right)=\left\{z_{\omega}=\left(x_{\omega},s_{\omega}\right)\mid x_{\omega}\in\mathbb{X}_{\Omega},s_{\omega}\in\mathbb{S}_{\Omega}\right\} \\
\left(\mathbb{X}_{\Omega}\otimes\mathbb{S}_{\Omega}\right)=\left(\mathbb{X}_{\mathfrak{U}}\otimes\mathbb{S}_{\mathfrak{U}}\right)\cup\left(\mathbb{X}_{\mathfrak{A}}\otimes\mathbb{S}_{\mathfrak{A}}\right)=\left\{z_{\omega}=\left(x_{\omega},s_{\omega}\right)\mid x_{\omega}=\left(x_{\mathfrak{a}},x_{\mathfrak{u}}\right),s_{\omega}\in\left(s_{\mathfrak{a}},s_{\mathfrak{u}}\right)\right\} \\
\mathfrak{K}=\left(\mathfrak{D}\otimes\mathbb{Z}_{\mathfrak{e}}\otimes\Pi\right)=\left\{\mathfrak{k}=\left(\mathfrak{d},z_{\mathfrak{t}},\eta\right)\mid\mathfrak{d}\in\mathfrak{D},z_{\mathfrak{t}}\in\mathbb{Z}_{\mathfrak{e}},\eta\in\Pi\right\} \\
\mathbb{U}=\mathfrak{f}\left(\mathbb{X}_{\Omega_{\nu}}\otimes\mathbb{X}_{\mathfrak{e}_{\nu}}\right)=\mathfrak{f}\left(\mathbb{X}_{\Omega_{\nu}}\otimes\mathfrak{I}\left(\mathbb{S}_{\Omega_{\nu}}\right)\right),\ \text{with}\ \left(\mathbb{X}_{\Omega_{\nu}}\cap\mathfrak{I}\left(\mathbb{S}_{\Omega_{\nu}}\right)\right)\neq\varnothing
\end{array}\right\}(6.2.2.1.2)$$

We may unite the *space of imagination-uncertainty dualities* with the *space of problem-solution dualities* to define the structure of the (probability) possibility space as:

$$\mathfrak{P}=(\mathbb{U}\cup\mathfrak{R}) = (\mathbb{A}\otimes\mathbb{Z}_{\Omega}\otimes\mathfrak{G}\otimes\mathfrak{B}\otimes\mathfrak{D})\cup(\mathbb{Z}_{\mathfrak{e}}\otimes\mathbb{V}\otimes\Pi\otimes\mathfrak{D}\otimes\mathfrak{K})$$
$$\left\{\begin{matrix}\text{space of Imagination-Uncertainty}\\ \text{Dualities with Anticipations,}\\ \text{necessity an initial Conditions}\end{matrix}\right\}\cup\left\{\begin{matrix}\text{Space of problem-solution}\\ \text{Dualities with Expectations}\\ \text{Freedom an sufficient conditions}\end{matrix}\right\} \quad (6.2.2.1.3)$$

The space of the *actuals* $(\mathfrak{A})$ is simply the intersection of the ontological space and the space of the potentials of the form:

$$\mathfrak{A}=\Omega\cap\mathfrak{U}=\left\{(\mathbb{A}\otimes\mathbb{Z}_{\Omega}\otimes\mathfrak{P}\otimes\mathfrak{B}\otimes\mathfrak{D})\cup(\mathbb{Z}_{\mathfrak{e}}\otimes\mathbb{V}\otimes\Pi\otimes\mathfrak{D}\otimes\mathfrak{K})\right\}\cap\left\{(\mathbb{V}_{\mathfrak{u}}\otimes\Phi_{\mathfrak{u}}\otimes\mathfrak{T}_{\mathfrak{u}}\otimes\mathbb{Z}_{\mathfrak{u}})\cup(\mathbb{V}_{\mathfrak{a}}\otimes\Phi_{\mathfrak{a}}\otimes\mathfrak{T}_{\mathfrak{a}}\otimes\mathbb{Z}_{\mathfrak{a}})\right\} \quad (6.2.2.1.4).$$

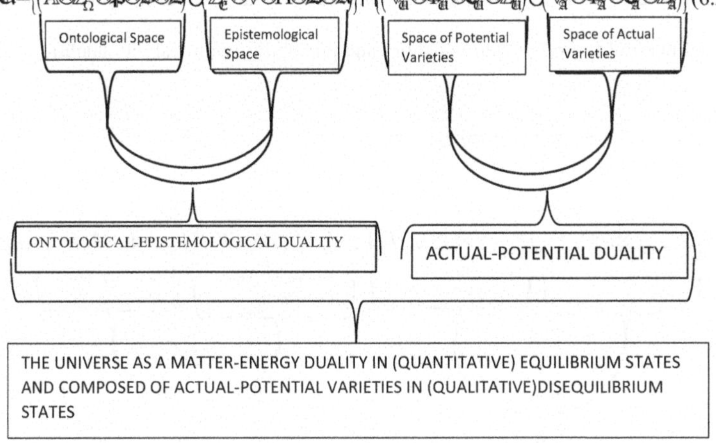

$$\left.\begin{array}{l}\Omega=(\mathfrak{A}\cup\mathfrak{U})=(\mathfrak{A}\otimes\mathfrak{U})=\text{ontological space}=\text{Space of all varieties}=\text{space of actual-potential dualities}\\(\mathfrak{A})=\Omega\cap\mathfrak{U}=\text{the space of actuals containing the space of realities}\\(\mathfrak{U})=\Omega\cap\mathfrak{A}=\text{the space of potentials containing the space of imaginations}\\(\Omega\otimes\mathfrak{E})=\text{Space of ontological-epistemological dualities containing the space of imagination-reality dualities}\\\mathfrak{O}=\left(\mathbb{X}_{\Omega}\otimes\mathbb{X}_{\mathfrak{E}}\right)=\left(\mathbb{X}_{\Omega}\otimes\mathfrak{I}\left(\mathbb{S}_{\Omega}\right)\right)\begin{cases}\text{Space of ontic-epistemic dualities containing ontological and epistemological}\\\text{characteristic dispositions where }\mathbb{X}_{\mathfrak{E}}=\mathfrak{I}\left(\mathbb{S}_{\Omega}\right)=\text{variety epistemic characteristic}\\\text{disposition, }\mathfrak{I}(\bullet)\text{ is signal decipher function.}\end{cases}\end{array}\right\}$$

The space of potential is one without an acquaintance but with space of imaginations containing the spaces of possibility-impossibility dualities and probability-improbability dualities and hence we have:

$$\mathfrak{U}=\mathfrak{A}\cap\Omega=\left\{\left(\mathfrak{G}\otimes\mathbb{Z}_{\Omega}\otimes\mathfrak{P}\otimes\mathfrak{B}\otimes\mathfrak{D}\right)\cup\left(\mathbb{Z}_{\mathfrak{E}}\otimes\mathbb{V}\otimes\Pi\otimes\mathfrak{D}\otimes\mathfrak{K}\right)\right\}\cap\left\{\left(\mathbb{V}_{\mathfrak{U}}\otimes\Phi_{\mathfrak{U}}\otimes\mathfrak{T}_{\mathfrak{U}}\otimes\mathbb{Z}_{\mathfrak{U}}\right)\cup\left(\mathbb{V}_{\mathfrak{A}}\otimes\Phi_{\mathfrak{A}}\otimes\mathfrak{T}_{\mathfrak{A}}\otimes\mathbb{Z}_{\mathfrak{A}}\right)\right\}\quad(6.2.2.1.4)$$

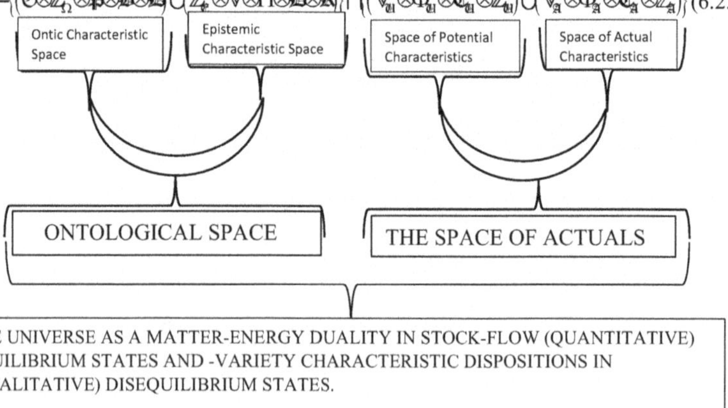

NOTE 6.2.2.1: Possibility and Possibility Space

A note on the interactive structures of the phenomena of possibility and possibility space on one hand and impossibility and impossibility space on the other hand will be useful for clarifications at the points of uncertainties which are divided into *uncertainties of anticipations (anticipation uncertainties)* at the level of possibility and *uncertainties of expectations* (*expectational uncertainties*) at the level of probability, where the uncertainties are generated from the ontic-epistemic differences in the space of ontic-epistemic dualities due to cognitive capacity limitations that generate variety ignorance over the space of acquaintance-description dualities $\mathfrak{Q} = (\mathbb{A} \otimes \mathbb{H}) = (\mathbb{A} \cup \mathbb{H})$ in a continuum and unity with fuzzy-dualistic conditionality $(\mathbb{A} \cap \mathbb{H}) \neq \varnothing$ where $(\mathbb{A})$ is the acquaintance space and $(\mathbb{H})$ is the space of descriptions . The possibility and impossibility present conceptual varieties in the space of social philosophical consciencism, where cognitive agents can perform telescopic future games with neuro-decision-choice actions as to what variety imaginations may enter the space of actuals which are natural constructs through the space of realities which in turn is a social construct under acquaintance and description. The possibility and impossibility spaces are the spaces for neuro-decision-choice connectivity between the telescopic present and telescopic future varieties. In other words, it is the game space of *what there is not* and *what there would be* from the space of the potentials as necessity to the space of imaginations with anticipation and hope to enter the space of realities contained in the space of actuals with expectational freedom and sufficient conditions of decision-choice outcomes. The possibility space houses necessity and anticipations of telescopic future varieties through functor actions in categorial conversions, where the anticipated varieties in the possibility space are expected to be on the path through the probability space to the space of realities contained in the space of actuals. The impossibility space on the other hand denies anticipations and varieties without variety expectations into the future under dualistic-polar conditions.

All the spaces in these discussions are constructed with fuzzy indicator functions and hence they are in intra-continua, inter-continua and in unity with give-and-take sharing modes. In the give-and-take sharing modes in dualistic-polar structures, there are contradictions and non-excluded middles, where the contingencies are seen in terms of neuro-decision-choice activities over the action space between the space of imaginations and the space of the actuals without reference to the defining probabilities. The possibility space may be used in neuro-decision-choice situations to create possible worlds for transformational varieties to be actualized for entries into the space of realities. This may be viewed as *transformational possible world* for planning and execution of to achieve goals, objectives, and visions. It may be used to create possible world for comparative varieties for logical analysis with existing varieties in the space of realities. It is comparative thought experiment with no transformative action and may be called *comparative possible worlds.* The differences between the transformation possible world and comparative world may be seen in terms their roles in telescopic past, telescopic present, and telescopic future. In the comparative epistemic world, there are no variety anticipations, expectations and categorial converters. In transformational possible world, there are variety anticipations, expectations, and conditions of converters.

The phenomena of possibility and the possibility space, impossibility, and impossibility space in knowing, information-knowledge production and decision-choice actions are always in relation to the future varieties and changes of perceptions on current varieties in the space of realities with cost-benefit balances. They encompass solution and solution space over the space of problem-solution dualities. The possibility space houses all the changes, transformations and the enveloping of telescopic future as conceptualized by cognitive agents. It houses visions, goals, objectives, conceptions, ideas, theories of existence, explanations and many such events relative to the principle of non-satiations. The impossibility space denies all the changes, transformations and the enveloping of telescopic future as conceptualized by cognitive agents. It denies visions, goals, objectives, conceptions, ideas, theories of existence, explanations and many such events relative to the principle of non-satiations. The possibility space provides a framework to design and implement actions to increate and out-create future realities in terms of *what there would be* and to connect to *what there is* as well as increase the information-knowledge accumulation as intellectual capital stocks through increasing successes of knowing over the space of identification-transformation dualities while the impossibility space denies the transformation and changes of varieties of the telescopic present over the space of existence-nonexistence dualities. The conceptions of varieties in the possibility space and impossibility space belong to the space of imaginations as a sub-space of potentials, and hence do not belong to the space of actuals under dualistic-polar conditions and fuzzy-conditionality restrictions. Given the concept and phenomenon of possibility, we now turn our attention to the concept of probability.

6.2.2.2 The Concepts and Phenomena of probability and Probability Space in the Extended Knowledge Square for Knowing and Information-knowledge production.

Let us examine the concepts and phenomena of probability and improbability with probability space, improbability space and the space probability-improbability dualities from figures (6.2.2.1- 6.2.2.3) relative to the concept and phenomenon of possibility with corresponding spaces and opposites, and how they help the understanding of the relationship between telescopic present and telescopic future relative to *what there is* and *what there would be* under knowing processes of variety outcomes of socio-natural processes over the space of certainty-uncertainty dualities with evidential support from the space of acquaintance in support of possibility. The evidential supports come from the space of empirical-axiomatic dualities of variety characteristic dispositions within the space of ontic-characteristic dualities. The concepts and phenomena of probability, improbability, probability space and improbability space are conceptual varieties within the space of acquaintance-description dualities with cognitive capacity limitations, where the telescopic future can only be conceived in terms cognitive projections of variety characteristics with some evidence but not necessary proofs of socio-natural processes over the spaces of identification problem-solution dualities and transformation problem-solution dualities under the conditions of knowing and information-knowledge production relative to the space of fundamental-applied dualities as one travels over the space of ignorance-knowledge dualities. Let us provide some working definitions that will help in the understanding of knowing as the production of intellectual investment flows, and information-knowledge accumulation as the production of intellectual capital stocks as well as the understanding of information-knowledge quality-control decisions over the spaces of acceptance-rejection dualities and credulity-doubt dualities. The definitions will also be presented as verbal ones and complemented by algebraic ones for increasing clarity.

DEFINITION 6.2.2.2.1A Probability, improbability, Probability Space, and Improbability Space (verbal)

Probability, just like *possibility,* is epistemic phenomena of varieties concerning the degree to which *what there would be* in the Sankofa-anoma traditions may enter into the space of actuals through the space of realities from the space of potentials through the possibility space contained in the space imaginations as distinct from *what there was* (past) and *what there is* (present) from the space of actuals to provide distinguishing characteristics of telescopic future varieties in the space of imaginations, as conceived by cognitive agents, based on the epistemic characteristics of telescopic past varieties and telescopic present varieties in an infinite information-time horizon within the system of information-time dualities in the space of realities from the infinite conditions of ontic characteristics of potential existence in relation to the infinite conditions of actual existence in conceptualizing the cognitive space of the future unknown varieties which contains variety expectations, freedom, and sufficient conditions from the characteristic dispositions of *what there was* and *what there is* as input restrictions in the space of neuro-decision-choice actions with further restrictions by cognitive capacity limitations over the space of acquaintance-description dualities.

The conditions of lack of degrees to enter the space of actuals, zero expectation, lack of freedom and complete insufficient conditions define the epistemic phenomenon of *improbability* in the cognitive space of knowing and information-knowledge production in relation to *what there would not be* relative to the space of actuals. The collection of all the varieties of *what there would be* with positive degree of migration to the space of actuals is the *probability space* which is connected to the sub-space of the epistemological space and the collection of all of *what there would not be* with zero degrees of migrating into the space of actuals is the *improbability space* which is also connected to the epistemological space through the space of imaginations. The union and the Cartesian product of probability and improbability spaces under a dualistic-polar structure define the space of probability-improbability dualities with fuzzy-conditionality restrictions.

DEFINITION 6.2.2.2.1B: Probability Space and Impossibility Space (Algebraic)

The probability space $(\mathfrak{b} \in \mathfrak{B})$ is a collection of all the telescopic future varieties considered as *what there would be* by cognitive agents with characteristic disposition $\left(\mathbb{X}_{(\mathfrak{p}_v)}\right)$, which is a union of the spaces of information-decision activities $(\mathfrak{d} \in \mathfrak{D})$ and space of imaginations $\mathfrak{G}$ such that $(\mathfrak{G} \otimes \mathfrak{D}) = (\mathfrak{G} \cup \mathfrak{D})$, with characteristic disposition $\left(\mathbb{X}_{(\mathfrak{G} \otimes \mathfrak{D})}\right)$, problem-solution dualities $\left(\rho \in \mathfrak{R} = (\mathfrak{H} \otimes \mathfrak{S})\right)$ with a characteristic disposition $\left(\mathbb{X}_{(\mathfrak{R})}\right)$, where the space of decision-choice activities involves sequential interactions of acquaintance space, information-knowledge processing space for *ignorance reduction* passing through the spaces of variety ignorance-knowledge dualities and information-knowledge-certainty for increasing *doubt reduction*. Similarly, the improbability space $(\mathfrak{h} \in \mathfrak{H})$ is the collection of all the telescopic future varieties considered by cognitive agents as not capable of *there would be* with zero degree of transformation from the space of imagination to the spaces of realities and actual. Algebraically, let $(\mathbb{A})$ represent the acquaintance space with a generic element $(\alpha \in \mathbb{A})$, $(\mathbb{V})$ is the space of varieties with a generic element $(\nu \in \mathbb{V})$, and characteristic disposition $(\mathbb{X}_{\mathbb{V}})$, $(\mathfrak{P})$ is the possibility space with a generic element $(\mathfrak{p} \in \mathfrak{P})$ and characteristic disposition $\left(\mathbb{X}_{(\mathfrak{p})}\right)$, $(\mathbb{Z}_\Omega)$ is the ontological information with a generic element $(z_\omega \in \mathbb{Z}_\Omega)$, $(\mathbb{Z}_\mathfrak{E})$ is the epistemological information with a generic element $(z_\mathfrak{e} \in \mathbb{Z}_\mathfrak{E})$, $(\mathfrak{K})$ is the information-knowledge space with a generic element $(\mathfrak{k} \in \mathfrak{K})$, $\mathbb{U} = \mathfrak{f}\left(\mathbb{X}_{\Omega_v} \otimes \mathbb{X}_{\mathfrak{E}_v}\right) = \mathfrak{f}\left(\mathbb{X}_{\Omega_v} \otimes \mathfrak{I}\left(\mathbb{S}_{\Omega_v}\right)\right)$ is the space of certainty-uncertainty dualities with a generic element $(\upsilon \in \mathbb{U})$, (Π) is the space of paradigms of thought for neuro-decision-choice actions with generic element $(\eta \in \Pi)$ on information. Given these representations of spaces and sub-spaces, the algebraic structure of the probability space is abstracted from the space of imaginations as a sub-space of epistemological space, represented as a Cartesian product with a complex system of unions, intersections and containments of all the relevant spaces in the production-consumption processes of intellectual investment flows and accumulation of intellectual capital stocks, where all information $(z \in \mathbb{Z})$ is defined as characteristic-signal disposition $(\mathbb{Z} = \mathbb{X} \otimes \mathbb{S})$ with $(\mathbb{X})$ as the content of the information and $(\mathbb{S})$ as the message of the representation, and communication over the space of source-destination dualities, where the Cartesian product and the union of possibility and probability spaces present the space of certainty-uncertainty dualities with probability space in the form:

$$\left.\begin{array}{l}
\mathfrak{B} = \left(\mathbb{Z}_{\mathfrak{e}} \otimes \mathfrak{G} \otimes \mathfrak{D}\right) = \left\{\mathfrak{b} = \left(z_{\mathfrak{e}}, \mathfrak{g}, \mathfrak{d}\right) \mid z_{\mathfrak{e}} \in \mathbb{Z}_{\mathfrak{e}}, \mathfrak{g} \in \mathfrak{G}, \mathfrak{d} \in \mathfrak{D}\right\} \\
\text{where the space of certainty-uncertainty dualities is specified as} \\
\mathbb{U} = \left(\mathbb{A} \otimes \mathbb{Z}_{\Omega} \otimes \mathfrak{B} \otimes \mathfrak{D}\right) = \left\{\upsilon = \left(\alpha, z_{\omega}, \mathfrak{b}, \mathfrak{d}\right) \mid \alpha \in \mathbb{A}, z_{\omega} \in \mathbb{Z}_{\Omega}, , \mathfrak{b} \in \mathfrak{B}, \mathfrak{d} \in \mathfrak{D}\right\} = \mathfrak{k}\left(\mathbb{X}_{\Omega,} \otimes \mathfrak{I}\left(\mathbb{S}_{\Omega,}\right)\right) \\
\mathbb{A} = \left(\mathbb{V} \otimes \mathbb{Z}_{\Omega} \otimes \Pi \otimes \mathfrak{D}\right) = \left\{\alpha = \left(\nu, z_{\omega} \eta, \mathfrak{d}\right) \mid \nu \in \mathbb{V}, z_{\omega} \in \mathbb{Z}_{\Omega}, \eta \in \Pi, \mathfrak{d} \in \mathfrak{D}\right\} \\
\mathfrak{E} = \left(\mathbb{U} \cup \mathfrak{R}\right) = \left\{\mathfrak{e} = \left(\upsilon, \rho\right) \mid \upsilon \in \mathbb{U}, \rho \in \mathfrak{R}\right\} \\
\mathfrak{P} = \left(\Omega \otimes \mathbb{V} \otimes \mathbb{U} \otimes \mathfrak{E} \otimes \mathfrak{G} \otimes \mathfrak{D} \otimes \mathbb{Z}_{\mathfrak{e}}\right) = \left\{\mathfrak{p} = \left(\omega, \nu, \mu, \mathfrak{d}, z_{\mathfrak{e}}, \mathfrak{g}\right) \mid \omega \in \Omega, \mathfrak{e} \in \mathfrak{E}, \nu \in \mathbb{V}, \mathfrak{g} \in \mathfrak{G}, \mathfrak{d} \in \mathfrak{D}, z_{\mathfrak{e}} \in \mathbb{Z}_{\mathfrak{e}}\right\} \\
\mathfrak{D} = \left(\mathbb{V} \otimes \Pi \otimes \mathbb{Z}_{\mathfrak{e}} \otimes \mathfrak{D} \otimes \mathfrak{K}\right) = \left\{\mathfrak{d} = \left(\nu, \eta, z_{\mathfrak{e}}, \mathfrak{d}, \mathfrak{k}\right) \mid \nu \in \mathbb{V}, \eta \in \Pi \otimes \mathbb{Z}_{\mathfrak{e}}, \mathfrak{d} \in \mathfrak{D}, \mathfrak{k} \in \mathfrak{K}\right\} \\
\Omega = \mathfrak{A} \otimes \mathfrak{U} = \mathfrak{A} \cup \mathfrak{U}, \text{ with } \mathfrak{A} \cap \mathfrak{U} \neq \varnothing, \mathfrak{u} \in \mathfrak{U}, \mathfrak{a} \in \mathfrak{A} \\
\mathfrak{G} = \mathfrak{B} \cup \mathfrak{P} = \mathfrak{B} \otimes \mathfrak{P} \text{ with } \mathfrak{B} \cap \mathfrak{P} \neq \varnothing, \mathfrak{b} \in \mathfrak{B}, \mathfrak{p} \in \mathfrak{P}, \left(\mathfrak{b}, \mathfrak{p}\right) \in \left(\mathfrak{B} \otimes \mathfrak{P}\right),
\end{array}\right\} \quad (6.2.2.2.1)$$

Similarly, we have:

$$\left.\begin{array}{l}
\mathfrak{C} = (\mathbb{U} \cup \mathfrak{R}) = \{\mathfrak{c} = (\upsilon, \rho) \mid \upsilon \in \mathbb{U}, \rho \in \mathfrak{R}\}, \\
\text{where the space of problem-solution dualities is specified as:} \\
\mathfrak{R} = (\mathbb{Z}_{\mathfrak{C}} \otimes \mathbb{V} \otimes \Pi \otimes \mathfrak{D} \otimes \mathfrak{K}) = \{\rho = (z_{\mathfrak{c}}, \nu, \eta, \mathfrak{d}, \mathfrak{k}) \mid z_{\mathfrak{c}} \in \mathbb{Z}_{\mathfrak{C}}, \nu \in \mathbb{V}, \eta \in \Pi, \mathfrak{d} \in \mathfrak{D}, \mathfrak{k} \in \mathfrak{K}\} \\
\mathbb{Z}_{\mathfrak{C}} = (\mathbb{X}_{\mathfrak{C}} \otimes \mathbb{S}_{\mathfrak{C}}) = \{z_{\mathfrak{c}} = (x_{\mathfrak{c}}, s_{\mathfrak{c}}) \mid x_{\mathfrak{c}} \in \mathbb{X}_{\mathfrak{C}}, s_{\mathfrak{c}} \in \mathbb{S}_{\mathfrak{C}}\} \\
\mathbb{Z}_{\Omega} = (\mathbb{X}_{\Omega} \otimes \mathbb{S}_{\Omega}) = \{z_{\omega} = (x_{\omega}, s_{\omega}) \mid x_{\omega} \in \mathbb{X}_{\Omega}, s_{\omega} \in \mathbb{S}_{\Omega}\} \\
(\mathbb{X}_{\Omega} \otimes \mathbb{S}_{\Omega}) = (\mathbb{X}_{\mathfrak{U}} \otimes \mathbb{S}_{\mathfrak{U}}) \cup (\mathbb{X}_{\mathfrak{A}} \otimes \mathbb{S}_{\mathfrak{A}}) = \{z_{\omega} = (x_{\omega}, s_{\omega}) \mid x_{\omega} = (x_{\mathfrak{a}}, x_{\mathfrak{u}}), s_{\omega} \in (s_{\mathfrak{a}}, s_{\mathfrak{u}})\} \\
\mathfrak{K} = (\mathfrak{D} \otimes \mathbb{Z}_{\mathfrak{C}} \otimes \Pi) = \{\mathfrak{k} = (\mathfrak{d}, z_{\mathfrak{c}}, \eta) \mid \mathfrak{d} \in \mathfrak{D}, z_{\mathfrak{c}} \in \mathbb{Z}_{\mathfrak{C}}, \eta \in \Pi\} \\
\mathbb{U} = \mathfrak{f}(\mathbb{X}_{\Omega_v} \otimes \mathbb{X}_{\mathfrak{C}_v}) = \mathfrak{f}(\mathbb{X}_{\Omega_v} \otimes \mathfrak{I}(\mathbb{S}_{\Omega_v})), (\mathbb{X}_{\Omega_v} \cap \mathfrak{I}(\mathbb{S}_{\Omega_v})) \neq \varnothing
\end{array}\right\} \quad (6.2.2.2.2)$$

We may unite the *space of imagination-uncertainty dualities* with the *space of problem-solution dualities* to define the structure of the probability space as:

$$\mathfrak{B} = (\mathbb{U} \cup \mathfrak{R}) = \underbrace{(\mathbb{A} \otimes \mathbb{Z}_{\Omega} \otimes \mathfrak{G} \otimes \mathfrak{P} \otimes \mathfrak{D})}_{\left\{\begin{array}{l}\text{space of Imagination-}\\ \text{UncertaintyDualities}\end{array}\right\}} \cup \underbrace{(\mathbb{Z}_{\mathfrak{C}} \otimes \mathbb{V} \otimes \Pi \otimes \mathfrak{D} \otimes \mathfrak{K})}_{\left\{\begin{array}{l}\text{Space of problem-solution}\\ \text{Dualities}\end{array}\right\}} \quad (6.2.2.3)$$

The space of the *actuals* is simply the intersection of the ontological space and the epistemological space of the form:

$$\mathfrak{A} = \mathfrak{U} \cap \Omega = \{(\mathbb{A} \otimes \mathbb{Z}_{\Omega} \otimes \mathfrak{P} \otimes \mathfrak{B} \otimes \mathfrak{D}) \cup (\mathbb{Z}_{\mathfrak{C}} \otimes \mathbb{V} \otimes \Pi \otimes \mathfrak{D} \otimes \mathfrak{K})\} \cap \{(\mathbb{V}_{\mathfrak{U}} \otimes \Phi_{\mathfrak{U}} \otimes \mathfrak{T}_{\mathfrak{U}} \otimes \mathbb{Z}_{\mathfrak{U}}) \cup (\mathbb{V}_{\mathfrak{A}} \otimes \Phi_{\mathfrak{A}} \otimes \mathfrak{T}_{\mathfrak{A}} \otimes \mathbb{Z}_{\mathfrak{A}})\} \quad (6.2.2.2.4)$$

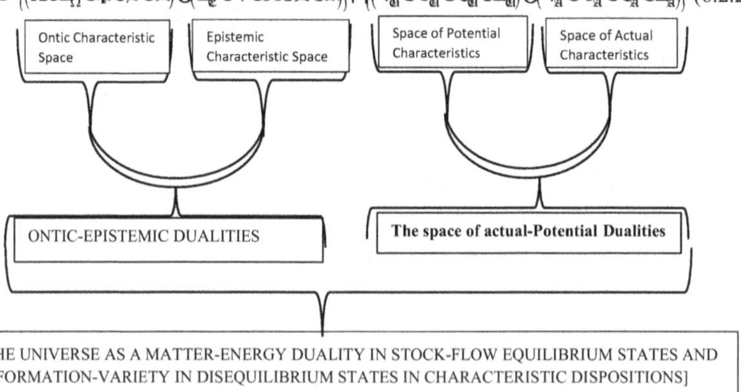

NOTE 6.2.2.2: Probability and Probability Space

A note on the interactive structures of the phenomena of probability and probability space on one hand and improbability and improbability space on the other hand will be useful for clarification the relationality of the concept and phenomenon of the space of probability-improbability dualities and the space of possibility-impossibility dualities as contained in the space of imaginations relative to the space ignorance defined by the space of variety epistemic distances. The probability space and the improbability space present a system of conceptual varieties in the space of philosophical consciencism, where cognitive agents can perform telescopic future games with neuro-decision-choice actions on what variety possibility in the space of imaginations may enter into the space of actuals through the space of realities as contained in the epistemological space, where the game in the space of imaginations is betting on outcome of future varieties and the reward successful variety actualization under the principle of non-satiation. The probability and improbability spaces jointly with the possibility and impossibility spaces constitute the neuro-decision-choice reflections on the system of certainty-uncertainty and doubt-surety spaces and sub-spaces within the space of imaginations for neuro-decision-choice connectivity between telescopic present and telescopic future varieties. Generally, the present conditions in the space of realities may be taken as known with small degree of ignorance. The future conditions within the space of potentials exist with substantial degree of ignorance with functor action results of ignorance.

In other words, it is the game space of *what there is not* with degrees of presumption and *what there would be* with degrees of presumption from the space of the potential relative to necessity and doubt to the space of imaginations relative to anticipation and hope to enter the space of actual through the space of reality with expectational freedom and sufficiency. The probability space houses freedoms and expectations of telescopic future varieties, from which the varieties with greater benefits over costs in the probability space are moved from the expected path to the space of realities contained in the space of actuals. The improbability space denies the existence of freedom and sufficiency of neuro-decision-choice of varieties without variety expectations into the future varieties. All the spaces are constructed and linked with fuzzy indicator functions in inter-continua and in unity with give-and-take sharing modes and non-excluded middles with contradictions, where contingencies with degrees of presumption are seen in terms of neuro-decision-choice activities over the action space between the space of imaginations contained in the space of potential, and the space of the realities as a sub-space of the space of the actuals with references to possibilities.

The phenomena of probability and improbability with corresponding spaces are always relative to the existence of possibility and the possibility space, impossibility, and impossibility space in knowing, information-knowledge production and neuro-decision-choice actions. These phenomena are always in relation to the future varieties and changes of perceptions on current varieties in the space of realities as seen in the space of variety cost-benefit distributions. They encompass solutions and solution space over the space of problem-solution dualities. Epistemically, the probability space, jointly with the possibility space houses all the changes, transformations and the enveloping of telescopic future as conceptualized by cognitive agents with cognitive capacity limitations relative to knowing and knowledge production over the spaces of ignorance-knowledge dualities, identification-transformation dualities, and fundamental-applied dualities in relation to the space of imaginations.

The space of imaginations houses visions, goals, objectives, conceptions, ideas, theories of existence, explanations, solutions, and many such telescopic future events relative to the principle of non-satiations connecting forecasting-discounting processes in the Sankofa-anoma continuum.

The improbability space denies all the changes, transformations and the enveloping of telescopic future as conceptualized by cognitive agents in the space of ignorance-knowledge dualities in the sense that there are variety possibilities that are improbable due to cognitive capacity and resource limitations in implementing the elements needed for their categorial converters (functors). It denies presumptions of visions, goals, objectives, solutions, conceptions, ideas, theories of existence, explanations and many such possible events relative to the principle of infeasibility. The probability space provides a framework for degrees of hope and freedom to implement actions for committing intellectual and non-intellectual capital resources to increate and out-create future realities in terms of *what there would be* relative to *what there is* as well as increase the future information-knowledge accumulation as the production of intellectual capital stocks through increasing degrees of successes of knowing over the space of identification-transformation dualities while the improbability space houses the denials of the changes of varieties over the space of identification-transformation dualities within the space of telescopic present-future dualities .

The conceptions of varieties in the probability space, improbability space and the space of probability-improbability dualities contained in and supported by the possibility space, impossibility space and the space of possibility-impossibility dualities jointly belong to the space of imaginations as a sub-space of the space of potentials. The conceptions of variety outcomes from the space of imaginations through the space of possibility-probability dualities and impossibility-improbability dualities belong to the space of realities as a sub-space of actuals under dualistic-polar conditions and fuzzy-conditionality restrictions. It is useful to note that the space of probability-improbability dualities is contained the space of possibility and hence an improbable event just like probable event is also possible event equipped with categorial convertor (functor), while possible events are equipped with functor implementors, the improbable events are not equipped with functor implementors. The possibility space is equipped with functor space. The impossibility space is not equipped with a functor space. The functor space generates the space of anticipations while the space of functor implementors generates the space of expectations. In this respect, the space of possibility-probability dualities generates the spaces of anticipation-expectation, necessity-freedom dualities, and initialization-sufficiency dualities within the generational social philosophical consciencism.

CHAPTER 7

REFLECTIVE ANALYTICS ON POSSIBILITY, POSSIBILITY SPACE, IMPOSSIBILITY, IMPOSSIBILITY SPACE, POSSIBLE AND IMPOSSIBLE WORLDS

ABSTRACT: CHAPTER 7

Chapter 7 follows chapter 6 to reflect the concepts of possibility, probability space, impossibility space and their relationships to possible and impossible worlds relative to the telescopic conditions over the space of identification-transformation dualities. The space of imaginations is shown as a derived space from the space of the potential and them shown to contain the possibility and probability spaces, which then give rise to the concepts and philosophy of possible and impossible worlds. The concepts and phenomena of possibility and possibility are defined and explicated with comparative differences and similarities differently from the existing ones. Their similarities and commonness of the relationships to the concepts and phenomena of possible and impossible worlds are discussed relative to the spaces of true-false and information-knowledge dualities. The chapter is concluded with examples of dualistic-polar membership functions that may be used under dualistic-polar conditions with the use of the fuzzy paradigm of thought.

7.1: Comparative Analytics of Possibility, Possibility Space, Probability, and Probability space

Analytically, what are the differences and similarities between the concept and phenomenon of possibility and the concept and phenomenon of probability? Similarly, what are the differences and similarities between the concept and phenomena of the possibility space and

K. K. Dompere, *The Theory of Epistemic Fields*, Studies in Systems, Decision and Control 498, https://doi.org/10.1007/978-3-031-42470-0_7

the concept and phenomenon of the probability space? Both possibility and probability are in relation to the concept and phenomenon of imagination relative to the concept and phenomenon of potential? The concept and phenomenon of imagination arise from the concept and phenomenon of the space of variety certainty-uncertainty dualities containing the concepts and phenomena of certainty and uncertainty and corresponding spaces which are contained in the space of certainty-uncertainty dualities as functor mappings of the concepts and phenomena of variety epistemic distance and the space of epistemic distances. The variety epistemic distance and the space of variety epistemic distances are derived from the space of ontic-epistemic dualities that houses both knowledge and ignorance defined by the ontic and epistemic characteristic dispositions, where the development of variety epistemic characteristic dispositions are constrained by cognitive capacity limitations of all forms over the space of acquaintance-description dualities which is derived from the concept and the phenomenon of ontological-epistemological duality. The degrees of cognitive capacity limitations produce the degrees of variety ignorance that establish the foundations of the concepts and phenomena of the space of certainty-uncertainty from which cognitive speculations emerge to give rise to the space of imaginations to compensate for the degrees of variety ignorance over the space of identification-transformation dualities. The cognitive activities over the space of imaginations are attempts to overcome cognitive capacity limitations to reduce variety epistemic distances to increase the quality of information-knowledge varieties associated with neuro-decision-choice activities to increase the degrees of variety successes and reduce the degrees of failures in creating future realities relative to present with the information-knowledge disequilibrium dynamics. The space of imaginations even though is derived from the potential space is linked to the space of actual in the dualistic-polar conditions of the space of actual-potential dualities.

There are important differences between possibility and probability and their corresponding spaces. Within the concept and phenomena of possibility we have the necessity, anticipation, initial conditions in terms of variety identifications, misinformation, disinformation, goals, objectives, visions, dreams, apparitions, preferences, conceptions, explanatory theories relative to initial conditions, prescriptive theories relative to anticipations, cognitive designs relative to plans and the concepts of future varieties in the sense that future realities from the space of identification-transformation dualities reside in the concept and phenomenon of possibility under the evolving social philosophical consciencism that contains national personalities that generate different types of national exceptionalism over the space of neuro-decision-choice actions in the social space in the broad general sense of human behaviors determined by the acquired information-knowledge conditions for conceptualizing the space of imaginations in terms of telescopic future.

Within the concept and phenomena of probability, we have the freedom, expectations, likelihood, sufficient conditions in terms of variety transformations in the sense that actualized future realities reside in the concept and phenomenon of possibility, success, failure, resource capacity limitations, risk, implementational hesitations and the concepts of future realities under social philosophical consciencism. The point here is that possibility is not the same as probability over the spaces of ignorance-knowledge dualities, identification-transformation dualities, and neuro-decision-choice actions. The possibility space houses the conceptions of telescopic future varieties without implementations and not in relation to resource conditions. The probability space houses the implementational actions in relation to resource limitations and risk over the spaces of success-failure dualities and cost-benefit dualities. The possibility space presents a dynamic model of human variety existential concepts contained in the space of imaginations, the size of which is constrained by forms of cognitive capacity limitations in conceptions under the evolving social philosophical consciencism. The possibility space is conceptualized without its relationship to the spaces of cost-benefit dualities and success-failure dualities.

The probability space presents a dynamic transformational system as an extension from the possibility space to connect the space of imaginations to the space of realities in terms of outcomes, where the success-failure dynamics are under creativity and resource limitations and the ruling social philosophical consciencism. It is over the probability space that detection of anomalies occurs, where cognitive conditionality specifies the areas of acceptable claims over the space of ignorance-knowledge dualities under the principle of reductionism which gives rise to data analytics, the theory of statics and general theory of entropy, and where applications of explanatory and prescriptive theories on varieties are exercised within the space of fundamental-applied dualities. It is at this point in knowing with increasing difficulties of ignorance reductions and knowledge enhancements that the concepts and phenomena of randomness, random variable, stochasticity, and stochastic variable are introduced over the probability space with some confusion in terms of the essential epistemic differences within the space of randomness-stochasticity dualities, where its connectivity to the possibility space is not defined or specified. It may then be assumed that the relationships between possibilistic variables and probabilistic variables in neuro-decision-choice actions are of conformal mappings that translate the probabilistic-possibilistic conditions to the stochastic and random variables. In this respect, the concepts and phenomena of randomness, stochasticity and statistical analytics are introduced to allow the definitions and stated boundaries of applicable claims in neuro-decision-choice actions that relate the conditions of the space of variety ontic-epistemic dualities from which the space of certainty-uncertainty dualities is derived, where the definitions are couched in probabilistic conditionality.

The set of intellectual anomalies arises within the space imaginations contained in the space of imagination-reality dualities. The telescopic future depends on what are imagined and what are not imagined in terms of possibilities and probabilities. It is within the space of imaginations containing possibility and probability spaces that intellectual constructs over the space of identification-transformation dualities must be examined where mathematical representations must be viewed as structures that inherit the cognitive capacity limitations. The belief that the sufficiently advanced intellect, in principle to mathematically, can prove everything that is true under mathematical correspondence of physical laws, must be examined over the space of imaginations. The Gödel incompleteness theorems, the Turing's proof of non-computable values, the formulation of quantum theory, chaos, and other developments are better understood in the space of imaginations in terms of cognitive capacity limitations of what we can theoretically represent, prove, compute, and predict in the family of ordinary languages (FOL) and the family of absolute languages (FAL). The belief rests on the idea that mathematical representations of physical systems are efficient enough to make the epistemic characteristic dispositions equal to the ontic characteristic dispositions.

7.2 THE POSSIBILITY SPACE, POSSIBILITY-IMPOSSIBILITY DUALITIES AND THE POSSIBLE AND IMPOSSIBLE WORLDS

Let us reflect on the phenomena of knowing and information-knowledge production as they relate to the concepts of possible world, impossible world, and the space of possible-world-impossible-world dualities from the understanding of possibility, probability and corresponding dualities and the corresponding spaces as they also relate to the spaces of variety acceptance-rejection dualities and ignorance-knowledge dualities. The phenomena of possible worlds and impossible worlds will be placed within the epistemic structure of the theory of knowledge square, where knowing by cognitive agents proceeds from potential space to possibility space contained in the space of imaginations, from the possibility space to probability space also contained in the space of imaginations and from the probability space to the space of the actuals that houses the space of realities. Let us also keep in mind of the relational interconnectivity with a system of give-and-take-sharing modes of the spaces and sub-spaces. The interconnectivity is such that the space of possibility-probability dualities is a sub-space contained in the space of imaginations as contained in the space of imagination-reality dualities in which cognitive agents operate over the space of ontological-epistemological dualities.

The space of possibility is contained in the space of imaginations and not in the space of realities even though the conditions of the space of realities affect the space of imaginations through the system of relational continua and give-and-take sharing modes within the spatial interconnectedness under the general principle of opposites. The possibility space and the possible worlds will then be related to variety natural transformations over the space of actual-potential dualities, where varieties are transformed from potentials to the actuals, and actuals are transformed to the potentials under ontological processes without going through the possibility and probability spaces contained in the space of imaginations due to the information-knowledge equality conditions as have been discussed. In the previous chapters, it was indicated that the information-knowledge equality simply indicates no risk, error, anticipation, and expectations in the ontological variety processes, where variety characteristic dispositions are the same as variety signal dispositions in the intra-ontological communications. In other words, there are no cognitive capacity limitations, resource limitations, ignorance, uncertainty, imagination, and the space of imaginations and hence no possibility and probability connecting the space of potential to the space of actuals which is directly connected to the space of potential as they are presented in the in the diagram of the knowledge square in Fig 6.2.2.1. This natural advantage of direct actual-potential connectivity is not available to cognitive agents operating with cognitive capacity limitations and resource capacity limitation over the epistemological space linked to the epistemic field with corresponding relational and information fields as a derived identities from the ontological space. The epistemological space is where cognitive agents, under cognitive capacity limitations and resource limitations, must move from space of potentials through the space of imaginations containing possibility and probability space in connecting the space of realities and the space of actuals is then connected to the space of potentials.

The cognitive model of knowing and knowledge production in terms of intellectual investment-capital accumulation processes over the space of input-output dualities is such that the understanding of concepts of possible worlds, impossible worlds and possible-world-impossible-world duality with corresponding spaces as they relate to knowing and knowledge productions must be seen within the concepts of ontological space, epistemological space and space of ontological-epistemological dualities under socio-natural dynamics as guarded by the universal principle of opposites over the matter-energy-information system with neutrality of time as the fourth dimension of universal existence generating variety and categorial give-and-take-sharing modes of relation. Our knowing processes and information-knowledge accumulation are developed around four basic pillars of the house of the universe of the ontological space and the house of knowledge of the epistemological space to understand ourselves and the universe in which cognitive and decision-choice agents reside and function, also as ontological elements, where the house of knowledge is a minuscular and unfinished structure at any moment of time within the house of the universe containing the space of finite-infinite dualities. It is useful to point out that the epistemological space as conceived is a small subspace within the ontological space, where the space of ontological-epistemological dualities presents the idea that every ontological variety may have a corresponding epistemological variety in the epistemological space but not every epistemological variety has an existence in the ontological space due to cognitive capacity limitations of many forms. In terms of information representation, the space of ontological-epistemological duality has a corresponding space of ontic-epistemic characteristic dualities, where the ontic characteristic dispositions relate to the perfect identities of ontological varieties and the epistemic characteristic dispositions relate to the imperfect identities of epistemological varieties. Corresponding to the space of variety ontological-epistemological dualities, there is the space of ontic-epistemic dualities of variety characteristic dispositions and corresponding to it is the space of perfect-imperfect dualities of variety characteristic dispositions that generates the space of epistemic distances which then gives rise to the space of certainty-uncertainty dualities to give rise to the space of imaginations that houses doubt-surety, success-failure, risk-benefit dualities and other such neuro-decision-choice dualities contained in the identification-transformation dualities.

Let us now examine the concepts of possibility, impossibility, possible world, and impossible world within the space of ontological-epistemological dualities with a clear understanding that the existence of cognitive capacity limitations can produce intended and unintended *anomalies* in the space of knowing and knowledge production processes over the space ignorance-knowledge dualities. The anomalies arise in all areas of knowing and hence are part of the factors that unite all areas of information-knowledge production, accumulations, and stocks over the space of fundamental-applied dualities. Anomalies in knowing are varieties in the epistemological space that are part of the information-knowledge accumulation system with self-excitement and self-correction. They do not occur in the ontological space, where there are no self-correction processes but continual variety transformations in accordance with the laws of intra-ontological and inter-ontological transformations.

7.2.1 Possibility Space, Possible Worlds, Impossible Worlds and Ontological Space

Let us examine the concepts of possibility, impossibility, Possible world, and impossible world in relations to conditions over the ontological space $\left(\Omega=\left(\mathfrak{U}\cup\mathfrak{A}\right)\right)$. The ontological space is the primary identity space relative to the knowing process, information-knowledge production, information-knowledge accumulations and decision-choice actions over the space of problem-solution dualities in relation to the ontological space which is a union between the space of actuals $(\mathfrak{A})$ and the space of potential $(\mathfrak{U})$ with corresponding characteristic-signal dispositions under the principle of opposites. The space of the actuals and the space of the potentials combine to form the space of actual-potential dualities $\left(\mathfrak{U}\cup\mathfrak{A}\right)=\left(\mathfrak{U}\otimes\mathfrak{A}\right)$ with relational continuum $\left(\mathfrak{U}\cap\mathfrak{A}\right)\neq\varnothing$, unity $\left(\mathfrak{U}\cup\mathfrak{A}\right)$ and dualistic-polar structures that are connected to the epistemological space as a derived identity from the primary identity of the ontological space which are then combined to form the space of ontological-epistemological dualities $\left(\Omega\cup\mathfrak{E}\right)=\left(\Omega\otimes\mathfrak{E}\right)$ that houses the space of actual-potential dualities $\left\{\left(\Omega\cup\mathfrak{E}\right)=\left(\Omega\otimes\mathfrak{E}\right)\right\}\subseteq\left\{\left(\mathfrak{U}\cup\mathfrak{A}\right)=\left(\mathfrak{U}\otimes\mathfrak{A}\right)\right\}$. From the space of actual-potential dualities is abstracted the space of imagination-reality dualities $\left\{\left(\mathfrak{G}\cup\mathfrak{H}\right)=\left(\mathfrak{G}\otimes\mathfrak{H}\right)\mid\left(\mathfrak{G}\cap\mathfrak{H}\right)\neq\varnothing\right\}$, where the space of imaginations is contained in the space potentials $(\mathfrak{G} \subset \mathfrak{U})$ and the space of realities is contained in the space of actual $(\mathfrak{H} \subset \mathfrak{A})$. We must always keep in mind that both the space of imaginations and the space of realities are social constructs under the principle of imperfections with information-knowledge inequality conditions. Similarly, both the space of actuals and the space of potentials are natural constructs under the principle of perfections with information-knowledge equality conditions, Analytically, the space of imagination-reality dualities is conceptualized in the epistemological space as a cognitive construct from the spaces of actual and potential to help in the understanding of the construct of the epistemic field with relational and information fields over the space of identification-transformation dualities and its dynamics over the space of static-dynamic dualities relative to the space of ignorance-knowledge dualities contained in the space of fundamental-applied dualities.

By the epistemic construct, the space of actual-potential dualities is the primary identity of the space of ontological varieties with the space of ontic characteristic dispositions, while the space of imagination-reality dualities is the derived identity of the space of epistemological varieties with the space of epistemic characteristic dispositions. Let us always keep in focus the nature of information-knowledge conditions of equality and inequality under the input-output processes in the variety transformations and changes in relation to ontological and epistemological processes, where the epistemological processes are derived from the ontological processes as identities. The identities of the epistemological varieties are derived identities by neuro-decision-choice actions over the space of acquaintance-description dualities in the knowing as the production of intellectual investment flows and information-knowledge accumulation as the production of intellectual capital stocks. The space of actuals is the collection of all *what there was and* all *what there is* at any point in time while the space of potentials is the collection of all *what there would be* in the space of actual-potential dualities. The space of realities is the collection of claimed known epistemic varieties of *what there was* and *what there is* at any time point. It houses dissatisfaction, problems, the will to change, the principle of non-satiation and the need for more varieties over the space of quality-quantity dualities. The space of imaginations is the collection of idealized epistemic varieties of *what there would be* at any time point. It houses visions, goals, objectives, solutions ideas, anticipations, expectations, and dreams, as they relate to aspirations and aims under the principle of non-satiation.

At the ontological space, the identifications and transformations of varieties are inter-actual-potential processes which are basically intra-ontological processes with information-knowledge equality conditions. At the epistemological space, the identifications and transformations of varieties are inter-imagination-reality processes which are basically intra-epistemological processes and connected to the ontological processes with information-knowledge inequality conditions and cognitive capacity limitations that present uncertainties and risks of neuro-decision-choice actions over the action space. Information is the same as knowledge in the natural processes over the ontological space, where nature does not distinguish between information and knowledge. Information is not the same in the social-epistemic processes over the epistemological space. Alternatively stated, the ontic conditions are different from epistemic conditions giving rise to identity conflicts between the ontological characteristic dispositions and the epistemic characteristic dispositions, where the ontic conditions are the primary category, and the epistemic conditions are the derived category in the processes of knowing and information-knowledge production [R17.18] [R17.19].

The set of conditions of information-knowledge equality implies that there is no uncertainty in terms of vagueness, volume-limitations, mistakes and risk over the ontological transformations and variety outcomes. Here, at the ontological space, every variety is a primary and derived identity to itself and to other varieties within categories, and every category is a primary and derived identity to itself and to other categories within the epistemic field and the ontological space is an identity to itself and to the epistemological space. The ontological input-output processes internally produce knowledge as inputs and outcomes as outputs. They also suggest that ontologically, there is one universe at any moment of time and that the universe is the universe of *what there is*, as the identity with all its natural technologies and processes forming the initial conditions for transformations and changes at the relevant ontological decision-time points, where the universe is one in relational continuum and always in the matter-energy stock-flow equilibrium and in an information-knowledge stock-flow disequilibrium. In the development of knowing and information-knowledge accumulation, no new matter and no new energy are being created only new information-knowledge disequilibrium processes through continual variety transformations with the same matter-energy stock conditions.

What we have is a continual transformation from the new to the old and old to the new, from life to death and death to life and from inputs to outputs and from outputs to inputs in the transformational recycling under the principle of opposites in relational continua and unity producing variety negations of negations under parent-offspring process. The organic identification-transformation process is the set of conditions of conflicts and tensions as universally produced from the principle of opposites that generates external necessities and directions of transformations of ontological varieties with internal freedoms. The understanding of the ontological transformations through knowing and knowledge production will help us to understand the meanings of epistemological anomalies, as frameworks of understanding are sought through information acquisition and the developments of paradigms of thought. There is no possibility, impossibility, and no possible and impossible worlds over the ontological space. The possibility is in impossibility, where there is no probability and hence necessity is in the freedom and freedom is in necessity over the ontological space, where conditions of social behaviors deviate from conditions of natural behaviors of the same variety due to the conditions of degrees of information-knowledge inequality and equality as defined over the ontological and epistemological spaces with variety epistemic distances. Generally, ontological activities operate over the space of absoluteness, where values are absolutes, while epistemological activities operate over the space of relativity where values are relatives over the space of quality-quantity dualities.

The universe as an ontological space may be considered in a category of spaces of three information-time dualities of the space of information past-present dualities, the space of information present-future dualities and the space of information past-future dualities in relational continua and unity, where these time dualities are made up of varieties with their corresponding characteristic dispositions, where the space of the time dualities has a corresponding space of past-present-future information constructs that reveal the contents and multiplicity of contents in varieties. In these spaces of time dualities, each space of time dualities is unique relative to the corresponding information duality at any time reference point and may vary at various time point. Within the system of the space of time dualities, the universe may be seen in terms of *what there was*, *what there is* and *what there would be* in assisting the diversity and unity of our knowing and information-knowledge development. The system of *what the was* contains the conditions of the beginning to be known, the system of *what there is* contains the conditions of the making to be known, the system of *what there would be* contains the conditions of the end to be known and the system of relational continua and unity contains the beginning-end connectivity to be known. The universe as *what there was* (past), therefore, constitutes the initial conditions for the transformation to the universe of *what there is* (present). In other words, *what there was* constitutes the primary category while *what there is* constitutes a derived category. Similarly, the universe, *as what there is* (present), constitutes the initially foundational conditions for natural transformations into *what there would be* in terms of new universal forms with new contents. *What there is* serves as the primary category of *what would be* as a derived category under conditions of successor-predecessor processes into future forms, where the universe is seen in term of collection of characteristic contents of varieties, states, relationality, processes, and transformations of forms given the initial matter-energy-information relational conditions. Within the space of the beginning-end duality lie the past-present, present-future and past-future dualities in the space of finite-infinite dualities, where every infinity is an offspring to another infinity as the parent in the precursor-successor dualities, every present telescopic disposition is an offspring of the past telescopic disposition, as well as the parent of the future telescopic disposition and the past telescopic disposition is the parent to the present telescopic disposition and the future telescopic disposition. It is within the sequence of past, present future dispositions of the Sankofa tradition that one may examine the concepts and phenomena of possible world, impossible world and the possible world-impossible world dualities.

The concepts of possible world and impossible world and possible-world-impossible-world dualities with ontological activities have no meaning over the ontological space and hence we must proceed to the epistemological space, where cognitive activities are formed and practiced relative to the conditions of the system of time-information dualities. From the ontological space to the epistemological space, the epistemic question in knowing, and our understanding is what are possible world and impossible world and how do we define them to give their identities and knowabilities and difference in the space of ignorance-knowledge dualities relative to knowing as the production of intellectual investment flows and information-knowledge accumulation as the production of intellectual capital stocks? This epistemic question belongs to the epistemological space and cognition of neuro-decision-choice agents, where information-knowledge disparity resulting from cognitive capacity limitations to create epistemic vacuums for cognitive speculations, designs, dreams, visions, illusions, confusions, and conflicts over the space of acceptance-rejections dualities in dualistic-polar conditions in the space of imagination-knowledge dualities where cognitive speculations are derived from the space of certainty-uncertainty dualities under the principle of insufficient reasoning. The conflicts over the space of acceptance-rejection dualities are motivated by variety ontic-epistemic differences to generate conditions contained in the space of certainty-uncertainty dualities to form important imaginative activities over the space of thinking-reasoning dualities for dreams, designs, visions, goals, and objectives and other such futuristic desires to take place in the space of imaginations contained in the space of potentials. The imaginative thinking-reasoning process is composed of possibilistic thinking-reasoning and probabilistic thinking-reasoning processes which are interdependent. Similarly, variety illusions, delusions and other such holdings take place in the space of realities where variety characteristic dispositions are misidentified and placed in the social philosophical consciencism in which variety desirability is ranked by cost-benefit relative conditions.

The number of future universal world forms in the possibility space is infinite and unknown to cognitive agents except in the size of the space of imaginations, where the universe is seen in terms of contents of varieties, holding the matter-energy in a stock-flow equilibrium over the ontological space. Conceptually, there are infinite future forms over the space of possibility-impossibility dualities that ontological space may be composed of. Only one form, however, will be available as actualized for the replacement of the current form for knowing and knowledge development from the present to the past in the telescopic process over the space of present-past dualities. The universal form that will be presented as actualized will be influenced by the collective and individual activities of the ontological varieties in relational diversity, continua, and unity in the space of relational give-and-take sharing disequilibrium dynamics. The information-knowledge equality in ontological transformations with no uncertainty and risk structure is such that *what there is,* is always, and always transformed to *what there would* be since these are identities and primary categories.

Let us repeat that the concepts of possibility, impossibility, possible world, impossible world have no meaning over the ontological space to understand natural activities and transformations since these concepts find their meanings in the information-knowledge disparities over the space of varieties, but such information-knowledge disparity is not a property of the ontological space where nature operates. The ontological space presents to us perfect varieties in identities and existence defined in terms of ontic characteristic dispositions as the contents of the varieties as well as the senders' variety signal dispositions in the space of source-destination dualities. It is this condition of information-knowledge equality and disparity that forms the analytical distinction between the ontological space and the epistemological space as well as establishes a distinction between the space of absoluteness and the space of relativity which will be relevant to activities of cognitive agents over the epistemological space. The information-knowledge equality over the ontological space also establishes conditions for information-knowledge structures over the space of acquaintance in terms of assumptions and the development of paradigms of thought in cognitive thinking and reasoning with information-knowledge disparity to derive conclusions for acceptances and actions, and between the nature of classical paradigm of thought and the fuzzy paradigm of thought. It is the same conditions of information-knowledge disparity that also establish variety epistemic distances, generate conditions of ignorance, speculations containing anticipations, expectations, risk and different outcomes in the knowing process and neuro-decision-choice actions that equip the information-knowledge system with self-exiting, self-motivating and self-correcting properties toward perfection, increasing degrees of knowing and increasing information-knowledge accumulation from the elements in the space of relativity to the space of absoluteness.

Over the epistemological space, where all cognitive agents operate, there is information-knowledge disparity between the ontological varieties and epistemological varieties as result of cognitive capacity limitations of different forms with neuro-decision-choice agents. The cognitive capacity limitations of neuro-decision-choice agents produce imperfections of fuzziness and limited-volume incompleteness of representation at the space of acquaintance to generate conditions of speculations over the space of imaginations. anticipations over the possibility space and expectations of variety identities of present and future over the spaces of probability, doubt-surety dualities and decidability-undecidability dualities. Some aspects of this set of complicated relations in the space of knowing and information-knowledge production under the management and control by neuro-decision-choice processes have been discussed in one form or other in the previous monographs [R17.15] [R17.17[R17.18]. The discussions in this monograph are further extensions relative to knowing and information-knowledge production over the conceptual space of economic actions, where knowing is views as the production of intellectual investment flows and information-knowledge accumulation as the production of intellectual capital stocks. The intellectual investment flows and the intellectual capital stocks combine to generate information-knowledge stock-flow disequilibrium dynamics over the spaces of problem-solution dualities, question-answer dualities ignorance-knowledge dualities with relational connectivity to the space of imagination-reality dualities. The problems and questions in knowing reside in the space of realities while the solutions and answers reside in the space of imaginations.

7.2.2: Possible Worlds, Impossible Worlds, Space of Imaginations and the Epistemological Space: Conceptual Reflections

We shall now examine the concepts of possibility, impossibility, possible world, and impossible world in relations to conditions over the epistemological space relative to the knowing process and information-knowledge production. Analytically, things are different over the epistemological space, where based on the variety conditions of ontological identity, one can conceive of many possibilities and hence an infinite set of conceptual possible worlds from the initial ontological conditions through the elements in the *space of imaginations* constrained by cognitive capacity limitations which place finiteness on the set of conceivable possible worlds from any set of initial variety conditions around which one may speculate and imagine. In the process of knowing and information-knowledge control, we have identified ontological varieties as the identities while the epistemological varieties as the derived identities. The characteristic dispositions of the ontological varieties are the *ontic characteristic dispositions* in the state of perfections. The characteristic dispositions of the epistemological varieties are the *epistemic characteristic dispositions* in the state of imperfections. The variety epistemic characteristic dispositions may deviate from the variety ontic characteristic disposition for the same variety since the epistemic characteristic dispositions are abstracted from the ontological signal dispositions over the space of acquaintance-description dualities. The variety ontic-epistemic characteristic deviations may be associated with varieties that may appear in the space of imagination-reality dualities as they relate to many different phenomena that may arise in the space of anomalies over the space of error-correction dualities. Here, the ontological space is the sender of signal dispositions and the epistemological space is the receiver in the space of source-destination dualities, where the neuro-decision-choice agents in the epistemological space decipher the signals from the acquaintance space to form experiential information or the knowledge by acquaintance under a system of cognitive capacity limitations including vagueness and volume incompleteness to produce incomplete acquaintances and defective information and then process it with a paradigm of thought to obtain knowledge by description.

Under the principle of opposites, one may conceptualize the possible world and possible-world space, impossible world and impossible-world space and the space of possible-world-impossible-world dualities in an interrelated and interdependent definitional system. The interrelated and interdependent definitional system of these concepts may have no *knowing value* or *information-knowledge value* as it relates to ontological varieties and their behavior. It may also have no direct epistemic value as either intellectual investment or information-knowledge intellectual capital accumulation. Indirectly, the interrelations and interdependences of this system of possibilities may be referenced to solutions and answers in the spaces of problem-solution dualities and question-answer dualities. The varieties in the possibility space may be related to possible worlds and impossible worlds as conceptualized in the epistemological space and related to the neuro-decision-choice actions in the space of imaginations contained in the space of imagination-creativity dualities. We shall examine the conditions under which the possible and impossible worlds will be useful in neuro-decision-choice actions, knowing as the production of intellectual investment flows and information-knowledge accumulation viewed as production of intellectual capital stocks under paradigms of thought as well as how the possibility and possible worlds relate to information-knowledge discovery, human variety creations and mimicries of natural processes and technologies over the space of possibility-probability dualities.

The characteristics of variety possibility, variety probability and the corresponding spaces have been stated but let us have some reflections with necessity and freedom with corresponding dualities and spaces. What are the characteristic dispositions of the concepts of possibility, impossibility, possibility-impossibility duality, and the corresponding analytical spaces, and how they connect to the concepts and phenomena of possible world and possible world space? Similarly, what are the characteristic dispositions of the concepts of impossible world and impossible-world space, possible-world-impossible-world dualities, and the corresponding space over the spaces of characteristic-signal dualities which may be used to define and explicate them? What is the usefulness of the concepts and phenomena of possible world and impossible world in the space of human actions relative to the spaces of ignorance-knowledge dualities, identification-transformation dualities, and problem-solution dualities? To help answer these questions, some implicit questions and other questions that may arise, we shall proceed by viewing then in conceptual spaces of characteristics and categories that have been already defined. New ones will be introduced, defined, and explicated. The system that is being presented is an analytical model, mathematical representation, economic-theoretic structure, neuro-decision-choice framework and philosophical explications for the general understanding of knowing as the production of intellectual investment flows and information-knowledge accumulation as the production of intellectual capital stocks, where every neuro-decision-choice action is a transformation of varieties over the space of problem-solution dualities relative to changes in relation, property or both, and every transformation is an economic production under cost-benefit conditions and rationality relative to an element or some elements in the space of goals, objectives and visions.

Let the variety possible world be $(\mathfrak{w})$ and corresponding possible-world space be $(\mathfrak{W})$ with a generic element $(\mathfrak{w} \in \mathfrak{W})$ and the characteristic-signal disposition $\{\mathbb{Z}_{\mathfrak{W}} = (\mathbb{X}_{\mathfrak{W}} \otimes \mathbb{S}_{\mathfrak{W}})\}$., where $(\mathbb{X})$ is characteristic disposition and $(\mathbb{S})$ is signal disposition and all the subscripts relate to the spaces, categories or varieties. Similarly, let the variety impossible world be $(\mathfrak{v})$ and the corresponding impossible-world space be $(\mathfrak{V})$ with a generic element $(\mathfrak{v} \in \mathfrak{V})$ and the characteristic-signal disposition $\{\mathbb{Z}_{\mathfrak{V}} = (\mathbb{X}_{\mathfrak{V}} \otimes \mathbb{S}_{\mathfrak{V}})\}$. Furthermore, let the space of possible-world-impossible-world dualities be $\mathfrak{L} = (\mathfrak{W} \otimes \mathfrak{V}) = (\mathfrak{W} \cup \mathfrak{V})$ with a generic element $(\mathfrak{w}, \mathfrak{v}) \in (\mathfrak{W} \otimes \mathfrak{V})$ and the characteristic-signal disposition $\{\mathbb{Z}_{\mathfrak{L}} = (\mathbb{X}_{\mathfrak{L}} \otimes \mathbb{S}_{\mathfrak{L}}) = (\mathbb{Z}_{\mathfrak{W}} \cup \mathbb{Z}_{\mathfrak{V}})\}$ with dualistic-polar relational conditionality of $(\mathbb{X}_{\mathfrak{W}} \cap \mathbb{X}_{\mathfrak{V}}) \neq \varnothing$, where the condition $(\mathbb{Z}_{\mathfrak{W}} \cap \mathbb{Z}_{\mathfrak{V}}) \neq \varnothing$ is to maintain the relational continuum and unity with relational give-and-take sharing modes through their interconnectedness and intersectionality under the principle of opposites that also binds all the varieties and categories together in unity.

How do these concepts of possible world, impossible world and the corresponding spaces relate to possibility and possibility space $(\mathfrak{P})$ with a generic element $(\mathfrak{p} \in \mathfrak{P})$ and the characteristic-signal disposition $\mathbb{Z}_{\mathfrak{p}} = (\mathbb{X}_{\mathfrak{p}} \otimes \mathbb{S}_{\mathfrak{p}})$, and probability and probability space $(\mathfrak{B})$ with a generic element $(\mathfrak{b} \in \mathfrak{B})$ and characteristic-signal disposition $\mathbb{Z}_{\mathfrak{B}} = (\mathbb{X}_{\mathfrak{B}} \otimes \mathbb{S}_{\mathfrak{B}})$ as seen over the space of possibility-probability dualities $(\mathfrak{Z} = \mathfrak{P} \otimes \mathfrak{B})$ with a generic element $(\mathfrak{z} = (\mathfrak{p}, \mathfrak{b}) \in \mathfrak{Z} \subset \mathfrak{G})$ and the characteristic-signal disposition of the form $\mathbb{Z}_{\mathfrak{G}} \supset \mathbb{Z}_{\mathfrak{z}} = (\mathbb{X}_{\mathfrak{z}} \otimes \mathbb{S}_{\mathfrak{z}}) = (\mathbb{Z}_{\mathfrak{p}} \cup \mathbb{Z}_{\mathfrak{B}}) = (\mathbb{X}_{\mathfrak{p}} \otimes \mathbb{S}_{\mathfrak{p}}) \cup (\mathbb{X}_{\mathfrak{B}} \otimes \mathbb{S}_{\mathfrak{B}})$, the space of necessity-freedom dualities over the space of neuro-decision-choice actions under the principle of non-satiation, where $(\mathfrak{N})$ is the space of necessity with generic element $(\mathfrak{n} \in \mathfrak{N})$, the characteristic-signal disposition $\{\mathbb{Z}_{\mathfrak{N}} = (\mathbb{X}_{\mathfrak{N}} \otimes \mathbb{S}_{\mathfrak{N}})\}$ and $(\mathfrak{F})$ is the space of freedoms with the generic element $(\mathfrak{f} \in \mathfrak{F})$, the characteristic-signal disposition $\mathbb{Z}_{\mathfrak{F}} = (\mathbb{X}_{\mathfrak{F}} \otimes \mathbb{S}_{\mathfrak{F}})$ and with space of necessity-freedom dualities $(\mathfrak{P} = (\mathfrak{N} \otimes \mathfrak{F}) = (\mathfrak{N} \cup \mathfrak{F}) \subset \mathfrak{G})$ with a generic element $\mathfrak{p} \in \mathfrak{P} \Rightarrow (\mathfrak{n}, \mathfrak{f}) \in (\mathfrak{N} \otimes \mathfrak{F})$ and the characteristic-signal disposition of the form $\{\mathbb{Z}_{\mathfrak{G}} \supset \mathbb{Z}_{\mathfrak{P}} = (\mathbb{X}_{\mathfrak{P}} \otimes \mathbb{S}_{\mathfrak{P}})\}$ over the space of potentials $(\mathfrak{U})$, all of which have spaces of characteristic-signal dispositions as information to reveal their identities, where the characteristic dispositions of the varieties contain the contents and the signal dispositions are the messages that house the contents of the varieties within the space of source-destination dualities through the system of multiplicity of communications over the space of ontological-epistemological dualities.

The dividedness in creating the spaces and subspaces in the general epistemological space is induced by fuzzy processes, where every space, category or set has a corresponding membership characteristic function for assigning categorial belonging and thus defining the path of generic evolution over the space of problem-solution dualities. The membership characteristic functions are cognitively crafted by a system of neuro-decision-choice actions over the space of problem-solution dualities, where every neuro-decision-choice agent has a choice of representation and hence a subjective influence and where the collectivity of representation is obtained by fuzzy optimal aggregations. It is always useful to keep in mind that every individual phenomenon and the corresponding space are derived from a space of dualistic-polar structures under the principle of opposites with relational continuum and unity and with the fuzzy paradigm of thought within the epistemic field, as the information and relational fields alter in the response to variety identifications and transformations relative to the space of static-dynamic dualities.

The epistemic journey from the space of imaginations contained in the potential space, through the possibility space and probability space within the space of imaginations to the space of realities contained in the space of actuals must also pass through the space of certainty-uncertainty dualities defined over the space of qualitative-quantitative dualities with the possibility presenting *degrees of anticipations,* where subjectivity dominates objectivity and probability presenting *degrees of expectations,* where there are always conflicts between objectivity and subjectivity relative to dominance in the claims of probability measure of probabilistic conditionality. The degrees of anticipation and expectation must be combined to create hybrid values for establishing qualitative-quantitative dispositions of events over the space of anticipation-expectation dualities which must be related to the spaces of possibility-probability dualities containing the space of necessity-freedom dualities and contained in the space of imaginations as a sub-space of potentials, where the qualitative and quantitative uncertainties are conceptualized as contained in the space of imaginations as the sub-space of potentials. The concepts and phenomena of subjective and objective probabilities are neuro-decision-choice dependent, and the similarities and differences must be ascertained from the conditions in the space of acquaintance-description dualities. The size of the variety epistemic disposition relative to the size of the variety ontic characteristic disposition presents relative frequency as a measure of the degree of ignorance contained in the space of ontic-epistemic dualities where the space of ontic-epistemic dualities is conformally mapped onto the space of degrees in certainty-uncertainty dualities and as the measures of objective probability without establishing the processes of the variety epistemic distances. The linking of the objective probability to the volume-limitations in the acquaintance space is difficult do established. The concept of relative frequency has a limited area of justification in the knowing process, where the degree of frequency is related to epistemological data in quantity without quality at the space of acquaintance [R24.17] [R24.33] and measure the degrees of quantitative nonacquaintance in the space of exact-inexact dualities without qualitative factors which dominate socio-natural existence and occurrence.

The degree of ignorance, just like the degrees of knowledge, is cognitively determined through the neuro-decision-choice actions, while the acceptance of the knowledge as an input into other decision choice actions reflects the degree of ignorance as reflecting the degree of failure subjectively computed as the probability within the neuro decision-choice situation, where the concept of relative frequency makes no sense over the space of theory-application dualities. This is the approach used in [R24.5] [R24.27][R24.37], where the subjective probability measures the degree of cognitive ignorance in the fuzzy information space. In this respect, the whole development of decision theory needs some critical review and major restructuring. The development of the decision-choice theory must include the development of its information-knowledge inputs as part of comprehensive theory of production-consumption processes as well as input-output processes relative to supply-demand decision chains. The decision-choice theory is about human actions over the space of identification-transformation dualities relative to either changes in variety relation, changes in variety properties or both, where information-knowledge items are acquired as inputs for representations and computations over the space of quality-quantity dualities which allows processes to deal with subjective and qualitative characteristics with judicious judgements. The theory of decision under uncertainty and the expected utility theory must assume the existence of probability and probability measure as weighted value of probabilistic uncertainty with very little discussions on the possibilistic uncertainty. The meaning of uncertainty and its relationship with probability are epistemically assumed for representations and computation.

It is at this junction of epistemic journey from the space of variety ignorance to the space of variety knowledge under cognitive capacity limitations of cognitive agents that the debate and discussions of the concepts and phenomena of subjective probability , objective probability and the search for relevant measurements arise to assist in the decision-choice process over the space of certainty-uncertainty dualities, where probability theory is developed to present a constrained space of knowledge and neuro-decision-choice claims. In all theoretical discussions, it is neglected that the measures of the uncertainties are used to transform the possibilistic-probabilistic cost-benefit values in the space of imaginations to cost-benefit equivalences in the space of realities as information-knowledge inputs into neuro-decision-choice actions for the expected outcomes. The relationships among cost-benefit values and neuro-decision-choice actions over the space of certainty-uncertainty dualities constitute the foundation of the development of theories of decisions under uncertainty, where indirectly variety anticipations in the possibility space are connected to variety expectations in the probability space.

The epistemic problems in the development of theories of decisions under uncertainties which are also related to knowing and information-knowledge production for acceptance are that the concepts of problem, uncertainty and probability are not clearly defined and explicated such that most of these theoretical developments of decision under uncertainty are erected on undefined, poorly defined or analytically weak foundation as dealt with over the space of absoluteness with the classical paradigm of thought generating some paradoxes, where such paradoxes like those of Gödel's incompleteness and completeness theorems reflect the analytical weakness of classical paradigm of thought [R21.5][R21,15] leading to the rise of all kinds of concepts and phenomena such as cognitive uncertainty and behavioral uncertainty without connections to the spaces of quality-quantity dualities , cognitive capacity limitations and the foundations of uncertainty-certainty dualities, where qualitative and quantitative uncertainties can be examined to develop hybrid values and their effects on neuro-decision-choice where the Gödel incompleteness axiomatic formal systems may be examined over the space absoluteness-relativity dualities. Similarly, the solution to the principle of insufficient reasoning is abstracted from the thinking algorithm of optimization of mutually constraining opposite duals or opposite poles under fuzzy mathematical programing, where the dualistic-polar optima are always at the point of $\left(\frac{1}{2}\right) \in [0,1]$ for the optimal functioning of elements under the principle of opposites defining the universal forces of conflicts and harmonies.

In the process of examining the concept of quantitative uncertainty from the probability space, the probability set has an implicitly assumed condition of possibility set in the classical probabilistic reasoning, where qualitative uncertainty is assumed away and all events with positive probability are placed in the possibility space. In this respect, the concept of possibility shows up under impossible outcomes. In this process, the varieties from the potential space are epistemically moved to the space of imaginations and then to the space of possibility-impossibility dualities and then separated into possibility and impossibility spaces. The possibility space is equipped with necessity and initial conditions with either categorial convertors or functors that generate anticipation operators. The impossibility space is not equipped with necessity and initial conditions with categorial convertors and hence has no anticipation operator, where every probability set has an implicitly underlying possibility set. The possibilistic values of the elements in the probability set are assumed to be positive between zero and one and uniformly distributed to sum up to one under the classical paradigm and hence unimportant in the development of classical probabilistic values and reasoning in a way where the effects of the possibility space is dismantled from probabilistic outcomes in the space of thinking-reasoning dualities except all events with positive probability belong to the possibility space with equal possibilities and anticipation operator, and all events with zero probability without expectations are classified as impossible events without anticipation and hence belong to the impossibility space rather than improbable events with anticipation operator but without expectation operator in the possibility space.

An outcome in the space of imagination-reality dualities may belong to the space of possibility-impossibility dualities, where such an outcome may be mapped into the space of probability-improbability dualities. The space of probability-improbability dualities is contained in the possibility space, equipped with the space of anticipation operators while the impossibility space is equipped with the space of non-anticipation operators. The anticipation operators are generated by necessity, initial conditions and categorial convertors or functor operators. All varieties in the probability space are equipped with both anticipation and expectation operators, where the anticipation operator is primary, and the expectation operator is derived and hence with measures of possibility and probability. The expectation operators are generated by conditions of freedom, sufficiency, and resource availability. The varieties in the improbability space are equipped with anticipations but not with expectations as seen in implementation requirements. In this respect, a zero probability implies a lack of resource capacity to actualize the possible variety. The possibility space is equipped with the space of categorial convertors or functors, and the impossibility space is not equipped with the space of categorial convertors or functors.

The analytical dismantling of the effects of the space of possibility from the space of probability has complex implications over the whole knowing, information-knowledge production and their neuro-decision-choice actions over the space of acceptance-rejection dualities with epistemic conditionality from the nature of cognitive capacity limitations of neuro-decision-choice agents. The implication here is that the possibilistic and probabilistic reasonings are not combine in all neuro-decision-choice activities over the action space. The conditions over the space of uncertainty-certainty dualities as generated by the space of epistemic distances from the conditions of the space of ontic-epistemic dualities are presented only in the possibility space to the neglect of the supporting possibility space. It is, here, over the space of anticipation-expectation dualities in relation to the space of possibility-probability dualities that theoretical difficulties and zones of neuro-decision-choice conflicts also arise between the advocates of subjective and objective (personal) probabilities, quantification of uncertainties [R24.1-R24.37], quality control of information-knowledge structures and the whole development of entropy [R21.1], where possibility, probability, necessity, freedom. subjectivity and objectivity have no important simultaneity in in neuro-decision-choice activities over the action space of knowing and information-knowledge development relative to the space of fundamental knowledge, and variety development and transformations relative to the space of applied knowledge, where creative link must be made over the space of fundamental-applied knowledge dualities in variety production processes.

The system of zones of epistemic conflicts is due to the manner in which the ontological signal dispositions are decoded and the variety epistemic distances are conceptualized to create knowledge by acquaintance as an input into the space of thinking-reasoning dualities to abstract the knowledge by description as well as reconciled the probability measures with the use of hybrid fuzzy-stochastic numbers to capture fuzziness in terms of subjectivity, quality, vagueness, necessity and volume-limitations in terms of objectivity in the spaces of acquaintance and representations of the characteristic-signal dispositions that impose cognitive capacity limitations in knowing and the information-knowledge production under quality control over the spaces of possibility-probability, necessity-freedom, anticipation-expectation and constructionism-reductionism dualities with a defined mega-paradigm of thought and a system of meta-paradigms of thought , where the variations of the nature of possibility set, as necessity and initial conditions, affect the constructed probability set and outcomes at the space of decidability-undecidability dualities. It is useful to examine the nature and the roles of possibility, possibility space, probability. probability space in relation to necessity, anticipation, freedom, and expectation in the process of knowing, information-knowledge production, cognitive capacity limitations of cognitive agents and the general neuro-decision-choice activities over the action space including knowing and elements in the spaces of problem-solution dualities and question-answer dualities in relational diversity, continuum, give-and-take-sharing modes and universal unity under the principle of opposites.

The concept of possibility is a present-future telescopic process in the space of knowing regarding variety identifications and transformations. The space of present-future telescopic process houses imagination, anticipations, initial conditions, necessity, illusion, hallucination, delusion, visions, goals, objectives, possible worlds, impossible worlds, and the space of possible-impossible-world dualities. The space of possible-impossible worlds may generate characteristic dispositions of illusory varieties, delusionary varieties, hallucination varieties, lie varieties, creativity varieties, hope varieties and falsely derived varieties with false anticipation over the space of true-false dualities from the elements of the space of ontological-epistemological dualities to produce fundamental and applied knowledge systems over the space of input-output dualities as seen from the space of actual-potential dualities, where the space of fundamental-applied dualities is defined over the space of imagination-reality dualities relative to the epistemological space. It is these uncertainties projected from the space of ontic-epistemic dualities that generate uncertainties and risks of cognitive creations over the space of fundamental-applied and success-failure dualities. It is also useful to the understanding that the behavioral dynamics of the space of success-failure dualities are consistent with behavior of space of parent-offspring dualities cost-benefit dualities, life-death dualities, old-new dualities, end-benefit dualities in the space of opposites where the opposites reside in each other such that in costs do we find benefits and vice versa and in negative do we find positive and vice versa in the general space of varieties.

Given the space of imagination-reality dualities, how does one know that the varieties of illusions, delusions, and hallucination, contained in the space of imaginations are not past, current. or future information-knowledge items in the space of knowing? Similarly, how do the conditions of these phenomena relate to the conditions of anomalies such as misinformation, disinformation, propaganda, ill-posed problem, phantom problem, fake news, unsolvable problem, unsolved problem, alternative realities, alternative facts and paradoxes as one seeks the conditions of diversity and unity principles of knowing and information-knowledge system over the space of imagination-reality dualities that may also house illusions of falsehood, misleading, hallucination, hope, goal varieties contained in space of impossibility-possibility dualities keeping in mind that hope not only resides in hopelessness but also resides in illusions and dilutions in the space of ignorance-knowledge dualities?

The concepts and phenomena of possibility, probability, possible worlds, and epistemic anomalies make sense only in the epistemological space, where every epistemic variety is cognitively derived and neuro-decision-choice based, relative to the available information through conceptual games over the space of possibility-probability dualities. These concepts make no material sense over the ontological space where variety transformations are directly implemented over the space of actual-potential dualities. The phenomena of possibility and probability are cognitive in formation that relate to human state of confidence in the available information-knowledge varieties as inputs for neuro-decision-choice actions. What are the characteristic dispositions that present the identities, differences, and similarities of possible and impossible worlds? Let us keep in mind that the concepts and phenomena of possible world and impossible world are derivatives from the concepts of possibility and possibility space and hence their sources are traceable to the sources of possibility-impossibility dualities? In other words, how does one know in a conceptual system that a possible world is a possible world and impossible world is an impossible world? The answer to this question is found from the concept of information defined as characteristic-signal disposition that allows identity of a variety to be known and established. Let us also keep in mind that the variety characteristic-signal disposition is such that the characteristic disposition presents the contents of the information and the signal disposition presents the message of the contents in the space of source-destination dualities of sender-receiver processes, where the signal dispositions are from the senders at the sources that must be deciphered by the receivers at the destinations to abstract the content and identify the varieties.

In view of these discussions on the conceptual anomalies, let us keep in mind the system of information-time dualities of the spaces of past-present dualities, present-future dualities and past-future dualities and their relationships to telescopic enveloping from *what there was*, *what there is* and *what there would be* over the space of identification-transformation dualities. It is also useful to see the connectivity actions of cognitive agents on the past-present-future processes over appropriate spaces and sub-spaces of knowing in terms of the *asantrofi-anoma* problem-solution dualities. The definitions, explications, and discussions on the conceptual anomalies in imagination, reality and space of imagination-reality dualities will be taken up in another monograph where definitions will be given to the concept of possible, impossible, and possible-impossible world dualities [R8.16b].

7.2.3: Examples of the Graphs of Dualistic-Polar Membership Characteristic Functions for Negative and Positive Duals in the Fuzzy Space

It is useful at this point to suggest some examples of mathematical and graphical representations of the membership belonging functions of the negative dual is a decreasing function of the number of the negative characteristics such that

$$\mu_{\mathrm{B}}(x)=\left\{(x,\mu_{\mathrm{B}}(x),n)\,|\,x\in\mathrm{B},\, n\subset\mathbb{N},\, \mu_{\mathrm{B}}(x)\in[0,1],\frac{d\mu_{\mathrm{B}}(x)}{dx}<0\right\} \text{ Negative Dual} \qquad (7.2.3.1)$$

Similarly, the mathematical and graph representations of the membership functions of the positive dual is an increasing function of the number of the positive characteristics such that,

$$\mu_{\mathrm{A}}(x)=\left\{(x,\mu_{\mathrm{A}}(x),n)\,|\,x\in\mathrm{A},\, n\in\mathbb{N},\, \mu_{\mathrm{A}}(x)\in[0,1],\frac{d\mu_{\mathrm{A}}(x)}{dx}>0\right\} \text{ Positive Dual } (7.2.3.2)$$

The following functional and graphical forms are provided and adopted from Kaufman [R5.35] and may be extended and replaced with other dualities and variables for analytical needs.

GRAPHS AND REPRESENTATIONS OF MEMBERSHIP FUNCTIONS FOR NEGATIVE DUAL

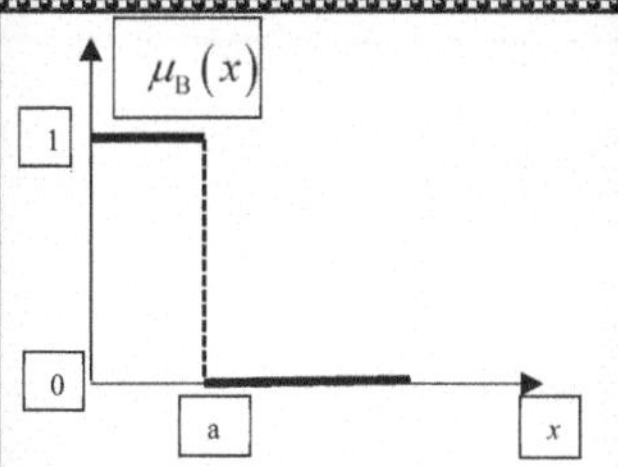

$$\mu_B(x)\begin{cases} =1 & \text{if } 0 \le x \le a \\ =0 & \text{if } x > a \end{cases} \quad \textbf{(7.2.3.3)}$$

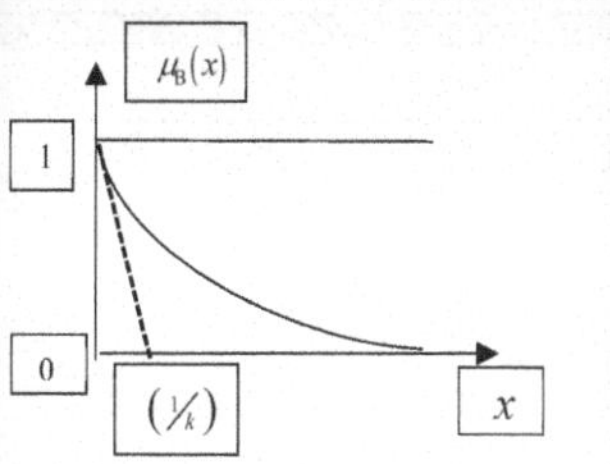

$$\mu_B(x) = e^{-kx}, \ \forall k > 0 \quad \textbf{(7.2.3.4)}$$

GRAPHS AND REPRESENTATIONS OF MEMBERSHIP FUNCTIONS FOR POSITIVE DUAL

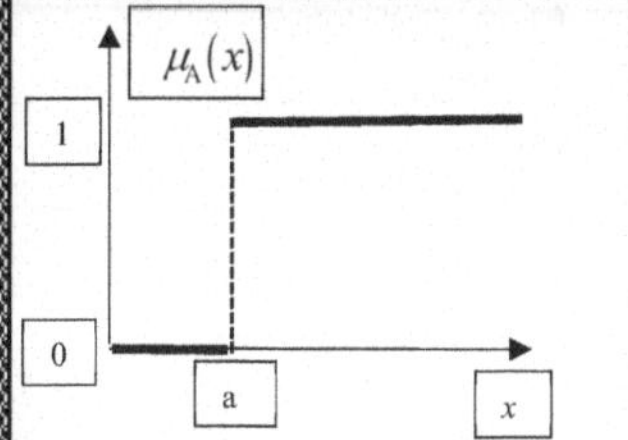

$$\mu_A(x)\begin{cases} =1, \text{ if } x \ge a \\ =0, \text{ if } 0 \le x < a \end{cases} \quad (7.2.3.9)$$

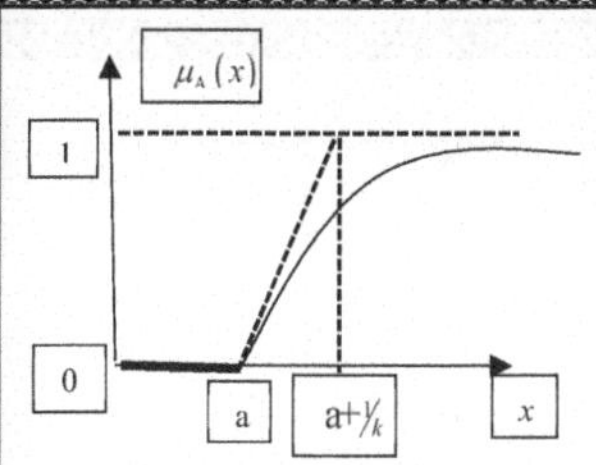

$$\mu_A(x)\begin{cases} =1-e^{-k(x-a)}, \text{ if } x \ge a, \\ =0, \text{ if } 0 \le x \le a, k > 0 \end{cases}$$

(7.2.3.10)

GRAPHS AND REPRESENTATIONS OF MEMBERSHIP FUNCTIONS FOR NEGATIVE DUAL

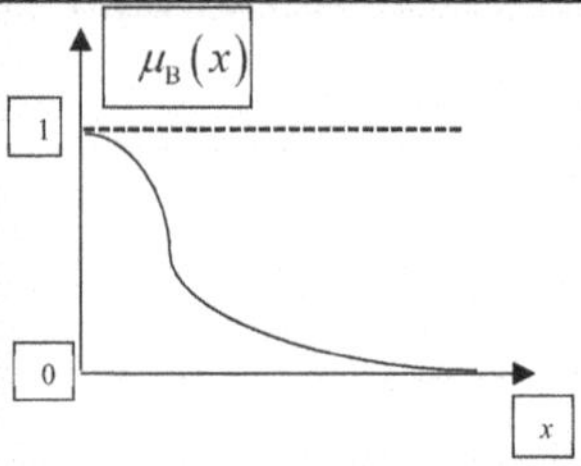

$$\mu_B(x)\left\{= e^{-kx^2}, \forall k > 0\right. \quad (7.2.3.5)$$

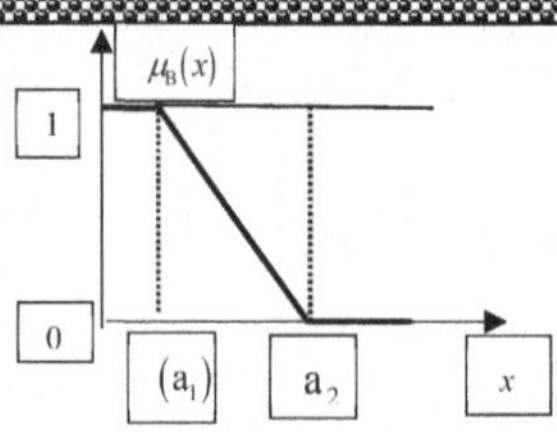

$$\mu_B(x) = \begin{cases} 1, & \text{if } 0 \le x \le a_1 \\ \dfrac{a_2 - x}{a_2 - a_1}, & \text{if } a_1 \le x \le a_2 \\ 0, & \text{if } x \ge a_2 \end{cases}$$

(7.2.3.6)

GRAPH AND REPRESENTATION OF MEMBERSHIP FUNCTIONS FOR POSITIVE DUAL

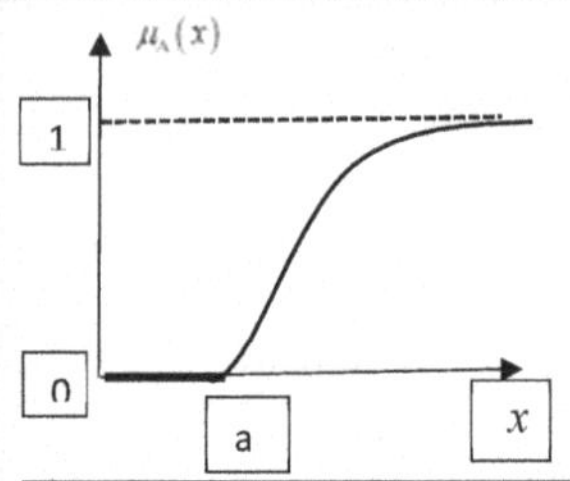

$$\mu_A(x)\begin{cases} = 1 - e^{-k(x-a)^2}, & \text{if } x \ge a, \\ = 0, & \text{if } 0 \le x \le a, k > 0 \end{cases}$$

(7.2.3.11)

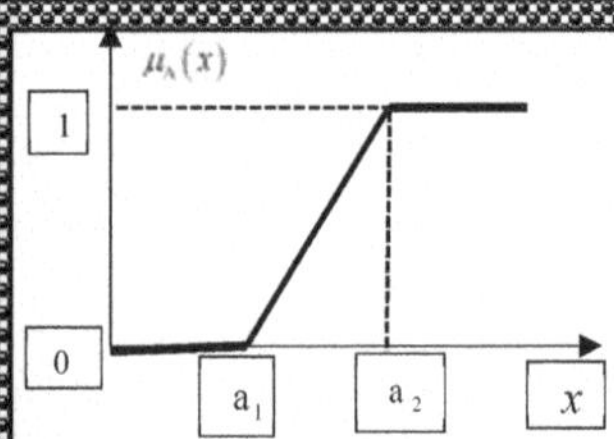

$$\mu_A(x) = \begin{cases} 1, & \text{if } x \ge a_2 \\ \dfrac{x - a_1}{a_2 - a_1}, & \text{if } a_1 \le x \le a_2 \\ 0, & \text{if } 0 \le x \ge a_1 \end{cases}$$

(7.2.3.12)

GRAPHS AND REPRESENTATIONS OF MEMBERSHIP FUNCTIONS FOR NEGATIVE DUAL

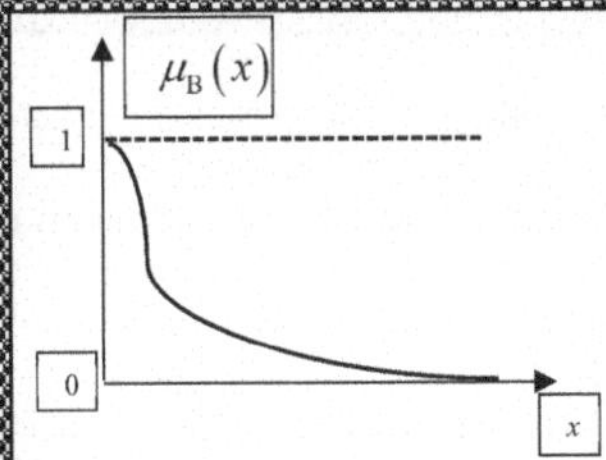

$$\mu_B(x)\left\{= \frac{1}{1+kx^2}, \forall k > 1\right.$$

(7.2.3.7)

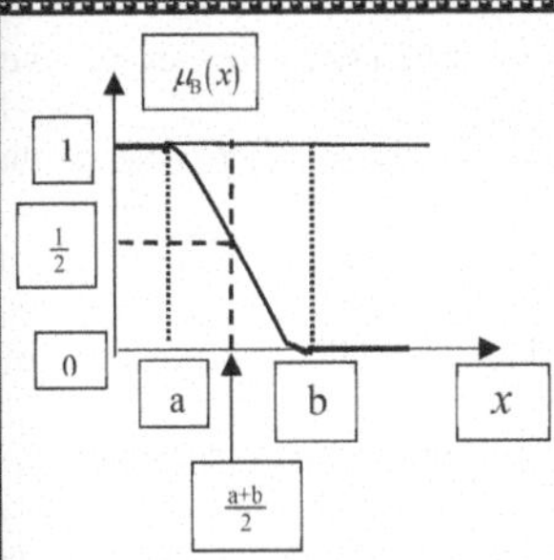

$$\mu_B(x) = \begin{cases} 1, & \text{if } 0 \le x \le a \\ \frac{1}{2} - \frac{1}{2}\sin\frac{\pi}{b+a}\left(x - \frac{a+b}{2}\right), & \text{if } a \le x \le b \\ 0, & \text{if } x \ge b \end{cases}$$

(7.2.3.8)

GRAPH AND REPRESENTATION OF MEMBERSHIP FUNCTIONS FOR POSITIVE DUAL

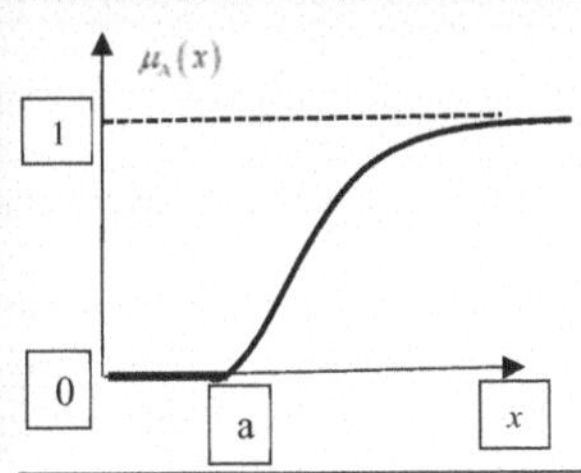

$$\mu_A(x)\begin{cases} = \frac{k(x-a)^2}{1+k(x-a)^2}, & \text{if } a \le x \le \infty \\ = 0, & \text{if } 0 \le x \le a, \quad k > 0 \end{cases}$$

(7.2.3.13)

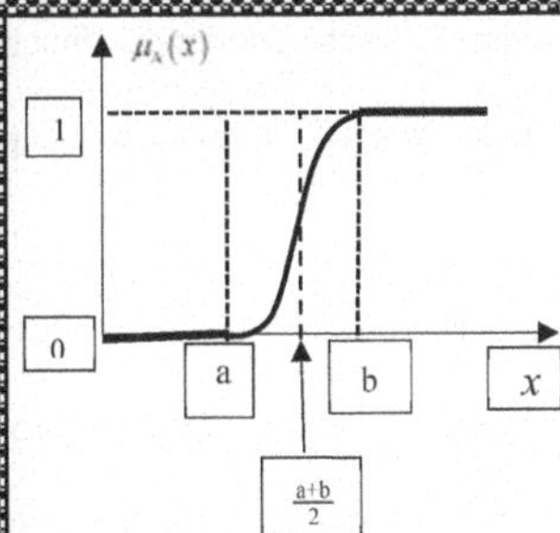

$$\mu_A(x) = \begin{cases} 1, & \text{if } x \ge b \\ \frac{1}{2} + \frac{1}{2}\sin\frac{\pi}{b-a}\left(x - \frac{a+b}{2}\right) & \text{if } a \le x \le b \\ 0, & \text{if } 0 \le x \le a \end{cases}$$

(7.2.3.14)

These are some of the fuzzy membership characteristic functions that may be used in theoretical and applied works in the analysis of the collaborative process of fuzzy paradigm of thought and information conditions of the principle of opposites with evolving duality and polarity in relational continuum and unity at static and dynamic states. One thing is clear. Equation (7.2.3.3) is not relevant in fuzzy reasoning over the space of relativity since all the continuous membership functions contain their end points which are one and zero with uniform distribution of belonging to the set. This representation is consistent with the classical paradigm under the principle of the excluded middle specified over the space of absoluteness. The question as to what functional form will be used depends on the nature of the variety, conditions of transformation and the questions to which answers are sought. The relational conditions of continuum and unity of the principle of opposites require that the continuous functional form must be used where every level negative-characteristic set that acts as cost has a supporting level positive-characteristic set that acts as a benefit. It is useful to have an understanding on the relationships among the space of relativity, the space of absoluteness, the principle of the excluded middle, the principle of relational continuum and unity, the principle of opposites, the classical paradigm of thought and the fuzzy paradigm of thought and the roles they play in human actions through the neuro-decision-choice actions that connect the past-present-future (Sankofa-anoma) telescopic processes. The formations of sets, groups, categories, and fields are undertaken with membership characteristic functions in selections, their placements and aggregation under fuzzy paradigm of thought.

The paradigms of thought are used to construct the membership characteristic functions in dealing with the general conditions over the space of imagination-reality dualities under cognitive capacity limitations of many forms that link the possibility space and the probability space and their relationships to probability and the development of fuzzy mathematics and non-fuzzy mathematics for the acquisition and processing of information over the space of exact-inexact dualities, subjective-objective dualities and vagueness-clarity dualities. There are other types of fuzzy numbers such as trapezoidal, decagonal, pentagonal, hexagonal numbers that are available to explore in the fuzzy paradigmatic approach to the theory epistemic field for critical analytics of formation of knowledge areas, departments within areas in the processes of knowing as the production of intellectual investment flows and information-knowledge accumulation as the production of intellectual capital stocks where these neuro-decision-choice actions are seen as economic activities over the space of input-output dualities and production-consumption dualities. The production of intellectual investment flows goes to update the intellectual capital stocks where the services of the intellectual capital stocks become inputs into the knowing processes.

EPILOUGUE

REFLECTIONS ON ORGANIZATIONAL ANALYTICS FOR FUTURE RESEARCH

ABSTRACT: Epilogue

The epilogue of this monograph provides a connectivity with the prologue and epistemic development of theories on knowing as social production of intellectual investment flows and information-knowledge accumulations as social production of intellectual capital stocks and projections of future actions on organizational analytics as central role of simultaneity of oneness, diversity, unity and disunity, where knowing and information knowledge accumulation are seen in terms stock-flow dynamics under neuro-decision-choice actions. The fuzzy rationality, principles of opposites with relational continua and unity, the theory of info-statics, identification problem-solution dualities, the theory of info-dynamics, transformation problem-solution dualities, the theory of problem-solution dualities as production-consumption dynamics over the spaces of input-output dualities and destruction-construction dualities, all are connected to the space of varieties as contained in the space of imagination-reality dualities as derived category by neuro-decision-choice actions from the space of actual-potential dualities. The connectivity systems of actions for variety identifications and transformations are done through the development of organizations and sub-organization containing institutions which are relevant for neuro-decision-choice actions and policies relative to dynamic categories of problem-solution dualities. The system of ideas and theories as outlined so far points to the development of the theory organizational analytics and the contents institutions within the processes of explanatory

K. K. Dompere, *The Theory of Epistemic Fields*, Studies in Systems, Decision
and Control 498, https://doi.org/10.1007/978-3-031-42470-0

and prescriptive sciences to give us pathways of understanding of parent-offspring dynamics of institutional analytics within the space of problem-solution dualities. We proceed with reflections on conclusions, discussions on organization of institutions and categories of the space of problem-solution dualities through economic concepts of production, consumption, and input-output processes of information-knowledge system. In this respect, an introduction to a prelude to the theory of neuro-decision-choice system is advanced for future development, where the natures of the roles of imagination, reality, belief, hope, courage, and creativity in the space of imagination-reality dualities are examined in the journeys over the space of ignorance-knowledge dualities.

I: REFLECTIONS ON CONCLUSIONS

The monograph on the development of a theory of epistemic fields supported with corresponding relational and information fields is concluded by connecting it with a framework for analyzing the role of certain mental dispositions of the neuro-decision-choice actions in knowing as the production of intellectual investment flows and information-knowledge accumulation as the production of intellectual capital stocks. The theoretical framework of the theory allows a simultaneous treatment of knowing and information-knowledge development over the space of fundamental-applied dualities under the dualistic-polar conditions, where the outputs of neuro-decision-choice activities over the space of fundamentals provide inputs into the neuro-decision-choice actions over the space of applications to give us an analytical framework for examining political economies in terms of information-knowledge intensities and rank them in degrees of information-knowledge intensities over the space of cross-sectional-timeseries dualities. The dualistic-polar relations of give-and-take sharing modes are such that the outputs of the neuro-decision-choice actions become inputs in the neuro-decision-choice actions over the space of the fundamentals. By placing the emphasis of knowing and information-knowledge acquisition through the theory of problem-solution dualities and polarities, the connections to the general economic theory of production-consumption dualities in understanding the stock-flow conditions of information-knowledge disequilibrium dynamics, principles of diversity and unity of knowing and science, and the fuzzy paradigm of thought, are logically opened to us in explaining the formation of categories, categorial conversions and the role of neuro-decision-choice actions as a dominant element in defining the meaning of life, essence of life and the different organizations of social set-ups in disequilibrium dynamics.

The conditions of the principles of diversity and unity of knowing point to the direction for the establishment of two interdependent sets of information-knowledge systems for fundamental and applied knowing and knowledge and corresponding to them are two sets of institutions for the development and maintenance of the social organization for neuro-decision-choice activities over the space of production-consumption dualities. One set of institutions is the institutional group for neuro-decision-choice actions on the space of fundamental problem-solution dualities to produce the fundamental information-knowledge system as dynamic intellectual socks. The set of institutions for neuro-decision-choice activities is the institutional group for neuro-decision-choice actions on the space of applied problem-solution dualities to produce the applied information-knowledge system. This approach of viewing the social organism in terms of information-knowledge systems allows one to see the dualistic-polar conditions of institutional diversity and unity, where the institutional diversities are directed to deal with interdependent categories of problem-solution dualities in the fundamental and applied neuro-decision-choice actions, where the unity is directed to pull together categorial diversities for neuro-decision-choice actions on the space of categorial unity of problem-solution dualities into instruments for managing the controls and the directions of variety transformations of the social organism.

The unity, further, allows us to investigate and understand the elemental neuro-decision-choice behaviors of individual-community dualities (*anoma-kokone-kone problem*) over the space of problem-solution dualities, where an individual is simultaneously a cost and a benefit to the community and the community is simultaneously a cost and a benefit to the individual to maintain the principle of relational cost-benefit duality. In the respect of human general production activities in all sectors of the social set-up, there are individual skills and abilities and collective skills and abilities in decision-choice actions over the space of the problem-solution dualities with differential outcomes and results. The unity finds expressions in organizations and sub-organizations that are used to exploit the categorial diversities in the general neuro-decision-choice actions over the space of problem-solution dualities. Diversities find expressions in the distribution of the socio-natural characteristic dispositions over the space of varieties. Unity in the intellectual space is seen as relaxing the individual cognitive capacity limitations and expanding the contributory zone of individual actions while simultaneously expanding the collective intelligence in knowing relative to social rationality over the space of problem-solution dualities. At the level of the space of problem-solution dualities, the concept of diversity finds expressions and meanings in *categorial specializations* in skills and know-hows from the understanding of some elements in the space of natural process-technological dualities while the concept of unity finds expressions and meanings in progressively inter-categorial supports , dependencies and inter-changes of varieties to bring about types of categorial specializations into the relational give-and-take sharing modes to maintain the conditions of internal and external mutual existence and destructions in the space of variety construction-destruction dualities, where over the epistemological space the elements in the destruction-construction dualities obey the social law of cost-benefit dynamics in continuity. The debate on capitalism, socialism, and communism relative to the space of individual-collective duality is not simply on the ideology of their basic social collective and collaborative works, (*anoma-kokone-kone problem*). Implicit in the debate is the ideology of cost-benefit distribution as the organization of collective work and effort (*asantrofi-anoma problem*) and the decision-choice distributive power over the spaces of justice-injustice dualities and unity-diversity dualities (*the funtum-mereku-denkem-mereku problem*) in the space of production-consumption dualities.

Over the ontological space, the elements in the space of destruction-construction dualities obey the laws of continuum, ecological stability, and parent-offspring dynamics. In all these processes in the infinite universe, given the matter-energy stock-flow equilibrium as establishing the permanency of the universe, and information-knowledge stock-flow disequilibrium as established by the continual variety transformations within the matter-energy statics and information-knowledge dynamics, nothing is created without something being destroyed, and nothing is destroyed without something being created, there is no real benefit without real cost, there is no knowledge without ignorance, there is no change without conflict, there is no conflict without tension, there is no tension without force, there is no force without energy and there is no energy without matter in the space of static-dynamic dualities as defined over the space of quality-quantity dualities. The permanency of the universe in the knowing process finds definition and expressions within the matter-energy existence which provides input-output varieties for its self-maintenance, while change in the universal existence finds definition and expressions within the space of variety disequilibrium dynamics, where information-knowledge existence provides input-output characteristic dispositions for its continual disequilibrium dynamics to maintain the spaces of varieties and the information-knowledge stock-flows, while changes in ignorance find definitions and expressions over the space of expanding varieties in the space of destruction-construction dualities over the space of oneness-dividedness dualities and polarities. This epistemic structure under the principle of opposites governs all planetary systems, galaxies and multiplicities of "solar systems" or exoplanetary systems in the infinite universe with mater-energy equilibrium, variety disequilibrium dynamics and information-knowledge disequilibrium dynamics with exoplanetary and multiplicities of relational continua and unity, where the universe constitutes an unbreakable matter-energy existence in relational and information continua and unity. It is important to understand the relationship between the conditions of matter-energy stock-flow equilibrium and the conditions of matter-energy variety transformation disequilibrium where matter-energy system continually transforms itself into variety of forms in disequilibrium states. The matter-energy stock-flow equilibrium means that the universe is not expanding. The variety transformation disequilibrium means that the system of varieties is continually expanding such that on the planet earth, volcanoes, earthquakes, and other natural events create new matter-energy varieties and information by transformations changing qualitative characteristics without disturbing the general principle of matter-energy equilibrium, where matter-energy flows are always zero to maintain the same matter-energy in quantitative equilibrium state . The input-output dynamics within matter-energy equilibrium states, where matter cannot be destroyed or created, mean that the is always quantitative equilibrium and qualitative disequilibrium with continual matter-energy variety transformation in the space of relation-property dualities in disequilibrium states. The changes in variety property induce corresponding changes in variety relation and vice versa. In this respect the concept of expansion of the universe involves decomposition and composition of the same matter-energy stock into forms without new flows. Non-expansion of universe involves constancy of matter-energy stock with zero matter-energy flows where there is no matter-energy destruction and creation but continual variety transformations in the space of input-output and relation-property dualities.

II: ORGANIZATIONS AND INSTITUTIONS OVER PROBLEM-SOLUTION DUALITIES

The conditions of the theory of epistemic fields with supporting relational and information fields provide an important logical framework for institutional developments and efficient social organization over the paths of social disequilibrium dynamics and problem-solution disequilibrium dynamics, where every variety solution generates variety problems in an unending processes defining the enveloping paths of social development, rise and fall of empires, the rise and decline of nations and the rise of different types of human social set-ups in relation to the social philosophical consciencism that has taken hold of the generation at relevant time. The theory of the epistemic fields also provides a framework for the development of departments of knowing showing the rise of interdisciplinary collaborations and integrations of knowing, learning, teaching and research as new problem-solution dualities arise within the space of problem-solution dualities. Social organizations are developed to pull together diverse institutions within the social set-ups in relations to goals, objectives and visions as collectively conceived in the space of imaginations. Institutions exist as vehicles through which knowing is developed, problems are identified, solutions are found, and actions are executed over the space of problem-solution dualities. Institutions are developed and tailored to solve problems while needs for new institutions arise as new problem-solution dualities appear as the offspring of the old problem-solution dualities over the epistemic field that houses the categories and departments of knowing and information-knowledge production.

The production of information-knowledge intellectual stocks supported by knowing as the production of intellectual investment flows must be organized, where the production effectiveness will depend on institutions and continual institutional creations and innovation to service the multiplicity of complex neuro-decision-choice actions. The institutions through which individual-community neuro-decision-choice actions are organized for fundamental problem-solution dualities to produce the fundamental information-knowledge system are housed in the *educational sector* which derives its power of creation from the governmental power system operating with rules and regulations as socially established in the legal structure and maintained by a system of enforcement processes by the political processes. The units of the education system derive their individual and collective legitimacies from the legal structure within the space of socially acceptable-rejection dualities. Similarly, the institutions through which individual-community neuro-decision-choice actions are organized for applied problem-solution dualities to produce applied information-knowledge system are housed in the *business sector* which also derives its social power of creation from the governmental power system operating with rules and regulations as socially established in the legal structure and maintained by a system of enforcement processes by the same political process. The information-knowledge stock-flow disequilibrium processes take place in the economic structure guided by the conditions of political structure that use the conditions of legal structure for controls and management with implicit inclusions of both the political and legal decision-choice activities in the economic structure relative to the space of input-output dualities.

The character of any organization depends on the creative interactions of leadership and institutions through which neuro-decision-choice actions are implemented while the strength of social institutions depends on the leadership, its personality, and neuro-decision-choice actions over the space of thinking-reasoning dualities as applied over the space of problem-solution dualities relative to the space of social goals, objectives, and visions. The outcomes of the social neuro-decision-choice actions and the rise of new social problem-solution dualities depend on information-knowledge input-output varieties, the logic of thought, the collective personality of the leadership and the administration staff, their integrity with individual and collective actions that are derived from the social philosophical consciencism from which negative and positive behavior may be ascertained and translated into the elements in the space of cost-benefit dualities.

The social unity of the progress of knowing and information-knowledge dynamics in the space of stability requires a continual organization and reorganization of institutional elements, casting new institutions within the space of construction-destruction dualities of cognitive diversities in the social space of input-output resource distribution and to develop organizational laws and rules over the space of fairness-unfairness dualities relative to the elements in the space of categories of problem-solution dualities and the management of intra-relationships and inter-relationships of contents and social varieties. The organizational laws and rules of fairness, equity and equality are reducible to the conditions over the space of real cost-benefit dualities. There are three important mega-sectorial organizations contained in the basic institutional structures of economics, politics, and law. These mega-organizations are *educational organization*, *business organization* and *governmental organization* involved in knowing and information-knowledge production in support of management and control of the social set-up. The educational organizational sector produces the fundamental knowledge, broadly defined, as inputs. The business organizational sector produces applied knowledge by using the fundamental knowledge as input to generate outputs. The governmental organizational sector produces rules and laws to coordinate and manage the organs with inputs from the fundamental and applied knowledge systems as well as provide helps and directions to the organs whenever there are instabilities to maintain the information-knowledge stock-flow disequilibrium processes relative to thew production-consumption disequilibrium processes also relative to the input-output disequilibrium processes. Each mega-sector is directed to deal with a family of categories of problem-solution dualities which constrains the individual and collective neuro-decision-choice actions over the space of individual-collective dualities in relation to the general space of problem-solution dualities and information-knowledge processes. The life blood of the socioeconomic system is information-knowledge flows, where the effectiveness of the information-knowledge flows is constrained by available transmission technologies over the space source-destination dualities.

The system of variety transformations is such that every variety in it is simultaneous an output from a primary input as well as an input into a process to generate an output as derived. In other words, the system of varieties is infinitely matter-energy closed under transformation and identification. Every social progress depends on the ability to produce and use information-knowledge inputs, preference ordering of the leadership, its personality, decision-choice actions, the people collective support under systems of organizations with their institutional configurations over the space of problem-solution dualities, where variety solutions are always set against variety problems to generate new variety problems in the space of chain problem-solution interconnectedness, where the problem-solution processes mimic input-output dynamics. Under the organizational considerations, the epilogue is used to discuss the frameworks for organizing the social setup to provide the best institutional arrangements to serve the information-knowledge needs for individual and social neuro-decision-choice actions over the space of problem-solution dualities to bring about optimal paths of destruction-replacement processes over the intergenerational enveloping for executing the infinite production-consumption and input-output processes over the fundamental and applied information-knowledge systems as a system of a family of variety production-consumption dynamics.

III PRODUCTION, CONSUMPTION, AND INPUT-OUTPUT PROCESSES OF INFORMATION-KNOWLEDGE SYSTEM

The governmental neuro-decision-choice actions on the space of politico-legal problem-solution dualities will affect both the efficiencies of all activities in education and business sectors and their interdependencies towards the path of stock-flow disequilibrium dynamics of the general information-knowledge system through their effects on the boundaries of neuro-decision-choice behavior in the space of individual-collective dualities and variety transformations defined by changes in the characteristic-signal dispositions. Both the fundamental and applied information-knowledge systems are socio-economic productions, the results of which are generated by individual and collective neuro-decision-choice actions working in diversity and unity over the economic structure to become socio-economic varieties of elements in the space of socio-economic problem-solution dualities guided by cost-benefit processes as viewed from the contents of social philosophical consciencism. The concepts and phenomena of problem and solution are defined in terms of relative cost-benefit dominance, and the governmental social policies as controls and management of the social system are to find the path of greater social benefits over social costs with the continual production of information-knowledge structures, where the information-knowledge production is relative to real production.

The educational production of socioeconomic varieties involves the creation of the fundamental information-knowledge systems as outputs which become inputs into the production of applied information-knowledge systems. The business production of socioeconomic varieties involves the creation of applied information-knowledge systems as outputs which then become inputs into the production of the fundamental information-knowledge systems and variety production-consumption systems in the space of changes of real-nominal dualities. The information-knowledge outputs become inputs into the governmental decision-choice processing system, the outcomes of which become inputs into educational, business, and governmental activities in the action space of variety identification-transformation dualities. The efficiencies of the neuro-decision-choice actions in the educational production and the business production depend on the general creative understanding and interdependent use of the education sector, business sector and judicious creation of rules, laws, regulations and policies of the decision-choice activities from the governmental sector to control and manage the social organism with elements from the intellectual capital accumulation given the population structure since these rules, laws, regulations and policies are constraints on the action space of thoughts and practices in the space of all preferences. It is here, in the space of production-consumption dualities of individual and social existence, that governmental governance in the framework of social organism may work to either facilitate or retard individual and social progress over the space of input-output dualities relative to the neuro-decision-choice actions within the space of problem-solution dualities and polarities in reference to research, teaching, learning, knowing and information-knowledge disequilibrium processes as they affect the behaviors of economic, Political and legal structures in human pursuit of happiness locked in the space of necessity-freedom dualities as generated by the space of possibility-probability dualities from the past to the present ang to the future.

The critical understanding of knowing and stock-flow dynamics of information-knowledge production is also the critical understanding of theory and applied economic processes in the space of input-output dynamics with information-knowledge stock-flow disequilibrium dynamics under the spaces of variety problem-solution dualities and destruction-construction dualities. The dynamics of the spaces of problem-solution and destruction-construction dualities are relative to the spaces of tangible-intangible dualities and paradigms of thought. The paradigms of thought harbor the created logics of actions in developing thinking algorithms which become guides to reasoning over the space of variety problem-solution dualities, setting the variety solutions against the variety problems relative to the individual and collective visions with goal-objective elements in the space of goals and objectives under the social and inter-generational preference orderings that are defined over the space of cost-benefit dualities.

All neuro-decision-choice actions are variety productions through transformations over the space of quality-quantity dualities, where varieties give way as their costs rise and in place are substituted with varieties with greater benefits. The construction-destruction dynamics are such that variety costs may become variety benefits as time moves on and information-knowledge expands with variety destructions and constructions. It is also here that thinking, reasoning and research must examine the relationships between info-statics and info-dynamics within the information-knowledge disequilibrium process on one hand and thermostatics and thermodynamics within matter-energy equilibrium process on other hand relative to the space of identification-transformation dualities containing cognition and all human actions.

The space of variety problem-solution dualities may be related to our understanding of medical sciences and the development of a theory of medical decisions based on the principle of opposites, fuzzy information-knowledge structure, and fuzzy paradigm of thought. Medical decisions are about neuro-decision-choice actions on the sub-space of variety problem-solution dualities that relate to conditions of human health. The basic structure of the theory relates to the idea that the human body is conceived as a macro-duality which is composed of two systems of micro-dualities, where each micro-duality in each of the system is under continual process of destruction-construction duality of deterioration-rejuvenation dynamics. Furthermore, the micro-dualities may also be grouped under a subsystem of *key aggregate parts*, and a subsystem of supporting *micro-parts*, all exist in dualities. Each micro-duality is composed of a negative micro-dual with corresponding characteristic disposition and a positive micro-dual with a corresponding characteristic disposition creating internal micro-conflicts in destruction-construction processes in give-and-take sharing modes for mutual existence and non-existence. The internal negative micro-dual is in a destructive and deterioration mode of the part and the internal positive dual is in a construction and rejuvenation mode of the same part.

The collection of all the negative micro-duals constitutes the negative macro-dual, while the collection of all the positive micro-duals constitutes the positive macro-dual in the space of life-death dualities. The positive macro-dual and the negative macro-dual constitute the duality of human individuality with deferential distributions of strengths, endowments, energies, and immune system of the macro-dualities to establish a unique set of characteristics in response to the common socio-natural environment. The concepts of negative and positive used to qualify dual of duality must be distinguished from applicational test of infections where negative and positive are indications of the status of infectious entry into the body. The understanding of the difference of the use of these concepts is important fuzzy decision-choice analytics. This is part of the theory of recycling where old forms give way to new forms and new forms give way to old forms in the space of production-consumption dualities in variety disequilibrium dynamics. Here, the human body is viewed as an instrument for activities over the space of input-output and production-consumption duality relative to the spaces of life-death and new-old dualities in existential dynamics.

The unique set of characteristics may be shared by different peoples to constitute a medical group. Different sets of medical characteristics will allow the population to be partitioned into categories with similar genetic characteristics (coding) in their responses to the common socio-natural environment. The elements in the genetic categories have similar fuzzy socio-natural responses to similar illness, deceases, viruses, and other identifiable ailments. Each genetic category has differential responds to same medical treatments as well as pharmaceutical treatments in that one shoe-fit-all medical practices must be abolished under information-knowledge rationality. It is this possible genetic partition of the population that suggests the path for the development of a *theory of targeted medical practices* and neuro-decision-choice actions over the space of medical problem-solution dualities. The input-output processes of the neuro-decision-choice actions are generated by fuzzy-stochastic information-knowledge systems on a system of dualities with relational continuum and unity under give-and-take sharing processes constrained by cost-benefit conditions over the space of relativity with the use of fuzzy-paradigm of thought to abstract cost-benefit balances for variety destruction-replacement actions over the space of problem-solution dualities. The development of a theory of targeted medical neuro-decision-choice actions over the space of medical problem-solution dualities with fuzzy-stochastic information and the use of fuzzy paradigm may have some difficulties but it is hoped to revolutionize medical and pharmaceutical practices over the space of real cost-benefit dualities, where side effects are real costs and health improvements are real benefits under the theoretical structure of neuro-decision-choice system. I hope that the development of this theoretical framework will be carried on in another monograph. It is useful at this point to outline the possible framework of a prelude to the theoretical framework of neuro-decision-choice system.

IV: A PRELUDE TO THE THEORY OF NEURO-DECISION-CHOICE SYSTEMS

The general explanatory and prescriptive processes of all human actions must be developed as the general theory of neuro-decision-choice System that will explain, predict and prescribe human actions over the space of problem-solution dualities defined in the space of cost-benefit dualities, where the problem-solution dynamics are directed toward achieving greater net benefits by balancing dualistic-polar conflicts in dualities and polarities through a multiplicity of institutions within a social organization as controlled and managed. The general theory of neuro-decision-choice system will affect national and international knowing and information-knowledge production systems broadly defined to include the theory and applications over the space of thinking-reasoning dualities. From this general theoretic position, shall we understand diversities and unity of human decision-choice activities of our social universe and problem-solution dynamics under the principle of opposites for continual dualistic-polar transformations of war-peace dualities, production mimicries of natural dualities, love-hate dualities, life-death dualities and many others dualistic-polar structures, where human neuro-decision-choice actions are constrained by conditions of cognitive capacity limitations, limitativeness and limitationality in relation to information and paradigm of thought, and where social tensions are generated by distributional conflicts of cost-benefit conditions of social outcomes over the space of input-output dualities, where matter and energy may be interchangeably cost and benefit in the space of identification-transformation dualities and in the space of knowing and information-knowledge accumulation under principles of categorial conversions, functor actions and social philosophical consciencism. The general understanding is between the interactive conditions of the space of *identification-transformation dualities* and the conditions of the space of *cost-benefit dualities* relative to elements in the space of goals, objectives. and visions under the *principle of non-satiation* with a preference scheme.

DEFINITION IVA: Neuro-decision-Choice and Neuro-decision-choice space

Let the general *neuro-decision-choice space* be represented by $(\mathfrak{D})$ with a generic element $(\mathfrak{d} \in \mathfrak{D})$, the general problem space be $(\mathfrak{P})$ with a generic element $(\mathfrak{p} \in \mathfrak{P})$, a general solution space $(\mathfrak{S})$ with a generic element $(\mathfrak{s} \in \mathfrak{S})$ and neurological space $(\mathfrak{N})$ with a generic element $(\mathfrak{n} \in \mathfrak{N})$, all conceived in the space of characteristic dispositions then the general *neuro-decision-choice space* is defined as the union of the general problem space, the general solution space and the general neurological space such that $\{(\mathfrak{D} = \mathfrak{P} \cup \mathfrak{S} \cup \mathfrak{N})\} = \{(\mathfrak{P} \otimes \mathfrak{N}) \cup (\mathfrak{S} \otimes \mathfrak{N})\}$ and if $(\mathfrak{p} \in \mathfrak{P} \mid \mathfrak{N})$ then the solution belongs to the intersection of the problem space and solution space in the form $\{(\mathfrak{s} \in \mathfrak{P} \cap \mathfrak{S}) \mid \mathfrak{n} \in \mathfrak{N}\}$. Similarly, if $(\mathfrak{s} \in \mathfrak{S})$ then the problem belongs to the intersection of the problem space and the solution space such that $\{(\mathfrak{p} \in \mathfrak{P} \cap \mathfrak{S}) \mid \mathfrak{n} \in \mathfrak{N}\}$. In this respect, given the space of real cost-benefit relativity conditions $(\mathfrak{r} \in \mathfrak{R})$ the *neuro-decision-choice space* is analytically $\mathfrak{D} = \{\mathfrak{d} = (\mathfrak{p}, \mathfrak{s}, \mathfrak{n} \mid \mathfrak{r}) \mid \mathfrak{p} \in \mathfrak{P}, \mathfrak{s} \in \mathfrak{S}, \mathfrak{n} \in \mathfrak{N}, \mathfrak{r} \in \mathfrak{R}\}$ and the space of problem-solution dualities $(\mathcal{R})$ is analytically the union of *problem space* $(\mathfrak{P})$ and the *solution space* $(\mathfrak{S})$ which is $(\mathfrak{R} = (\mathfrak{P} \cup \mathfrak{S}) = (\mathfrak{P} \otimes \mathfrak{S}) \subseteq \mathfrak{C})$ with a fuzzy conditionality of the form $(\mathfrak{P} \cap \mathfrak{S} \neq \varnothing)$ which is a non-zero intersectionality and $(\mathfrak{R})$ is the *relative cost-benefit space.*

NOTE IVA: DEFINITION OF NEURO-DECISION-CHOICE ACTION

A decision-choice activity is a mental action over the problem-solution space to reconcile the conflicts over the real cost and real benefit information structures as revealed by the inherent negative-positive characteristic dispositions in alternative actual and potential varieties to solve the ranking and selection problems in the cost-benefit and criterion spaces to bring about identifications and transformations and new information outputs. It is also a cognitive action in the problem-solution space which encompasses all aspects of human existence over the actual and potential spaces in the ontological space through the epistemological space. The mental actions are derived through a complex neurological system from the neurological space that allows comparative analytics of qualitative and quantitative characteristics for assessments and ranking of varieties relative to a preference scheme under the evolving social philosophical consciencism. The collection of all the neuro-decision-choice actions on negative-positive dualities from the space of cost-benefit dualities constitutes the general neuro-decision-choice space which is a union of the *general problem space, the general solution space* and the *neurological space* given the *criterion* and *vision spaces*. The problems, solutions, preferences, and decision-choice activities are all defined in the space of varieties where each variety is simultaneously a cost and a benefit, and where the relative cost-benefit information becomes an information input in the space of variety assessments and ranking under individual, collective and social preference schemes with decision-choice activities under deferential cognitive capacity limitations and the sociopsychological principle of non-satiation over the space of varieties.

The neuro-decision-choice space is not only a collection of varieties of decision-choice actions over the space of imagination-reality dualities and polarities as contained in the space of actual-potential dualities, but it is also a collection of action processes that define the path of social histories under individual and collective cognitive capacity limitations. The questions that arise are simply what cognitive capacity limitations are, and what are their manifestations and how can they be reduced or eliminated to increase the efficiencies of decision-choice outcomes relative to either the goal-objective elements or visions, where the establishments of visions and goal-objective elements belong to the space of problem-solution dualities. The cognitive capacity limitations are seen in terms of information-knowledge deficiencies due to differences in the space of ontic-epistemic dispositions and or paradigm deficiencies in thinking and reasoning over the space of problem-solution dualities with a set of goal-objective elements toward a vision. The information-knowledge deficiency is a neuro-decision input limitation, while paradigmatic deficiency is a neuro-decision processing limitation creating thinking and reasoning difficulties, errors, corrections over the space of success-failure dualities. The directional dispositions to these questions will be provided through definitions which will extend the vocabulary of the neuro-decision-choice language for exposition and communication.

DEFINITION IVB: Cognitive Capacity Limitation

A neuro-decision-choice system, $(\mathfrak{D})$ is said to be under cognitive capacity limitations if the expansion of the supporting information-knowledge structure in an *information field* is necessary and sufficient to obtain the desired goal-objective variety and changes of paradigm of thought are necessary and sufficient to arrive at the zone of correctness and absoluteness of the claims without epistemic fuzzy-stochastic conditionality.

DEFINITION IVC: Information-Knowledge Capacity Limitations

A neuro-decision-choice system, $(\mathfrak{D})$ under cognitive capacity limitations is said to be information-knowledge limitational (limitative) if the expansion of the supporting information structure from the information field is necessary (sufficient) to obtain the desired goal-objective variety toward a vision through a correct paradigmatic application.

DEFINITION IVD. Paradigmatic Capacity Limitations

A neuro-decision-choice system, $(\mathfrak{D})$ under cognitive capacity limitations is said to be *paradigmatic limitational (limitative)* if changes in the paradigm of thought for information-knowledge processing in decision-choice actions are necessary (sufficient) to arrive at the zone of correctness and absoluteness of the claims, where the paradigm of thought is equivalent to methodology containing techniques and methods.

A NOTE IVB: On Definitions of Cognitive Capacity Limitations

There are two ways of looking at the concept of cognitive limitations on the space of decision-choice actions. The two ways relate to the necessary and sufficient conditions of outcomes, where necessity relates to initial conditions and anticipations over the possibility space and sufficiency relates to freedom and expectations over the probability space while necessity-freedom dualities relate to possibility-probability dualities in decision-choice actions with outcomes constrained by fuzzy-stochastic conditionality in the space of relativity. The fuzzy-stochastic conditionality indicates the applicable and acceptable zone of claims with error conditionality established by fuzzy-stochastic conditionality with defined conditions of epistemic transversality conditions. The limitationality indicates requirements of necessity, and limitativeness indicates requirements for sufficiency in relations to inputs and processing to achieve outcomes toward a vision in the neuro-decision-choice system. In some situations, the neuro-decision-choice system may be information-knowledge complete but paradigm deficient. It may also be paradigmatically complete but information-knowledge deficient. It may be both information-knowledge and paradigm deficient. These conditions must be considered in the development of the theory of neuro-decision-choice systems as general theory of human actions with subjective judgements over the space of problem-solution dualities toward the understanding of the knowing processes and information-knowledge accumulations in the general stock-flow disequilibrium dynamics and matter-energy equilibrium states over the space of ontological-epistemological dualities. The knowing process and information-capital accumulation must be placed in the general space of economic production-consumption dualities and then make it clear the epistemic need to study the socio-natural variety dynamics, categorial-conversion dynamics, functor dynamics and the theory of dividedness in relation to the dynamics of intra-generational and inter-generation preferences guided by the contents of the social philosophical consciencism at generational points. The differences in the contents of social philosophical consciencism explain the intergenerational as well as inter-country differences in the success-failure dynamics over the space of problem-solution dualities.

The theory of human actions is the theory of neuro-decision-choice actions over the space of problem-solution dualities relative to the space of variety cost-benefit dualities, where every variety resides in the space of negative-positive dualities which is embedded in the space of cost-benefit dualities relative to individual and social preferences over the space of imagination-reality dualities. The space of cost-benefit dualities is the space of incentives for neuro-decision-choice actions over the space of identification-transformation dualities. All human actions over the space of static-dynamic dualities are seen in the two states of motions as changes in relations or stay in relations, and changes in properties or stay in properties induced by neuro-decision-choice activities under a system of individual and social preference ordering and real cost-benefit (pain-pleasure) conditions in duality and polarity. Both variety relations and properties within the space of static-dynamic dualities characterize all decision-choice activities of variety destruction-replacement processes. Variety identification is a change in relations between ignorance and knowledge that presents decision-choice actions in the space of variety identification problem-solution dualities while variety transformation is a change in property between ignorance and knowledge that presents a decision-choice actions in the space of variety transformation problem-solution dualities.

The space of variety identification-transformation structures of problem-solution dualities is central in knowing and information-knowledge accumulation under the neuro-decision-choice activities, where outcomes establish the enveloping of *past social history*, *current social history, and future social history* in all areas of human actions where history is seen within the space of success-failure dualities relative to the elements in the goal-objective space. The variety real cost-benefit dispositions provide conditions for variety ranking while the neurological structures relate the socio-natural preferences to variety real cost-benefit dispositions relative to the elements in the space of goals and objectives on the path to a vision in the space of absoluteness under the principle of non-satiation. The variety real cost-benefit dispositions incentivize the variety neuro-decision-choice actions in the space of relativity of dualistic-polar conditions over the space of imagination-reality dualities within the epistemological space.

In decision-choice processes, the negative dual and the positive dual reside in the space cost-benefit dualities relative to the elements in the space of goals and objectives which provide dualistic-polar conditions to establish individual and social visions. In analytical structure, we have individual (collective) negative (positive) dual, where the negative (positive) dual may be either cost or benefit dual relative to a variety in the space of goals, objectives, and visions. The conditions of the real cost-benefit duality point to the idea that new reflections must be brought to examine the way we conceptualize problems and solutions as well as decision-choice actions, theory of knowing and theory of knowledge. The concept of the real cost-benefit duality must be broadly interpreted where for example, failure-success duality is seen in the same light as pain-pleasure duality or gain-loss duality and many others. This approach allows us to view human decision-choice actions in terms of problem-solution interplays where problem is a net cost and solution is a net benefit. The development of neuro-decision-choice theory must, therefore, see all human decision-choice actions in terms of variety destruction-replacement processes under economic conditions of production-consumption dualities and reproduction of varieties from the variety space. This is also applicable to knowing, science, non-science, our understanding of knowledge and the theory of cognition in the areas of fundamentals and applications in the spaces of production-consumption and input-output dualities, where the intellectual capital stock is the composite sum of the available categories of knowing.

At the level of neuro-decision-choice actions, variety knowing is an intellectual investment process and information-knowledge accumulation is an intellectual capital-stock process which constitutes the accumulation of intellectual capital stocks about variety identities [R17.17], and the accumulations of technological transformation processes [R.17.16] in unity and diversity over the space of socio-economic activities. Every neuro-decision-choice activity is an action on variety problem-solution duality. Every problem-solution variety is either an identification problem-solution duality or a transformation problem-solution duality or both over the space of static-dynamic dualities, where neuro-decision-choice agents traverse through the space of ignorance-knowledge dualities. The identification problem is either a property-identification problem, relation-identification problem, or both. The transformation problem is either a property-change problem, relation-change problem, or both. The property-relation changes involve real cost-benefit comparative analytics over the space of imagination-reality duality as abstracted from the space of actual-potential dualities, where the actual variety problem-solution duality is compared to potential problem-solution dualities in terms of relative cost-benefit standing. The relational structure gives some ideas for the development of the theory of neuro-decision-choice systems that encompasses all areas of human decision-choice actions and actions over the space of variety problem-solution dualities under cost-benefit conditions as incentive over the spaces of neuro-decision-choice dynamics and success-failure dynamics. In the framework of neuro-decision-choice systems, if you are confronted with ignorance, which undoubtedly will, keep on thinking; if you meet difficulties, which undoubtedly will happen, keep on thinking; If you are overwhelmed by failures, which undoubtedly will occur, keep on thinking and if you want to reside in the space of necessity-freedom dualities, you must continue thinking with no stoppage because thinking is the life blood of anticipations in possibility, expectations in probability and successes of human actions in the life-death processes in the space of realities under the infinite process of human existence, where every problem has solutions in thinking and every selected solution generates new problems in thinking pointing to the development of algorithmic reasoning for the best selections.

We have provided definitions for some essential concepts used in the monograph wherever it is thought that there is some deficiency in clarity of their uses in technical communications. Over the space of understanding of decision-choice actions and problem-solution dualities, there are some confusions among the concepts of goals, objectives, mission, and vision. The concepts of goal, objective and goal-objective space have been defined. What are the concepts of mission and vision and what are their similarities and differences, and how do they relate to the concepts of goal, objective and knowing? It is also maintained that every definition must obey the principle of identity characteristic disposition relative to the space of real-nominal existence and maintain the lexical elasticity of substitutions over the space of synonym-antonym dualities [R4.13]. Let us refresh on definitional concept that may be helpful in a further development over the space of fundamental-applied dualities.

DEFINITION IVE: Vision, Vision Structure and Vision Space

Vision is a desirably imaginational organic variety as currently conceived to be an ultimate variety relative to current state, to which there is a sequential action drives to overcome a series of problem-solution dualities in its way by conducting a sequence of missions to conquer the goals and objectives that will lead to the vision as conceived. It is a conceptualization of variety in the potential space through the space of imaginations to be actualized in the space of actuals through the space of realities with the use of a series of decision-choice actions over the space of the problem-solution dualities relative to the space of imagination-reality dualities passing through the space of possibility-probability dualities. The collection of the set of goal-objective enablers and the vision is the *visionary structure* and the collection of all the vision structures constitutes the *vision space*.

DEFINITION IVF: The Spaces of the Actual and the Potential

The space of actuals (potentials) is a collection of varieties where the actual (potential) characteristic dispositions dominate the potential (actual) characteristic dispositions in dualistic-polar structures, where every actual variety has internal and external destructible capacity to be turned into a potential variety and every potential variety has internal and external constructible capacity to be turned into an actual variety through natural transformation processes and technologies such that each variety is seen as resides in the space of actual-potential dualities in a dualistic-polar continuum and unity.

NOTE IVC: On Definitions IVE &IVF

It is useful to keep in mind the definitional conditions of real and nominal varieties in the dualistic-polar space in terms of domination of characteristic dispositions as seen in the space of relativity and not in the space of absoluteness. To explain the epistemic relations among the concepts and phenomena of vision, goal, objective, mission, problem-solution duality, the brain and neuro-decision-choice action, it will be useful to visit my work on the theory of knowledge square that shows the complex relational structure from the potential to the possible, from the possible to the probable, from the probable to the actual and from the actual to the potential [R4.10]. At the level of socio-natural existence, varieties resides in the space of actual-potential dualities with dualistic-polar characteristic dispositions for internal conflicts and natural transformation processes over the space of the actual-potential dualities, where an actual variety is transformed by a destruction process and hence enters into the potential space and a potential variety changes characteristic disposition by a construction process into an actual and hence enters into the space of the actuals. The destruction of an actual variety, irrespective of how, is an input and a vacuum-creation process in a dualistic-polar setting to allow for a variety construction of a potential variety, as an output and a vacuum-replacement process, irrespective of how, over the space of the actuals. The natural transformation dynamics do not pass through the space of cost-benefit duality, where real opportunity cost and the real benefit have no analytical usefulness over the space of natural identification-transformation processes because there is no space of cost-benefit dualities in the ontological setting.

The natural transformation processes take place over the space of actual-potential dualities moving directly from actuals to potentials or from potentials to actuals through one dualistic-polar space, the elements of which are cognitively independent of the existence and actions of other varieties in the space of actual-potential dualities and hence the processes are independent of neuro-decision-choice actions of cognitive agents. The reason is that all-natural transformations and processes take place with perfect information inputs without uncertainties and risks since they are produced by nature and used by the same nature in the action processes. There are no cognitive capacity limitations in the sense that nature produces its own information, where the information is also the knowledge as the input into the natural decision-choice actions with no imperfections to produce other information-knowledge types as further inputs into the natural transformation processes. At the level of natural processes, information is knowledge; and knowledge is information with certainties and risk-lessness of the corresponding natural outcomes. There is no need for information-knowledge quality control for acceptances by nature as the identities relative to social existence. In general, uncertainties arise where and when there are disparities between information and knowledge and where there is an observational deficiencies over the space of acquaintance to create defective experiential information to create ontic-epistemic disparities in variety characteristic dispositions which may be amplified by paradigmatic conditions as one moves from the space of knowledge by acquaintance to the space of knowledge by descriptions [R17.16],[R17.17],[R17.18]. The concepts of imagination-reality duality, possibility-impossibility duality, possibility-probability duality, necessity-freedom duality, and the transversality conditions do not arise for nature operating over the ontological space.

The question of whether nature has a vision with goals and objectives is not our concern here. At the level of neuro-decision-choice actions, the visionary space is a family of visionary structures each of which is a sequence of inter-supportive variety characteristic dispositions established as a hierarchy of goals and objectives over the space of goals and objectives and then mapped onto the space of problem-solution dualities for sequential mission actions. The conditions of the principle of non-satiation point to the series of visionary structures without end and relate to greedy behavior depending on the practice of non-satiation behaviors over the social space. For the individual, the sequence of the visionary structures ends with finality of life while the sequence of generational visionary structures is endless with forward-backward telescopic processes with a continual creation of information-knowledge structures. In the process of knowing, it must be observed that methods and techniques are simply to help reconcile the variety problem-solution dualities in the visionary processes based on real cost-benefit conditions of outcomes of the imagination-reality processes relative to the space of cognitive perceptions of the outcomes of actual-potential transformation processes.

The information-knowledge structures from science, non-science, or any category of knowing are merely inputs into neuro-decision-choice actions in moving from one step to another, where decision-choice actions are impossible without information-knowledge inputs and where the outputs of decision-choice actions are information-knowledge items. In this respect, the information-knowledge processes are decision-choice activities through thinking-reasoning processes over the space of production-consumption dualities with or without markets. The whole knowing process is that the space of real cost-benefit dualities is contained in the space of imagination-reality dualities, where the space of cost-benefit dualities constitutes a dynamic incentive space with or without markets for its understanding and analytical utilization in line with dualistic-polar conditions. The space of cost-benefit duality is a concept and phenomenon that is defined in the social space with preference ordering within the space of relativity, giving a justification to the concept of the dual uses of varieties, but not defined in the space of absoluteness The cost-benefit characteristics may be reduced to monetary units by appropriate price constructions as have been discussed in [R5.14] [R13.15].

At the level of social knowing, the individual, collective and social neuro-decision-choice actions over the spaces of acquaintance and description create imperfect information-knowledge structures with fuzziness and volume incompleteness due to cognitive capacity limitations with the interpretations of qualitative and quantitative signal dispositions leading to various degrees of uncertainties, where outcomes pass through the complex space of doubt-surety dualities. The neuro-decision-choice processes over the space of actual-potential dualities are always indirect, moving through an abstracted space of imagination-reality dualities which contains sub-spaces that provide us with information-knowledge structures to examine socio-economic phenomena such as varieties in the spaces of success-failure dualities, justice-injustice duality and many such social dualistic-polar conditions with continual error-correction pfocesses between the duals and the poles. The existence of cognitive capacity limitations introduces imperfect information structures at the source of knowing as inputs into the variety knowing processes, general information-knowledge accumulation processes and all neuro-decision-choice actions at both the levels of individual, collective and social activities. The individual, collective and social neuro-decision-choice processes involving all endeavors must connect the actual and potential varieties. The connections are done through the space of possibility-probability dualities contained in the space of imagination-reality dualities.

At the point of knowing and variety creations in the epistemological space, the neuro-decision-choice actions proceed from the potential space to the space of imaginations, from the space of imaginations to the space of possibility space, and from the possibility space to the probability space contained in the space of imaginations and from the probability space to the space of varieties and then to the space of actual and from the space of actuals to the space of potentials in never-ending cyclical processes. The transformation from the space of actual to the space of potentials involves a process of destruction. The transformation from the space of potentials through the possible to the probable to the reality and actual involves construction by variety identification to establish necessity, anticipation, and transformation to establish freedom and expectations in the probability space and then to the actual. The neuro-decision-choice processes go through interconnected spaces of dualities of the spaces of potential-imagination dualities, the possibility-probability dualities, the probability-reality dualities, reality-actual dualities, and the actual-potential dualities in dealing with knowing, identification and transformation under dualistic-polar conditions of varieties over the space of matter-energy dualities.

The space of visions, goals and objectives are improper subspaces of the space of imaginations as abstracted from the potential space, where vision is related to an aspiration and necessity containing goal-objective varieties as problem-solution dualities that must be overcome with a series of missions if the vision is to be accomplished. A vision is an imagined organic variety to be actualized from the potential space. Any vision is a mental image conceived as a desirable variety that may or may not be reached by cognitive actions and journey to it. When a vision is defined, it remains as a mental image that is taken from the space of imaginations to the possibility space by establishing conditions of its actualization. These conditions are cognitively abstracted as characteristic dispositions from the potential space as the required problem-solution dualities that must be carefully overcome by the neuro-decision-choice actions. The required problem-solution dualities are the goals and objectives with their characteristic dispositions also as elements in the space of imaginations that stand as impediments that must be overcome on the way to reach the vision. The goals and objectives relative to the anticipated vision with information and paradigmatic constraints are moved into the probability space for missionary actions, while the vision remains in the possibility space with possibilistic and probabilistic uncertainties. The goals and objectives become probabilistic varieties under missionary actions with both possibilistic and probabilistic uncertainties or fuzzy-stochastic uncertainties.

Action analytics and organizational analytics are required over the space of possibility-probability dualities to establish variety vision and the supporting goals and objectives with well-defined characteristic dispositions. At the possibility space, a vision is a *necessity* defined by a possibilistic disposition with possibilistic uncertainties, where the necessity is related to anticipation. The goals, objectives and mission are established by *freedom* constrained by probabilistic uncertainties in the actualization processes given the possibilistic necessity and anticipation. Let us keep in mind that there is an intimate relationship between the space of possibility-probability dualities and those of necessity-freedom and anticipation-expectation dualities in all neuro-decision-choice processes. Possibility, necessity, and anticipation initialize the neuro-decision-choice process while probability, freedom and expectation carry on the actionable missions to the results of transformations as constrained by the resource space. The essential features of the theory of neuro-decision-choice system are attempts to present a general theory of decision-choice system that is applicable to all areas of human endeavors and provide channels to internalize the users into the process as well as encompass sub-theories involving the conditions of quality-quantity dualities and balances of the input-output structures, the corresponding cost-benefit dispositions and socio-economic transformations, where variations are due to information-knowledge specificities. Let us keep in mind that every human activity involves decision and choice where this decision-choice action is expressed over a variety in the space of problem-solution dualities setting solutions against the problems that lead to either a change in relation or a change in property or both. In this respect, the theory of knowing and the theory of information-knowledge accumulation may be seen in terms of relational connections among the elements in the potential space, space of imaginations, possibility space, the probability space and the spaces realities, and the actuals under individual, collective and social preferences. The connectivity links are shown in Figure IVA below.

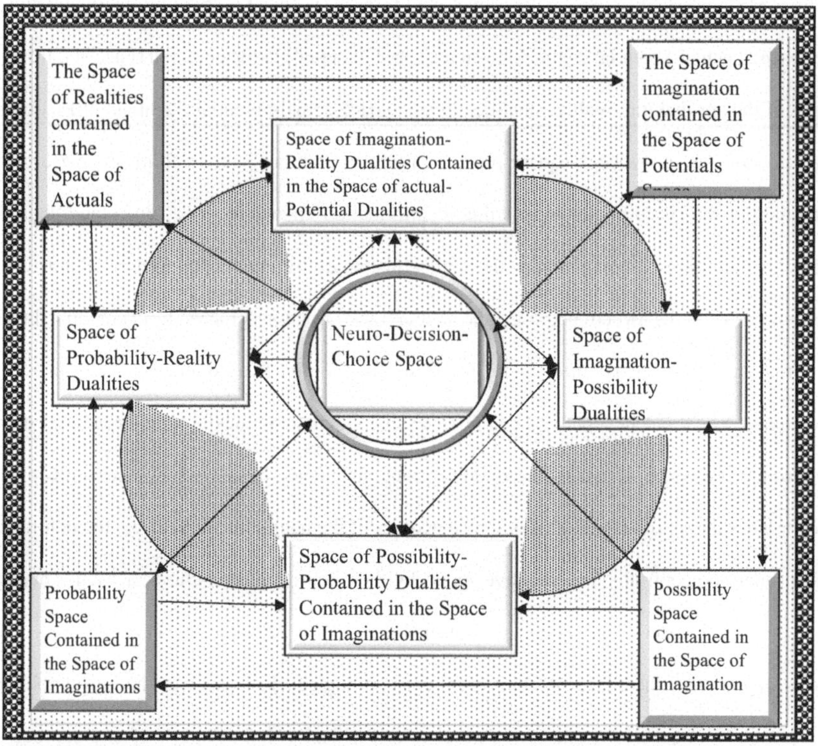

Figure IVA: A Cognitive Geometry of Dualities for Neuro-Decision-Choice Actions

The theory of neuro-decision-choice systems as may be conceived and constructed is to provide an explanatory links from the space of potentials through the space of imaginations to the possibility space as conceptually determined, from the space of imagination-possibility dualities to the space of possibility-probability dualities contained in the space of imaginations, from the possibility space to the probability space and from the space of possibility-probability dualities to the space of probability-reality dualities, and from probability space to space of realities contained in the space of the actuals, and from the space of probability-reality dualities to the space of actual-potential dualities, from the space of actuals to the potential space and from the space of the probability-reality dualities to the space actual-potential dualities in the implicit extension of the knowledge-square process providing indirect journey from the space of the potentials to the space of the actuals due to information-knowledge inequalities and cognitive capacity limitations as explaining the ontic-epistemic differences in variety characteristic dispositions. At the level of ontology, the natural transformations proceed from potential directly to actual and from actual directly to potential with continual accumulation of information-knowledge capital independently of cognitive agents. The natural transformation processes do not go through the space of imagination-reality space and hence bypass possibilistic-probabilistic uncertainties and corresponding risks. The outcomes of the natural processes over the ontological space cannot be controlled by cognitive agents by their neuro-decision-choice activities over the epistemological space. The choice of controllability is not available to neuro-decision-choice agents acting over the space of ontological-epistemological duality. Over the epistemological space, cognitive agents have one option which is how to deal with the impacts of natural processes of ontological outcomes with cognitive capacity limitations. The dealings with the impacts of ontological outcomes are far beyond the capacity of individual persons which gives rise of collective actions.

At level of society, the social transformations are carried by neuro-decision-choice agents through the dynamics of knowing and variety transformations, where the direct processes between the actual and potential are not possible. The indirect way is such that the neuro-decision-choice system proceeds from the potential space to derive a space of imaginations and then journey through the space of imaginations to proceed to the possibility space to establish the possibility set containing the conditions of necessity and anticipations then to the probability space to establish the probability set containing the conditions of the degrees of freedom and sufficiency and then to the space of actuals through the space of realities to actualize the potentials in the space of imaginations into the space of realities as the subspace of the space of actuals. The cognitive journey in the knowing and intellectual stock-flow system is through construction processes while the intellectual capital quality-control process is through the reduction processes over the space of constructionism-reductionism dualities. At the space of the realities, as a subspace of the space of actuals, the undesirable actualized realities are selected to enter the potential space by destruction and replaced by an actualized potential based on the *asantrofi-anoma* rationality in the space of cost-benefit dualities. This is the destruction-replacement processes of varieties under cost-benefit dualistic-polar conditions and preference schemes relative to the space of imagination-reality dualities through which cognitive agents operate in the space of actual-potential dualities.

Over the space of possibility-probability dualities arise several epistemic difficulties of concepts regarding their status in the space of knowing and information-knowledge production as intellectual investment-capital accumulation processes. These concepts include imagination, belief, vision, dream, hope, will, reality and their relationships to the knowing process and information-knowledge accumulation as the information-knowledge stock-flow disequilibrium dynamics. The most disturbing thing for the uses of these concepts in thinking and reasoning is the use of the concept of belief to define knowledge as "justified true belief" (JTB) without any connection to information which is not defined and sometimes taken to be the same thing as knowledge as well as taken to be the same as either fact and evidence in the confused interdependencies violating laws in the theory definitions, lexicography, lexical elasticity and variety identity over the space of real-nominal dualities. Again, what is reality and how does reality differ from actual in the space of knowing to understand *what there was, what there is* and *what there would be*? How do these interchangeable concepts meet conditions of identity as well as help in communications in the space of source-destination dualities to avoid misunderstanding and confusions of all forms over the space of synonym-antonym dualities? The classical definition of knowledge as justified true belief must be critically examined. How does knowledge as a justified true belief (JTB) relates to the concept and phenomenon of information as well as abstract meanings of the concepts of fact, data, and evidence in knowing and neuro-decision-choice processes?

The epistemic direction suggested here in defining information is through the principle of characteristic disposition as the content of information and the principle of signal disposition as message of information in communication and the processes of knowing, where variety knowledge is a derivative of variety information, the characteristic disposition of which is always less than or equal to the variety characteristic disposition in the space of primary-derived dualities. In the traditional epistemology, the concept of "justified true belief" seems to have no analytical meaning. What are the intrinsic meanings of the sub-concepts of belief, true and justified and how are they related in the space of quality-quantity dualities? These concepts are defined in in the space of vagueness with shades of meaning that belong to fuzzy set characterizations with fuzzy conditionality abstracted from the space of relativity and not in the space of absoluteness for the use of the classical paradigm of thought. The concept of "justified true belief" belongs to a phantom space if it is not related to some form of information as seen in the space of matter-energy dualistic-polar structure. As it stands, the justified true belief is disconnected from information and the principle of variety identity that makes it possible to construct dictionary of terms or glossary of terms.

The resolution of the phantom nature of the classical definition of knowledge is resolved by relating the concept of knowledge to the concept of information defined as characteristic-signal disposition. In this respect, justification will be related to comparative characteristic dispositions as defining either a nominal or real variety. The implication of the characteristic deposition relates to the theory of definitions, where distinctions are made between nominal definition and real definition. The principle of definitions is specified over the space of relativity under fuzziness that allows subjective interpretation in propositions and lexical elasticity of linguistic meanings. In fact, ideas must be examined within the space of imagination-reality dualities and to show how ideas relate to information and knowledge. Part of the complex conceptual relations are discussed in this monograph requiring more discussions.

V: METHODOLOGY, INFORMATION AND PROBLEM-SOLUTION DUALITIES

The set of interactive conditions of problem-solution dualities, information and methodology has been discussed in the monograph. There are several important points that must be kept in mind. The variety problem-solution dualities are made available indirectly to cognitive agents because of their interpretations of variety signal dispositions that reveal their characteristic dispositions at the level of acquaintance. The variety characteristic-signal dispositions play important roles at the epistemological space, where there are complex interactions between possibility and possibility, necessity, and freedom in the space of neuro-decision-choice actions. First, they form the experiential information also called knowledge by acquaintance, at the level of acquaintance, as the first step of knowing. The next step is to relate the experiential information as inputs for cognitive processing, where such experiential information may have different subjective interpretations with a use of a paradigm of thought. The need to have consistency and collective agreements leads to the development of a language and a paradigm of thought with rules for collective use to claim agreeable conclusions and examine conflicting claims. The nature of the paradigm depends on the conditions of vagueness and completeness of the experiential information inputs as part of the knowing process. The selected paradigm of thought will guide thinking and its use in reasoning with all information structures over the space of the problem-solution dualities under the existing social philosophical consciencism. The destruction-construction processes relate to neuro-decision-choice actions over the space of input-output dualities, where the variety problem-solution replacement dynamics will also be affected in the paradigmatic framework. The social philosophical consciencism is a composite aggregate of everything in the society including language, culture, proven and unproven information-knowledge elements, beliefs, preference scheme, decision-choice methods, and many social elements, all of which support the national personality, exceptionalism, identity, way of life, and existence.

The paradigm of thought defines the framework of all decision-choice actions, nature of information representation, operations on qualitative and qualitative structures in static and dynamic domains, the nature of thinking and processing of qualitative and quantitative characteristics, mathematics, logic and boundaries of pure and applied mathematics, where the applied mathematics must be interpreted in a broad general way in the representations of algorithms of thinking in support of reasoning. The computability-un-computability duality of mathematical representations of problem-solution duality will depend on Brouwer-Russell vagueness conditions, Turing computability conditions and Gödel completeness conditions of the paradigm of thought and whether one is operating in space of relativity or space of absoluteness. The paradigm of thought that one uses as the guide to thinking is important in all epistemic designs over the space of the problem-solution dualities, variety solutions and the subsequent variety problems that the solutions may generate. The nature of the paradigm of thought may constrain progress of knowing and scientific discoveries.

One important element in the principle of opposite in relation to knowing is that paradigms of thought must be able to establish a dynamic connectivity in parent-offspring process of problem-solution dualities where relationality is an essential element in changes in relations and properties on the basis of internal or external forces, and that our qualitative and quantitative mathematics must reveal these relations if contradictions, paradoxes and ill-posed problems are to be avoided. Let us keep in mind that the central element in all human actions is the mind transformed to deal with multiplicity of different challenges in socio-natural environments which are presented as problem-solution dualities that constrain needs, goals, wants, desires, visions, and progresses in all attempts to balance the forces of duals of dualities and poles of polarities in all socio-natural relations under variety of preference relations induced by neurological systems of individuality and collectivity in the conflict and cooperative zones of preferences. It is the same mind that develops information structures as inputs into decision-choice actions, the results of which become outputs and then transformed into inputs by the same mind for further decision-choice actions over the space of problem-solution dualities in the infinite processes of human existence over the space of cross-sectional-timeseries dualities. The space of the neuro-decision-choice actions, its interactions with the space of problem-solution dualities and the space of source-destination dualities in a system of variety dynamics over the space of identification-transformation dualities constitute the information-knowledge economy which is then translated into variety economy. Challenges relate to difficulties in finding solutions to problems or difficulties in locating conditions of problems over the space of cost-benefit dualities, where such difficulties may be viewed as secondary problems.

Nothing is known by cognitive agents in the universal system except in the space of variety acquaintance and with variety signal dispositions that are sent by variety characteristic dispositions linking the ontological space to the epistemological space for neuro-decision-choice recording and interpretation for variety knowing as an intellectual investment under flow conditions and intellectual accumulations under stock conditions. The space of acquaintances may be a multiplicity with or without awareness where perception of our acquaintances may not reveal all the variety characteristic dispositions due to our cognitive capacity limitations that generate vagueness and volume incompleteness to create qualitative and quantitative uncertainties over the space of possibility-probability dualities, where the theory of knowledge may be seen as the theory of the theory of intellectual capital accumulation under stock conditions over the space of doubt-surety dualities for which the conditions of entropy may help to resolve the dualistic-polar conflicts with neuro-decision-choice actions in the space of doubt-surety dualities of varieties, where variety sureties are set against variety doubts or vice versa with an entropic conditionality.

In viewing the neuro-decision-choice processes over the socio-natural varieties, several questions arise as to 1) what is information? 2) What is knowledge and its relationship to information? 3) What is data and its relationship to information and knowledge? 4) What is fact and its relationships to data, information, and knowledge? 5) what is evidence and how is it related to information, knowledge, data, fact, and varieties? 6) What is the relational connectivity of these concepts to each other, to thinking, reasoning and neuro-decision-choice input-output processes over the spaces of variety problem-solution dualities and variety production-consumption dualities? 7) What are the differences between information-knowledge production and information-knowledge transmission? These questions are epistemic challenges to traditional theories of knowledge, where the theories must relate to all areas of knowing and the traditional theory of information that concentrate on forms of storage, datamatics, communication and applications. The most important thing in conceptualizing the system of knowing, knowledge and unity of science is the question of how knowing and knowledge are related to each other and to the decision-choice actions over the space of problem-solution dualities as a collection of categorial problem-solution dualities into epistemic fields with relational and information fields. The discussions in this epilogue are the points of departure and entry points for future research on knowing and information-knowledge production over the space of fundamental-applied dualities.

VI: RELECTIONS ON CERTAINTY-UNCERTAINTY DUALITIES NEURO-DECISION-CHOICE ACTIONS AND PROBLEM-SOLUTION DUALITIES

In this epilogue for concluding the monography on the theory of epistemic fields, we have offered some short reflections of future works on organizations and institutions over problem-solution dualities, production, consumption, and input-output processes of information-knowledge system as well as a prelude to the theory of neuro-decision-choice systems on methodology, information and problem-solution dualities as seen in the space of ontological-epistemological dualities. Here, information constitutes an input-output in the neuro-decision-choice actions while the paradigm of thought constitutes the methodological instrument for the input processing to abstract outputs. As it has been presented, much is known about the quality of the input-output structure in terms of doubt, surety, success, failure, and many others over the outcome space. The quality of the input-output structures is defined in the space of information transmitted over the space of source-destination dualities contained in the space of variety certainty-uncertainty dualities which is defined by characteristic dispositions as the contents and signal dispositions as the message. Discussions and defining elements have been given on the concepts and phenomena of possibility and probability without explicit connectivity to the space of certainty-uncertainty dualities. There is a system of well-developed theories of probability, where the probabilities are associated with different degrees of *likelihood* without a clear definition and explication of what constitutes likelihood and how this likelihood appears in cognition composed of knowing, thinking, and decision-choice actions and human limitations. There is not a system of well-developed theories of uncertainty and how the concepts of certainty and uncertainty appear in human cognitions and actions. This epistemic gap may be closed with the developments of interdependent theories of uncertainty, imagination, possibility, and probability in the space of human actions.

In the literature on decision and choice, over all areas of cognitive actions, there is a section devoted to decision under uncertainty without the definition and explication of uncertainty. The concepts and phenomena of certainty and uncertainty are assumed as known, where their uses over the space of theory-application dualities are carried on within the space of possibility-probability dualities. The concepts and phenomena of certainty and uncertainty take on the structures and the forms of possibility, impossibility, probability improbability and corresponding spaces, where probability theory has become the theory of uncertainty such that the cognitive character within the space of quality-quantity dualities is lost projecting the epistemic debate over the space of subjective-objective and epistemic-non-epistemic dualities with excluded middle in non-dualistic-polar relational continuum and unity. The conceptual relations between the connectivity of certainty-uncertainty space and the possibility-probability space are not well defined and explicated. Over the space of qualitative-quantitative dualities of variety outcomes one may emphasize either quantitative disposition or qualitative disposition or seek a creative combination of the quality-quantity dispositions whether one is examining epistemic conditions or aleatory situations. Irrespective of how one views the situations in the space of human actions, there is a need to develop a theory uncertainty and certainty, where the concepts and phenomena of both certainty and uncertainty are clearly defined and explicate in human knowing and actions and how they are related to ignorance, creativity, innovation discovery and other qualitative human elements.

BLURB

THE THEORY OF EPISTEMIC FIELDS IN THE ROBLEM-SOLUTION DUALITIES AND POLARITIES:

Fuzzy-Semantic Foundations of Interdisciplinary Categories for the Diversity and the Unity of Knowing and the Unity of Science

The monograph is about the development of the theory of epistemic fields with the corresponding relational and information fields as a framework for the understanding of strategies and tactics of theory of knowing as production of intellectual investment flows and the theory of knowledge accumulation as the production of intellectual capital stocks in systems of factories and departments. Epistemic analytics of information fields, relational fields, categorial dividedness, characteristic-signal fields, diversity, and unity of the space of problem-solution dualities in knowing and information-knowledge accumulation are developed within this framework. Comparative analytics of classical and fuzzy indicator functions are discussed for the construct of categorial dividedness as factories within the space of problem-solution dualities for decompositions and compositions relative to institutional formations. The concepts and the roles of thinking and reasoning with curiosity, creativity, hope, Ill-posed problems, phantom problems, unsolved problems, misinformation, disinformation, fake news and courage are introduced, defined and analyzed on the cognitive journeys over the space of ignorance-knowledge dualities, where dualistic-polar conflicts between duals in the space of ignorance-knowledge dualities are resolved with the instruments of fuzzy optimization, the result of which is used to induced the zones of ignorance, the zones of knowledge and the zones of contentions. A complete development of the set of connecting paths of spaces and sub-spaces are provided, where all varieties, categories and spaces reside in dualistic-polar structures with knowledge stock views as a single tree with same roots, one

K. K. Dompere, *The Theory of Epistemic Fields*, Studies in Systems, Decision
and Control 498, https://doi.org/10.1007/978-3-031-42470-0

trunk, many branches and fruit cocktail. The ontological space contains the space of actual-potential dualities as the primary category of knowing and the epistemological space contains the space of imagination-reality dualities as the derived category of knowing within the space of primary-derived dualities. The space of potentials contains the space of imaginations which contains the sub-spaces of possibility-impossibility, probability-improbability, and possibility-probability dualities with corresponding spaces of necessity-freedom and anticipation-expectation dualities leading to the conception of the possible-world-impossible-world dualities in the space of semantic-non-semantic dualities. This monograph is also a continuation of the sequence of my works on the theories of paradigms of thought, rationality, *info-statics*, *info-dynamics, entropy, problem-solution dualities* in self-contained mathematics and philosophy, and their relational connectivity to information, language, knowing, knowledge, cognitive practices relative to nominalism and the space of construction-reduction dualities over the spaces of fundamental-applied and production-consumption, input-output and cost-benefit dualities.

..

THE THEORY OF EPISTEMIC FIELDS

INTERDISCIPLINARY AND MULTIDISCIPLINARY REFERENCES

R 1. CATEGORY AND FUNCTOR THEORIES IN MATHEMATICS, LOGIC, AND SCIENCES

[R1.1] Arzi-Gonczaworski, Z., "Perceive This as That — Analogies, Artificial Perception, and Category Theory", *Annals of Mathematics and Artificial Intelligence*, 26 (1), 1999, pp. 215–252.

[R1.2] Awodey, S., "Structure in Mathematics and Logic: A Categorical Perspective," *Philosophia Mathematica,* Vol. 3. 1996, pp. 209-237.

[R1.3] Baez, J. & Dolan, J., "Categorification", *Higher Category Theory* (Contemporary Mathematics, Volume 230), Ezra Getzler and Mikhail Kapranov (eds.), Providence: AMS, 1998, pp.1–36.

[R1.4] Baez, J. & May, P. J., *Towards Higher Category Theory*, Berlin: 2010,Springer.

[R1.5] Bain, J., "Category-theoretic Structure and Radical Ontic Structural Realism", *Synthese*, Vol. 190, 2013, pp.1621–1635.

[R1.6] Bell, J. L., "Category Theory and the Foundations of Mathematics," *British Journal of Science,* Vol. 32, 1981, pp. 349-358.

[R1.7] Bell, J. L., "Categories, Toposes and Sets," *Synthese,* Vol. 51, 1982, pp. 393-337

[R1.8a] Bell, J. L, 1995, "Infinitesimals and the Continuum", *Mathematical Intelligencer*, Vol, 17 (2), pp. 55–57.

[R1.8b] Bergman, S. *The Kernel Function and Conformal Mapping.* New York: Amer. Math. Soc., 1950

[R1.9] Black, M., *The Nature of Mathematics,* Totowa, N.J., Littlefield, Adams, and Co., 1965.

[R1.10] Blass, A., "The Interaction between Category and Set Theory," *Mathematical*

K. K. Dompere, *The Theory of Epistemic Fields*, Studies in Systems, Decision and Control 498, https://doi.org/10.1007/978-3-031-42470-0

Applications of Category Theory, Vol. 30, 1984, pp. 5-29.

[R1.11] Brown, B., and J Woods (eds.), *Logical Consequence; Rival Approaches and New Studies in exact Philosophy: Logic, Mathematics and Science*, Vol. II Oxford, Hermes, 2000.

[R1.12] Buchsbaum, D.A., "Exact Categories and Duality", *Transactions of the American Mathematical Society*, 80 (1), 1955, pp. 1–34.

[R1.13a] Carathéodory, C. *Conformal Representation.* New York: Dover, 1998.

[R1.13b] Domany, J. L.et al., *Models of Neural Networks III,* New York, Springer, 1996

[R1.14] Feferman, S., "Categorical Foundations and Foundations of Category Theory," in R. Butts (ed.), *Logic, Foundations of Mathematics and Computability*, Boston, Mass., Reidel, 1977, pp.149-169.

[R1.15] Freyd, P., *Abelian Categories. An Introduction to the Theory of Functors*, New York: Harper & Row. 1964.

[R1.16] Freyd, P., "The Theories of Functors and Models". *Theories of Models*, Amsterdam: North Holland, 1965, pp.107–120.

[R1.17] Freyd, P., Friedman, H. & Scedrov, A., "Lindembaum Algebras of Intuitionistic Theories and Free Categories", *Annals of Pure and Applied Logic*, Vol. 35 (2), 1987, 167–172.

[R1.18] Glimcher, P. W., *Decisions, Uncertainty, and the Brain: The Science of Neoroeconomics,* Cambridge, Mass., MIT Press, 2004

[R1.19] Gray, J.W. (ed.) *Mathematical Applications of Category Theory* (American Mathematical Society Meeting 89th Denver Colo. 1983)., Providence, R.I., American Mathematical Society, 1984.

[R1.20] Healy, M. J., & Caudell, T. P., "Ontologies and Worlds in Category Theory: Implications for Neural Systems",*Axiomathes*, Vol. 16 (1–2), 2006, pp.165–214.

[R1.21a] Hellman, G., "Does Category Theory Provide a Framework for Mathematical

Structuralism?", *Philosophia Mathematica*, Vol. 11 (2), 2003, pp.129–157.

[R1.21b] Hellman, G, "A Small Complete Category", *Annals of Pure and Applied Logic*, Vol. 40 (2), 1988, pp. 135–165.

[R.1.22] Johansson, Ingvar, *Ontological Investigations: An Inquiry into the Categories of Nature, Man, and Society,* New York, Routledge, 1989.

[R1.23] Kamps, K. H., D. Pumplun, and W. Tholen (eds.) *Category Theory: Proceedings of the International Conference*, Gummersbach, July 6-10, New York, Springer, 1982.

[R1.24] Kan, D. M., "Adjoint Functors", *Transactions of the American Mathematical Society*, Vol. 87, 1958, pp. 294–329.

[R1.25] Kythe, P. K. Computational Conformal Mapping. Boston, MA: Birkhäuser, 1998.

[R1.26] Lal, R. and Teh, N., "Categorical Generalization and Physical Structuralism", *British Journal for the Philosophy of Science*, Vol. 68 (1), 2017: pp. 213–251.

[R1.27a] Lambek, J. and Scott, P.J., "Intuitionistic Type Theory and Foundations", *Journal of Philosophical Logic*, Vol.10 (1), 1981: pp. 101–115.

[R1.27b] Lambek, J. and Scott, P.J, "New Proofs of Some Intuitionistic Principles", *Zeitschrift für Mathematische Logik und Grundlagen der Mathematik*, 29 (6), 1983, pp.493–504.

[R1.28] Landry, E., "Category Theory: The Language of Mathematics", *Philosophy of Science*, Vol.66 (3) (Supplement) 1999, pp. S14–S27.

[R1.29] Lawvere, F. W., "Functorial Semantics of Algebraic Theories", *Proceedings of the National Academy of Sciences U.S.A.*, Vol 50, 1963, pp. 869–872.

[R1.30] Lawvere, F. W., "Foundations and Applications: Axiomatization and Education. New Programs and Open Problems in the Foundation of Mathematics", *Bulletin of Symbolic Logic*, 9 (2), 2003: pp. 213–224.

[R1.31] Landry, E., Category Theory: The Language of Mathematics," *Philosophy of*

Science, Vol. 66, (Supplement), pp. S14-S27.

[R1.32] Landry E. and J.P Marquis, "Categories in Context: Historical, Foundational and Philosophical," *Philiosophia Mathematica*, Vol. 13, 2005, pp. 1-43.

[R1.33] Linnebo, O. & Pettigrew, R., "Category Theory as an Autonomous Foundation," *Philosophia Mathematica*, Vol. 19 (3), 2011, pp. 227–254.

[R1.34] Mac Lane, S., "Dualities for Groups", *Bulletin of the American Mathematical Society*, Vol. 56, 1950, pp, 485–516.

[R1.35] Makkai, M. & Paré, R., *Accessible Categories: The Foundations of Categorical Model Theory*, Contemporary Mathematics Vol.104, Providence: AMS1989, .

[R1.36] Makkai, M. & Reyes, G., *First-Order Categorical Logic*, Springer Lecture Notes in Mathematics Vol. 611, New York: Springer,1977.

[R1.37] Makkai, M. & Reyes, G, "Completeness Results for Intuitionistic and Modal Logic in a Categorical Setting", *Annals of Pure and Applied Logic*, Vol. 72 (1), 1995, pp. 25–101.

[R1.38] Marquis, J. –P., "Three Kinds of Universals in Mathematics," in B. Brown, and J. Woods (eds.), *Logical Consequence; Rival Approaches and New Studies in exact Philosophy: Logic, Mathematics and Science*, Vol. II Oxford, Hermes, 2000, pp. 191-212.

[R1.39] McLarty, C., "Category Theory in Real Time," *Philosophia Mathematica,* Vol. 2, 1994, pp. 36-44.

[R1.40] McLarty, C., "Learning from Questions on Categorical Foundations," *Philosophia Mathematica,* Vol.13, 2005, pp. 44-60.

[R1.41]Moon, P. and Spencer, D. E. "Conformal Transformations." in *Field Theory Handbook, Including Coordinate Systems, Differential Equations, and Their Solutions, 2nd ed.* New York: Springer-Verlag, pp. 49-76, 1988.

[R1.42] Nehari, Z., <u>*Conformal Mapping.*</u> New York: Dover, 1982.

[R1.43] Pareigis, B., *Categories and Functors*, New York: Academic Press, 1970 .

[R1.44] Peruzzi, A., "The Meaning of Category Theory for 21st Century Philosophy", *Axiomathes*, Vol. 16 (4), 2006 pp. 424–459.

[R1.45] Peruzzi, A., "Category Theory and the Search for Universals: A Very Short Guide for Philosophers", in *Modern Logic 1850–1950, East and West*, F. Abeles & M. Muller (eds.), Basel: Birkhauser, 2016, pp. 159–182.

[R1.46] Rodin, A., *Axiomatic Method and Category Theory* (Synthese Library: Volume 364), New York: Springer, 2014.

[R1.47] Spivak, D., *Category Theory for the Sciences*, Cambridge: MIT Press, 2014.

[R1.48] Taylor, P., "Intuitionistic sets and Ordinals", *Journal of Symbolic Logic*, Vol. 61, 1996, pp.705–744.

[R1.49] **Ross**, Don, *Economic theory and Cognitive Science; Microexplanation*, Cambridge, Mass., MIT Press, 2005.

[R1.50] Rodabaugh, S. et. al., (eds.), *Application of Category Theory to Fuzzy Subsets*, Boston, Mass., Kluwer, 1992.

[R1.51] Sieradski, Allan, J., An Introduction to Topology and Homotopy, PWS-KENT Pub. Boston,1992.

[R1.52] Spivak, D., *Category Theory for the Sciences*, Cambridge: MIT Press, 2014.

[R1.53] Taylor, J.G. (ed.), *Mathematical Approaches to Neural Networks*, New York North-Holland, 1993.

[R1.54] Taylor, P., "Intuitionistic sets and Ordinals", *Journal of Symbolic Logic*,Vol. 61, 1996, pp. 705–744.

[R1.55] Van Benthem, J. et al.(eds.), *The Age of Alternative Logics: Assessing Philosophy of Logic and Mathematics Today,* New York, Springer, 2006.

R2. CONCEPTS OF INFORMATION, FUZZY PROBABILITY, FUZZY RANDOM VARIABLE AND RANDOM FUZZY VARIABLE

[R2.1] Bandemer, H., "From Fuzzy Data to Functional Relations," *Mathematical Modelling*, Vol. 6, 1987, pp. 419-426.

[R2.2] Bandemer, H. et. al., *Fuzzy Data Analysis*, Boston, Mass, Kluwer, 1992.

[R2.3] Kruse, R. et. al., *Statistics with Vague Data,* Dordrecht, D. Reidel Pub. Co., 1987.

[R2.4] El Rayes, A.B. et. al., "Generalized Possibility Measures," *Information Sciences,* Vol. 79, 1994, pp. 201-222.

[R2.5] Dumitrescu, D., "Entropy of a Fuzzy Process," *Fuzzy Sets and Systems*, Vol. 55, #2, 1993, pp. 169-177.

[R2.6] Delgado, M. et. al., "On the Concept of Possibility-Probability Consistency," *Fuzzy Sets and Systems*, Vol. 21, #3, 1987, pp. 311-318.

[R2.7] Devi, B.B. et. al., "Estimation of Fuzzy Memberships from Histograms," *Information Sciences*, Vol. 35, #1, 1985, pp. 43-59.

[R2.8] Dubois, D. et. al., "Fuzzy Sets, Probability and Measurement," *European Jour. of Operational Research*, Vol. 40, #2, 1989, pp. 135-154.

[R2.9] Fruhwirth-Schnatter, S., "On Statistical Inference for Fuzzy Data with Applications to Descriptive Statistics," *Fuzzy Sets and Systems*, Vol. 50, #2, 1992, pp. 143-165.

[R2.10] Gaines, B.R., "Fuzzy and Probability Uncertainty logics," *Information and Control*, Vol. 38, #2, 1978, pp. 154-169.

[R2.11] Geer, J.F. et. al., "Discord in Possibility Theory," *International Jour. of General Systems*, Vol. 19, 1991, pp. 119-132.

[R2.12] Geer, J.F. et. al., "A Mathematical Analysis of Information-Processing Transformation Between Probabilistic and Possibilistic Formulation of Uncertainty," *International Jour. of General Systems*, Vol. 20, #2, 1992, pp. 14-176.

[R2.13] Goodman, I.R. et. al., *Uncertainty Models for Knowledge Based Systems*, New York, North-Holland, 1985.

[R2.14] Grabish, M. et. al., *Fundamentals of Uncertainty Calculi with Application* *to Fuzzy Systems*, Boston, Mass., Kluwer, 1994.

[R2.15] Guan, J.W. et. al., *Evidence Theory and Its Applications*, Vol. 1, New York, North-Holland, 1991.

[R2.16] Guan, J.W. et. al., *Evidence Theory and Its Applications*, Vol. 2, New York, North-Holland, 1992.

[R2.17] Hisdal, E., Are Grades of Membership Probabilities?" *Fuzzy Sets and Systems*, Vol. 25, #3, 1988, pp. 349-356.

[R2.18] Höhle Ulrich , "A Mathematical Theory of Uncertainty," in R.R. Yager (ed.) *Fuzzy Set and Possibility Theory: Recent Developments*, New York, Pergamon, 1982, pp. 344 – 355.

[R2.19] Kacprzyk , Janusz and Mario Fedrizzi (eds.) *Combining Fuzzy Imprecision with Probabilistic Uncertainty in Decision Making*, New York, Plenum Press, 1992.

[R2.20] Kacprzyk, J. et. al., *Combining Fuzzy Imprecision with Probabilistic Uncertainty in Decision Making*, New York, Springer-Verlag, 1988.

[R2.21] Klir, G.J., "Where Do We Stand on Measures of Uncertainty, Ambignity, Fuzziness and the like?" *Fuzzy Sets and Systems*, Vol. 24, #2, 1987, pp. 141-160.

[R2.22] Klir, G.J. et. al., *Fuzzy Sets, Uncertainty and Information*, Englewood Cliff, Prentice Hall, 1988.

[R2.23] Klir, G. J. et. al., "Probability-Possibility Transformations: A Comparison," *Intern. Jour. of General Systems*, Vol. 21, #3, 1992, pp. 291-310.

[R2.24] Kosko, B., "Fuzziness vs Probability," *Intern. Jour. of General Systems*, Vol. 17, # (1-3) 1990, pp. 211-240.

[R2.25] Manton, K.G. et. al., *Statistical Applications Using Fuzzy Sets*, New York, John Wiley, 1994.

[R2.26] Meier, W., et. al., "Fuzzy Data Analysis: Methods and Industrial Applications," *Fuzzy Sets and Systems*, Vol. 61, #1, 1994, pp. 19-28.

[R2.27] Nakamura, A., et. al., "A logic for Fuzzy Data Analysis," *Fuzzy Sets and Systems,* vol. 39, #2, 1991, pp. 127-132.

[R2.28] Negoita, C.V. et. al., *Simulation, Knowledge-Based Computing and Fuzzy Statistics*, New York, Van Nostrand Reinhold, 1987.

[R2.29] Nguyen, H.T., "Random Sets and Belief Functions," *Jour. of Math. Analysis and Applications*, Vol. 65, #3, 1978, pp. 531-542.

[R2.30] Prade, H. et. al., "Representation and Combination of Uncertainty with belief Functions and Possibility Measures," *Comput. Intell.* ,Vol. 4, 1988, pp. 244-264.

[R2.31] Puri, M.L. et. al., "Fuzzy Random Variables," *Jour. of Mathematical Analysis and Applications*, Vol. 114, #2, 1986, pp. 409-422.

[R2.32] Rao, N.B. and A. Rashed, "Some Comments on Fuzzy Random Variables," *Fuzzy Sets and Systems*, Vol. 6, # 3, 1981, pp.285-292.

[R2.33] Sakawa, M. et. al., "Multiobjective Fuzzy linear Regression Analysis for Fuzzy Input-Output Data," *Fuzzy Sets and Systems*, Vol. 47, #2, 1992, pp. 173-182.

[R2.34] Schneider, M. et. al., "Properties of the Fuzzy Expected Values and the Fuzzy Expected Interval," *Fuzzy Sets and Systems*, Vol. 26, #3, 1988, pp. 373-385.

[R2.35] Stein, N.E. and K Talaki, "Convex Fuzzy Random Variables," *Fuzzy Sets and Systems*, Vol. 6, #3, 1981, pp. 271-284.

[R2.36] Sudkamp, T., "On Probability-Possibility Transformations," *Fuzzy Sets and Systems*, Vol. 51, #1, 1992, pp. 73-82.

[R2.37] Walley, P., *Statistical Reasoning with Imprecise Probabilities*, London Chapman and Hall, 1991.

[R2.38] Wang, G.Y. et. al., "The Theory of Fuzzy Stochastic Processes," *Fuzzy Sets and Systems*, Vol. 51, #2 1992, pp. 161-178.

[R2.39] Zadeh, L. A., "Probability Measure of Fuzzy Event," *Jour. of Math Analysis and Applications*, Vol. 23, 1968, pp. 421 – 427.

R3. EXACT SCIENCE, INEXACT SCIENCES, INFORMATION AND KNOWLEDGE

[R3.1] Achinstein, P., "The Problem of Theoretical Terms," in Brody, Baruch A. (Ed.) *Reading in the Philosophy of Science*, Englewood Cliffs, NJ. Prentice Hall, 1970.

[R3.2] Amo Afer, A. G., *The Absence of Sensation and the Faculty of Sense in the Human Mind and Their Presence in our Organic and Living Body, Dissertation and Other essays 1727-1749,* Halle Wittenberg, Jena, Martin Luther University Translation, 1968.

[R3.3] Beeson, M. J., *Foundations of Constructive Mathematics*, Berlin/New York, Springer,1985.

[R3.4] Benacerraf, P., "God, the Devil and Gödel," *Monist,* Vol. 51, 1967, pp.9-32.

[R3.5] Benecerraf, P and H. Putnam (eds.), *Philosophy of Mathematics: Selected Readings* , Cambridge, Cambridge University Press, 1983.

[R3.6] Black, Max, *The Nature of Mathematics,* Totowa, Littlefield, Adams and Co. 1965.

[R3.7] Blanche, R., *Contemporary Science and Rationalism*, Edinburgh, Oliver and Boyd, 1968.

[R3.8] Blanshard, Brand, *The Nature of Thought*, London Allen and Unwin, 1939.

[R3.9] Blauberg, I. V., V.N. Sadovsky and E.G. Yudin, Systems Theory: Philosophical and Methodological Problems, Moscow, Progress Publishers, 1977.

[R3.10] Braithwaite, R. B., *Scientific Explanation*, Cambridge, Cambridge University Press. 1955.

[R3.11] Brewer, M.B., and B. E. Collins (eds.) Scientific Inquiry and the Social Sciences, Washington, Jossey-Bass Pub, 1981

[R3.12a] Brody, Baruch A. (ed.), *Reading in the Philosophy of Science*, Englewood Cliffs, N.J., Prentice Hall, 1970.

[R3.12b] Brody, Baruch A., "Confirmation and Explanation," in Brody, Baruch A. (ed.) *Reading in the Philosophy of Science*, Englewood Cliffs, N.J., Prentice-Hall,1970, pp. 410-426.

[R3.13] Brouwer, L.E.J., "Intuitionism and Formalism", *Bull of American Math. Soc.*, Vol. 20, 1913, pp81-96.; Also, in Benecerraf, P. and H. Putnam (eds.), *Philosophy of Mathematics: Selected Readings* , Cambridge, Cambridge University Press, 1983. pp. 77-89.

[R3.14] Brouwer, L.E.J., "Consciousness, Philosophy, and Mathematics," in Benecerraf, P. and H. Putnam (eds.), *Philosophy of Mathematics: Selected Readings* , Cambridge, Cambridge University Press, 1983, pp.90-96.

[R3.15] Brouwer, L. E. J., *Collected Works, Vol. 1: Philosophy and Foundations of Mathematics* [A Heyting (ed.)], New York, Elsevier, 1975.

[R3.16] Campbell, Norman R., *What is Science?* New York, Dover, 1952.

[R3.17] Carnap, R., "Foundations of Logic and Mathematics," in *International Encyclopedia of Unified Science*, Chicago, Univ. of Chicago,1939, pp.143-211.

[R3.18] Carnap, Rudolf, "On Inductive Logic," *Philosophy of Science*, Vol. 12, 1945, pp. 72-97.

[R3.19] Carnap, Rudolf, "The Methodological Character of Theoretical Concepts," in Herbert Feigl and M. Scriven (eds.) *Minnesota Studies in the Philosophy of Science, Vol. I*, 1956, pp. 38-76.

[R3.20] Charles, David and Kathleen Lennon (eds.), *Reduction, Explanation, and Realism*, Oxford, Oxford Unive3rsity Press, 1992.

[R3.21] Cohen, Robert S. and Marx W. Wartofsky (eds.), *Methodological and Historical Essays in the Natural and Social Sciences*, Dordrecht, D. Reidel Publishing Co. 1974.

[R3.22] Dalen van, D. (ed.), *Brouwer's Cambridge Lectures on Intuitionism,* Cambridge, Cambridge University Press, 1981.

[R3.23] Davidson, Donald, *Truth and Meaning: Inquiries into Truth and Interpretation*, Oxford, Oxford University Press, 1984.

[R3.24] Davis, M., *Computability and Unsolvability*, New York, McGraw-Hill, 1958.

[R3.25] Denonn Lester E. (ed.), *The Wit and Wisdom of Bertrand Russell*, Boston, MA., The Beacon Press, 1951.

[R3.26] Dompere, Kofi K., *Fuzziness and Foundations of Exact and Inexact Sciences*, New York, Springer, 2013

[R3.27] Dummett, M., "The Philosophical Basis of Intuitionistic Logic," in Benecerraf, P. and H. Putnam (eds.), *Philosophy of Mathematics: Selected Readings* , Cambridge, Cambridge University Press, 1983, pp.97-129

[R3.28] Feigl, Herbert and M. Scriven (eds.), *Minnesota Studies in the Philosophy of Science,* Vol. I, 1956.

[R3.29] Garfinkel, Alan, *Forms of Explanation: Structures of Inquiry in Social Science,* New Haven, Conn., Yale University Press, 1981.

[R3.30] George, F. H., *Philosophical Foundations of Cybernetics*, Tunbridge Well, Great Britain, 1979.

[R3.31] Gillam, B., "Geometrical Illusions," *Scientific American*, January, 1980, pp.102-111.

[R3.32] Gödel , Kurt., "What is Cantor's Continuum Problem?" in Benecerraf, P. and H. Putnam (eds.), *Philosophy of Mathematics: Selected Readings*, Cambridge, Cambridge University Press, 1983, pp.470-486.

[R3.33] Gorsky, D.R., *Definition,* Moscow, Progress Publishers, 1974.

[R3.34] Gray, William and Nicholas D. Rizzo(eds.), *Unity Through Diversity*. New York, Gordon and Breach, 1973.

[R3.35] Hart, W. D. (ed.), *The Philosophy of Mathematics*, Oxford, Oxford University

Press, 1996.

[R3.36] Hartkamper, a and H.- Schmidt, Structure and Approximation in Physical Theories, New York, Plenum Press, 1981.

[R3.37] Hausman, David, M., *The Exact and Separate Science of Economics,* Cambridge, Cambridge University Press, 1992.

[R3.38] Helmer, Olaf and Nicholar Rescher, *On the Epistemology of the Inexact Sciences*, P-1513, Santa Monica, CA, Rand Corporation, October 13, 1958.

[R3.39] Hempel, C. G., "Studies in the Logic of Confirmation," *Mind,* Vol.54, Part I , 1945, pp. 1-26.

[R3.40] Hempel, Carl G., "The Theoretician's Dilemma," in Herbert Feigl and M. Scriven (eds.) *Minnesota Studies in the Philosophy of Science*, Vol.II, 1958, pp. 37-98.

[R3.41] Hempel, C. G. and P. Oppenheim, "Studies in the Logic of Explanation," *Philosophy of Science*, Vol. 15, 1948, pp. 135-175. [also, in Brody, Baruch A. (ed.) *Reading in the Philosophy of Science*, Englewood Cliffs, NJ., Prentice-Hall,1970, pp. 8-27.

[R3. 42] Heyting, A., *Intuitionism: An Introduction,* Amsterdam: North-Holland, 1971.

[R3.43] Hintikka, Jackko (ed.), *The Philosophy of Mathematics*, London, Oxford University Press, 1969.

[R3.44] Hockney D. et al. (eds.), *Contemporary Research in Philosophical Logic and Linguistic Semantics*, Dordrecht-Holland, Reidel Pub., Co. 1975.

[R3.45] Hoyninggen-Huene, Paul and F. M. Wuketits, (eds.), *Reductionism and Systems Theory in the Life Science: Some Problems and Perspectives*, Dordrencht, Kluwer Academic Pub. 1989.

[R3.46] Ilyenkov, E.V., *Dialectical Logic: Essays on Its History and Theory*, Moscow, Progress Publishers, 1977.

[R3.47] Kedrov, B. M., "Toward the Methodological Analysis of Scientific Discovery," *Soviet Studies in Philosophy*, Vol. 11962, pp. 45 – 65.

[R3.48] Kemeny, John G, and P Oppenheim, "On Reduction," in Brody, Baruch A. (ed.) *Reading in the Philosophy of Science*, Englewood Cliffs, NJ., Prentice-Hall,1970, pp. 307-318.

[R3.49] Klappholz, K., "Value Judgments of Economics," *British Jour. of Philosophy*,

Vol. 15, 1964, pp. 97-114.

[R 3.50] Kleene, S.C.,"On the Interpretation of Intuitionistic Number Theory," Journal of Symbolic Logic, Vol 10, 1945, pp. 109-124.

[R3.51] Kmita, Jerzy, "The Methodology of Science as a Theoretical Discipline," *Soviet Studies in Philosophy*, Spring, 1974, pp. 38 –49

[R3.52] Krupp, Sherman R., (ed.), *The Structure of Economic Science*, Englewood Cliff, N. J., Prentice-Hall, 1966.

[R3.53] Kuhn, T., *The Structure of Scientific Revolution,* Chicago, University of Chicago Press, 1970.

[R3.54] Kuhn, Thomas, "The Function of Dogma in Scientific Research," in Brody, Baruch A. (ed.) *Reading in the Philosophy of Science*, Englewood Cliffs, NJ., Prentice-Hall,1970 pp. 356-374.

[R3.55] Kuhn, Thomas, *The Essential Tension: Selected Studies in Scientific Tradition and Change,* Chicago, University of Chicago Press, 1979.

[R3.56] Lakatos, I. (ed.), *The Problem of Inductive Logic,* Amsterdam, North Holland,1968.

[R3.57] Lakatos, I., *Proofs and Refutations: The Logic of Mathematical Discovery*, Cambridge, Cambridge University Press, 1976.

[R3.58] Lakatos, I., *Mathematics, Science and Epistemology: Philosophical Papers Vol. 2*, edited by J. Worrall and G. Currie, Cambridge, Cambridge Univ. Press, 1978.

[R3.59] Lakatos, I., *The Methodology of Scientific Research Programmes*, Vol 1, New York, Cambridge University Press, 1978.

[R3.60] Lakatos, Imre and A. Musgrave (eds.), *Criticism and the Growth of Knowledge*, New York, Cambridge University Press, 1979. Holland, 1979, pp. 153 – 164.

[R3.61] Lawson, Tony, *Economics and Reality*, New York, Routledge, 1977.

[R3.62] Lenzen, Victor, "Procedures of Empirical Science," in Neurath, Otto et al. (eds.), *International Encyclopedia of Unified Science, Vol. 1 – 10,* Chicago, University of Chicago Press, 1955, pp. 280-338.

[R3.63] Levi, Isaac, "Must the Scientist make Value Judgments?" in Brody, Baruch A. (Ed.) *Reading in the Philosophy of Science*, Englewood Cliffs, NJ., Prentice-Hall,1970, pp.559-570.

[R3.64] Tse-tung, *Mao, On Practice and Contradiction*, in *Selected works of Mao Tse-tung*, Piking, 1937. Also, London, Revolutions, 2008.

[R3.65] Lewis, David, *Convention: A Philosophical Study,* Cambridge, Mass., Harvard University Press, 1969.

[R3.66] Mayer, Thomas, *Truth versus Precision in Economics*, London, Edward Elgar 1993.

[R3.67] Menger, Carl, *Investigations into the Method of the Social Sciences with Special Reference to Economics*, New York, New York University Press, 1985.

[R3.68] Mirowski, Philip (ed.), *The Reconstruction of Economic Theory*, Boston, Mass. Kluwer Nijhoff, 1986.

[R3.69] Mueller, Ian, *Philosophy of Mathematics and Deductive Structure in Euclid's Elements*, Cambridge, Mass., MIT Press, 1981.

[R3.70] Nagel, Ernest, "Review: Karl Niebyl, Modern Mathematics and Some Problems of Quantity, Quality, and Motion in Economic Analysis," *The Journal of Symbolic Logic*, 1940, p.74.

[R3.71] Nagel, E. et al. (ed.), *Logic, Methodology, and the Philosophy of Science*, Stanford, Stanford University Press 1962.

[R3.72] Narens, Louis, "A Theory of Belief for Scientific Refutations," *Synthese,* Vol.145, 2005, pp. 397-423.

[R3.73] Narskii, I. S., "On the Problem of Contradiction in Dialectical Logic," *Soviet Studies in Philosophy,* Vol. vi, #4 pp.3-10, 1965.

[R3.74] Neurath, Otto et al. (eds.), *International Encyclopedia of Unified Science, Vol. 1 – 10,* Chicago, University of Chicago Press, 1955.

[R3.75] Neurath Otto, "Unified Science as Encyclopedic," in Neurath, Otto et al. (eds.), *International Encyclopedia of Unified Science, Vol. 1 – 10,* Chicago, University of Chicago Press, 1955, pp.1-27.

[R3.76] Planck, Max, *Scientific Autobiography and Other Papers*, Westport, Conn. Greenwood 1971.

[R3.77] Planck, Max, "The Meaning and Limits of Exact Science," in Max Planck, *Scientific Autobiography and Other Papers,* Westport, Conn. Greenwood, 1971, pp. 80 -120.

[R3.78] Polanyi, Michael, "Genius in Science," in Robert S. Cohen, and Marx W. Wartofsky (eds.), *Methodological and Historical Essays in the Natural and Social Sciences*, Dordrecht, D. Reidel Publishing Co. 1974, pp.57-71.

[R3.79] Popper, Karl, *The Nature of Scientific Discovery*, New York, Harper and Row, 1968.

[R3.80] Putnam, Hilary., "Models and Reality," in Benecerraf, P. and H. Putnam (eds.), *Philosophy of Mathematics: Selected Readings* , Cambridge, Cambridge University Press, 1983. pp. 421- 444.

[R3.81] Reise, S., *The Universe of Meaning*, New York, The Philosophical Library, 1953.

[R3.82] Robinson, R., *Definition*, Oxford, clarendon Press, 1950.

[R3.83] Rudner, Richard, "The Scientist qua Scientist Makes Value Judgments," *Philosophy of Science*, Vol. 20, 1953, pp 1-6.

[R3.84] Russell, B., *Our Knowledge of the External World*, New York, Norton, 1929.

[R3.85] Russell, B., *Human Knowledge, Its Scope and Limits*, London, Allen and Unwin, 1948.

[R3.86] Russell, B., *Logic and Knowledge: Essays 1901-1950,* New York, Capricorn Books, 1971.

[R3.87] Russell, B., *An Inquiry into Meaning and Truth*, New York, Norton, 1940.

[R3.88] Russell, Bertrand, *Introduction to Mathematical Philosophy*, London, George Allen and Unwin, 1919.

[R3.89] Russell, Bertrand, *The Problems of Philosophy*, Oxford, Oxford University Press, 1978.

[R3.90] Rutkevih, M. N., "Evolution, Progress, and the Law of Dialectics," *Soviet Studies in Philosophy,* Vol. IV, #3, PP. 34-43, 1965.

[R3.91] Ruzavin, G. I., "On the Problem of the Interrelations of Modern Formal Logic and Mathematical Logic," *Soviet Studies in Philosophy*, Vol. 3, #1, 1964, pp.34- 44.

[R3.92] Scriven, Michael, "Explanations, Predictions, and Laws," in Brody, Baruch A. (ed.) *Reading in the Philosophy of Science, Englewood Cliffs*, NJ., Prentice- Hall, 1970, pp.88-104.

[R3.93] Sellars, Wilfrid, The Language of Theories," in Brody, Baruch A. (ed.) *Reading in the Philosophy of Science,* Englewood Cliffs, NJ., Prentice-Hall,1970 pp.343-353.

[R3.94a] Sterman, John, "The Growth of Knowledge: Testing a Theory of Scientific Revolutions with a Formal Model," *Technological Forecasting and Social Change*, Vol. 28, 1995, pp. 93-122.

[R3.94b] Tsereteli, S. B. "On the Concept of Dialectical Logic,", *Soviet Studies in Philosophy*, Vol. V, #2, 1966, pp. 15-21.

[R3.95] Tullock, Gordon, *The Organization of Inquiry,* Indianapolis, Indiana, Liberty Fund Inc. 1966.

[R3.96] Van Fraassen, B., *Introduction to Philosophy of Space and Time*, New York, Random House, 1970.

[R3.97] Veldman, W., "A Survey of Intuitionistic Descriptive Set Theory," in P.P. Petkov (ed.), Mathematical Logic: Proceedings of the Heyting Conference, New York, Plenum Press, 1990, pp. 155-174.

[R3.98] Vetrov, A. A., "Mathematical Logic and Modern Formal Logic," *Soviet Studies in Philosophy*, Vol. 3, #1, 1964 pp. 24 –33.

[R3.99] von Mises, Ludwig, *Epistemological Problems in Economics*, New York, New York University Press, 1981.

[R3.100] Wang, Hao, *Reflections on Kurt* Gödel, Cambridge, Mass. MIT Press,1987

[R3.101] Watkins, J. W. N., "The Paradoxes of Confirmation"," in Brody, Baruch A. (ed.) *Reading in the Philosophy of Science*, Englewood Cliffs, NJ., Prentice-Hall, 1970, pp. 433-438.

[R3.102] Whitehead, Alfred North, *Process and Reality*, New York, The Free Press, 1978.

[R3.103] Wittgenstein, Ludwig, *Tractatus Logico-philosophicus,* Atlantic Highlands, N.J., The Humanities Press Inc.1974.

[R3.104] Woodger, J. H., *The Axiomatic Method in Biology*, Cambridge, Cambridge University Press, 1937.

[R3.105] Zeman, Jiří′, "Information, Knowledge and Time," in Libor Kubát, and J. Zeman (eds.), *Entropy and Information in the Physical Sciences*, Amsterdam,

Elsevier, 1975, pp. 245-260.

R4. FUZZY LOGIC, INFORMATION AND KNOWLEDGE-PRODUCTION

[R4.1] Baldwin, J.F., "A New Approach to Approximate Reasoning Using a Fuzzy Logic," *Fuzzy Sets and Systems*, Vol. 2, #4, 1979, pp. 309-325.

[R4.2] Baldwin, J.F., "Fuzzy Logic and Fuzzy Reasoning," *Intern. J. Man-Machine Stud.*, Vol. 11, 1979, pp. 465-480.

[R4.3] Baldwin, J.F., "Fuzzy Logic and Its Application to Fuzzy Reasoning," in. M. M. Gupta et al. (eds.), *Advances in Fuzzy Set Theory and Applications*, New York, North-Holland, 1979, pp.96-115.

[R4.4] Baldwin, J.F. et al., "Fuzzy Relational Inference Language," *Fuzzy Sets and Systems*, Vol. 14, #2, 1984, pp. 155-174.

[R4.5] Baldsin, J. and B.W. Pilsworth, "Axiomatic Approach to Implication For Approximate Reasoning With Fuzzy Logic," *Fuzzy Sets and Systems*, Vol. 3, #2, 1980, pp. 193-219.

[R4.6] Baldwin, J.F. et. al., "The Resolution of Two Paradoxes by Approximate Reasoning Using A Fuzzy Logic," *Synthese*, Vol. 44, 1980, pp. 397-420.

[R4.7] Dompere, K. K., *Fuzzy Rationality: Methodological Critique and Unity of Classical, Bounded and Other Rationalities*, (Studies in Fuzziness and Soft Computing, vol.235) New York, Springer, 2009.

[R4.8] Dompere Kofi K., *Epistemic Foundations of Fuzziness*, (Studies in Fuzziness and Soft Computing, vol. 236), New York, Springer, 2009.

[R4.9] Dompere Kofi K., *Fuzziness and Approximate Reasoning: Epistemics on Uncertainty, Expectation and Risk in Rational Behavior*, (Studies in Fuzziness and Soft Computing, vol. 237), New York, Springer, 2009.

[R4.10] Dompere, Kofi K., *The Theory of the Knowledge Square: The Fuzzy Rational Foundations of Knowledge-Production Systems,* New York, Springer, 2013.

[R4.11] Dompere, Kofi K., "Cost-Benefit Analysis, Benefit Accounting and Fuzzy Decisions: Part I, Theory", *Fuzzy Sets and Systems*, Vol. 92, 1997, pp. 275-287.

[R4.12] Dompere, Kofi K., "The Theory of Social Cost and Costing for Cost-Benefit Analysis in a Fuzzy Decision Space", *Fuzzy Sets and Systems*. Vol. 76, 1995, pp. 1-24.

[R4.13] Dompere, Kofi K., *Fuzzy Rational Foundations of Exact and Inexact Sciences,* New

York, Springer, 2013.

[R4.14] Gaines, B.R., "Foundations of Fuzzy Reasoning," *Inter. Jour. of Man-Machine Studies*, Vol. 8, 1976, pp. 623-668.

[R4.15] Gaines, B.R., "Foundations of Fuzzy Reasoning," in Gupta, M.M. et. al. (eds.), *Fuzzy Information and Decision Processes*, New York North-Holland, 1982, pp. 19-75.

[R4.16] Gaines, B. R., "Precise Past, Fuzzy Future," *International Journal. Of Man-Machine Studies.*, Vol. 19, #1,1983, pp.117-134.

[R4.17] Giles, R., "Lukasiewics Logic and Fuzzy Set Theory," *Intern. J. Man-Machine Stud.*, Vol. 8, 1976, pp. 313-327.

[R4.18] Giles, R., "Formal System for Fuzzy Reasoning," *Fuzzy Sets and Systems*, Vol. 2, #3, 1979, pp. 233-257.

[R4.19] Ginsberg, M. L.(ed.), *Readings in Non-monotonic Reason*, Los Altos, Ca., Morgan Kaufman, 1987.

[R4.20] Goguen, J.A., "The Logic of Inexact Concepts," *Synthese*, Vol. 19, 1969, pp. 325-373.

[R4.21] Gottinger, H.W., "Towards a Fuzzy Reasoning in the Behavioral Science," *Cybernetica*, Vol. 16, #2, 1973, pp. 113-135.

[R4.22] Gupta, M.M. et. al., (eds.), *Approximate Reasoning In Decision Analysis,* North Holland, New York, 1982.

[R4.23] Höhle Ulrich . and E.P. Klement , *Non-Clasical Logics and their Applications to Fuzzy Subsets: A Handbook of the Mathematical Foundations of Fuzzy Set Theory*, Boston, Mass. Kluwer, 1995.

[R4.24] Kaipov, V., Kh. et. al., "Classification in Fuzzy Environments," in M. M. Gupta et al. (eds.), *Advances in Fuzzy Set Theory and Applications,* New York, North-Holland, 1979, pp. 119-124.

[R4.25] Kaufman, A., "Progress in Modeling of Human Reasoning of Fuzzy Logic" in M. M. Gupta et al. (eds.), *Fuzzy Information and Decision Process*, New York, North-Holland, 1982, pp. 11-17.

[R4.26] Lakoff, G., "Hedges: A Study in Meaning Criteria and the Logic of Fuzzy Concepts," *Jour. Philos. Logic,* Vol. 2, 1973, pp. 458-508.

[R4.27] Lee, R.C.T., "Fuzzy Logic and the Resolution Principle," *Jour. of Assoc. Comput.* Mach., Vol. 19, 1972, pp. 109-119.

[R4.28] LeFaivre, R.A., "The Representation of Fuzzy Knowledge", *Jour. of Cybernetics*, Vol. 4, 1974, pp. 57-66.

[R4.29] Negoita, C.V., "Representation Theorems for Fuzzy Concepts," *Kybernetes*, Vol. 4, 1975, pp. 169-174.

[R4.30] Nowakowska, M., "Methodological Problems of Measurements of Fuzzy Concepts in Social Sciences", *Behavioral Sciences*, Vol. 22, #2, 1977, pp. 107-115.

[R4.31] Skala, H.J., "On Many-Valued Logics, Fuzzy Sets, Fuzzy Logics and Their Applications," *Fuzzy Sets and Systems*, Vol. 1, #2, 1978, pp. 129-149.

[R4.32] Van Fraassen, B.C., "Comments: Lakoff's Fuzzy Propositional Logic," in D. Hockney et. al., (Eds.), *Contemporary Research in Philosophical Logic and Linguistic Semantics* Holland, Reild, 1975, pp. 273-277.

[R4.33] Yager, R.R. et. al. (eds.), An Introduction to Fuzzy Logic Applications in *Intelligent Systems*, Boston, Mass., Kluwer 1992.

[R4.34] Zadeh, L.A., "Quantitative Fuzzy Semantics," *Inform. Science*, Vol. 3, 1971, pp. 159-176.

[R4.35] Zadeh, L.A., "A Fuzzy Set Interpretation of Linguistic Hedges," *Jour. Cybernetics*, Vol. 2, 1972, pp. 4-34.

[R4.36] Zadeh, L.A, "The Concept of a Linguistic Variable and Its Application to Approximate Reasoning," in K.S. Fu et. al. (eds.), *Learning Systems and Intelligent Robots*, Plenum Press, New York, 1974, pp. 1-10.

[R4.37] Zadeh, L.A., et. al., (eds.), *Fuzzy Sets and Their Applications to Cognitive and Decision Processes*, New York, Academic Press, 1974.

[R4.38] Zadeh, L.A., "The Birth and Evolution of Fuzzy Logic," *Intern. Jour. of General Systems*, Vol. 17, #(2-3) 1990, pp. 95-105.

R5. FUZZY MATHEMATICS AND PARADIGM OF APPROXIMATE REASONING UNDER CONDITIONS OF INEXACTNESS AND VAGUENESS

[R5.1] Bellman, R.E., "Mathematics and Human Sciences," in J. Wilkinson et. al. (eds.), *The Dynamic Programming of Human Systems*, New York, MSS Information Corp., 1973, pp. 11-18.

[R5.2] Bellman, R.E and Glertz, M., "On the Analytic Formalism of the Theory of Fuzzy Sets," Information *Science*, Vol. 5, 1973, pp. 149-156.

[R5.3] Butnariu, D., Fixed Points for Fuzzy Mapping," *Fuzzy Sets and Systems*, Vol. 7, #2, pp. 191-207, 1982.

[R5.4] Butnariu, D., "Decompositions and Range for Additive Fuzzy Measures", *Fuzzy Sets and Systems*, Vol. 10, #2, pp. 135-155, 1983.

[R5.5] Chang, C.L., "Fuzzy Topological Spaces," *J. Math. Anal. and Applications*, Vol. 24, 1968, pp. 182-190.

[R5.6] Chang, S.S.L., "Fuzzy Mathematics, Man and His Environment", *IEEE Transactions on Systems, Man and Cybernetics*, SMC-2 1972, pp. 92-93,

[R5.7] Chang, S.S., "Fixed Point Theorems for Fuzzy Mappings," Fuzzy *Sets and Systems*, Vol. 17, 1985, pp. 181-187.

[R5.8] Chapin, E.W., "An Axiomatization of the Set Theory of Zadeh," *Notices, American Math. Society*, 687-02-4 754, 1971.

[R5.9] Chaudhury, A. K. and P. Das, "Some Results on Fuzzy Topology on Fuzzy Sets," *Fuzzy Sets and Systems,* Vol. 56, 1993, pp. 331-336.

[R5.10] Chitra, H., and P.V. Subrahmanyam, "Fuzzy Sets and Fixed Points," *Jour. of Mathematical Analysis and Application*, Vol. 124, 1987, pp. 584-590.

[R5.11] Czogala, J. et. al., Fuzzy Relation Equations on a Finite Set," *Fuzzy Sets and Systems,* Vol. 7, #1,1982. pp. 89-101.

[R5.12] DiNola, A. et. al., (eds.), *The Mathematics of Fuzzy Systems*, Koln, Verlag TUV Rheinland, 1986.

[R5.13] Dompere, Kofi K., *Cost-Benefit Analysis and the Theory of Fuzzy Decisions: Identification and Measurement Theory* (Series: Studies in Fuzziness and Soft Computing, Vol. 158), Berlin, Heidelberg, Springer, 2004.

[R5.14] Dompere, Kofi K., *Cost-Benefit Analysis and the Theory of Fuzzy Decisions: Fuzzy Value Theory* (Series: Studies in Fuzziness and Soft Computing, Vol.160), Berling, Heidelberg, Springer, 2004.

[R5.15] Dubois, D. and H. Prade, *Fuzzy Sets and Systems*, New York, Academic Press, 1980

.

[R5.16] Dubois, "Fuzzy Real Algebra: Some Results," *Fuzzy Sets and Systems*, Vol. 2, #4, pp. 327-348, 1979.

[R5.17] Dubois, D. and H. Prade, Possibility Theory: An Approach to Computerized Processing of Uncertainty, New York, Plenum Press, 1988

[R5.18] Dubois, D. and H. Prade , "Gradual Inference Rules in Approximate Reasoning," *Information Sciences*, Vol. 61(1-2), 1992, pp. 103-122.

[R5.19] Dubois, D. and H. Prade, "On the Combination of Evidence in various Mathematical Frameworks," In: Flamm. J. and T. Luisi, (eds.), *Reliability Data Collection and Analysis*. Kluwer, Boston, 1992, pp. 213-241.

[R5.20] Dubois, D. and H. Prade, "Fuzzy Sets and Probability: Misunderstanding, Bridges and Gaps." *Proc. Second IEEE Intern. Conf. on Fuzzy Systems*, San Francisco, 1993, pp. 1059-1068.

[R5.21] Dubois, D. and H. Prade "A Survey of Belief Revision and Updating Rules in Various Uncertainty Models," *Intern. J. of Intelligent Systems*, Vol. 9, #1, 1994, pp. 61-100.

[R5.22] Filev, D.P. et. al., "A Generalized Defuzzification Method via Bag Distributions," *Intern. Jour. of Intelligent Systems*, Vol. 6, #7, 1991, pp. 687-697.

[R5.23] Goetschel, R. Jr., et. al., "Topological Properties of Fuzzy Number," *Fuzzy Sets and Systems*, Vol. 10, #1, pp. 87-99, 1983.

[R5.24] Goodman, I.R., "Fuzzy Sets as Equivalence Classes of Random Sets" inYager, R.R. (ed.), *Fuzzy Set and Possibility Theory: Recent Development*, New York, Pergamon Press 1992. pp. 327-343.

[R5.25] Gupta, M.M. et. al., (eds.), *Fuzzy Automata and Decision Processes*, New York, North-Holland, 1977.

[R5.26] Gupta, M.M. and E. Sanchez (eds.), *Fuzzy Information and Decision Processes*, New York, North-Holland, 1982.

[R5.27] Higashi, M. and G.J. Klir, "On measure of fuzziness and fuzzy complements," Intern. *J. of General Systems*, Vol. 8 #3, 1982, pp. 169-180.

[R5.28] Higashi, M. and G.J. Klir, "Measures of uncertainty and information based on possibility distributions," *International Journal of General Systems*, Vol. 9 #1, 1983, pp. 43-58.

[R5.29] Higashi, M. and G.J. Klir, "On the notion of distance representing information closeness: Possibility and probability distributions," *Intern. J. of General Systems*, Vol. 9 #2, 1983, pp. 103-115.

[R5.30] Higashi, M. and G.J. Klir , "Resolution of finite fuzzy relation equations," *Fuzzy Sets and Systems*, Vol. 13, #1,1984, pp. 65-82.

[R5.31] Higashi, M. and G.J. Klir , "Identification of fuzzy relation systems," *IEEE Trans. on Systems, Man, and Cybernetics*, Vol.14 #2, 1984, pp. 349-355.

[R5.32] Jin-wen, Z., "A Unified Treatment of Fuzzy Set Theory and Boolean Valued Set theory: Fuzzy Set Structures and Normal Fuzzy Set Structures," *Jour. Math. Anal. and Applications*, Vol. 76, #1, 1980, pp. 197-301.

[R5.33] Kandel, A. and W.J. Byatt, "Fuzzy Processes," *Fuzzy Sets and Systems*, Vol. 4, #2, 1980, pp. 117-152.

[R5.34] Kaufmann, A. and M.M. Gupta, *Introduction to fuzzy arithmetic: Theory and applications*, New York Van Nostrand Rheinhold,. 1991.

[R5.35] Kaufmann, A., *Introduction to the Theory of Fuzzy Subsets*, Vol. 1, New York, Academic Press, 1975.

[R5.36] Klement, E.P. and W. Schwyhla, "Correspondence Between Fuzzy Measures and Classical Measures," *Fuzzy Sets and Systems,* Vol. 7, #1, 1982.pp. 57-70.

[R5.37] Klir, George and Bo Yuan, *Fuzzy Sets and Fuzzy Logic*, Upper Saddle River, NJ Prentice Hall, 1995.

[R5.38] Kruse, R. et al., *Foundations of Fuzzy Systems*, New York, John Wiley and Sons, 1994.

[R5.39] Lasker, G.E. (ed.), *Applied Systems and Cybernetics, Vol. VI: Fuzzy Sets and Systems*, Pergamon Press, New York, 1981.

[R5.40] Lientz, B.P., "On Time Dependent Fuzzy Sets", *Inform, Science*, Vol. 4, 1972, pp. 367-376.

[R5.41] Lowen, R., "Fuzzy Uniform Spaces," *Jour. Math. Anal. Appl.*, Vol. 82, #21981, pp. 367-376.

[R5.42] Michalek, J., "Fuzzy Topologies," *Kybernetika*, Vol. 11, 1975, pp. 345-354.

[R5.43] Negoita, C.V. et. al., *Applications of Fuzzy Sets to Systems Analysis*, Wiley and Sons, New York, 1975.

[R5.44] Negoita, C.V., "Representation Theorems for Fuzzy Concepts," *Kybernetes*, Vol. 4, 1975, pp. 169-174.

[R5.45] Negoita, C.V. et. al., "On the State Equation of Fuzzy Systems," *Kybernetes*, Vol. 4, 1975, pp. 231-214.

[R5.46] Netto, A.B., "Fuzzy Classes," *Notices, American Mathematical Society,* Vol.68T-H28, 1968, pp.945.

[R5.47] Pedrycz, W., "Fuzzy Relational Equations with Generalized Connectives and Their Applications," *Fuzzy Sets and Systems*, Vol. 10, #2, 1983, pp. 185-201.

[R5.48] Raha, S. et. al., "Analogy Between Approximate Reasoning and the Method of Interpolation," *Fuzzy Sets and Systems*, Vol. 51, #3, 1992, pp. 259-266.

[R5.49] Ralescu, D., "Toward a General Theory of Fuzzy Variables," *Jour. of Math. Analysis and Applications*, Vol. 86, #1, 1982, pp. 176-193.

[R5.50] Rodabaugh, S.E., "Fuzzy Arithmetic and Fuzzy Topology," in G.E. Lasker, (ed.), *Applied Systems and Cybernetics, Vol. VI: Fuzzy Sets and Systems*, Pergamon Press, New York, 1981 pp. 2803-2807.

[R5.51] Rosenfeld, A., "Fuzzy Groups," *Jour. Math. Anal. Appln.*, Vol. 35, 1971, pp. 512-517.

[R5.52] Ruspini, E.H., "Recent Developments in Mathematical Classification Using Fuzzy Sets," in G.E. Lasker, (ed.), *Applied Systems and Cybernetics, Vol. VI: Fuzzy Sets and Systems*, Pergamon Press, New York, 1981. pp. 2785-2790.

[R5.53] Santos, E.S., "Fuzzy Algorithms," *Inform. and Control*, Vol. 17, 1970, pp. 326-339.

[R5.54] Stein, N.E. and K Talaki, "Convex Fuzzy Random Variables," *Fuzzy Sets and Systems*, Vol. 6, #3, 1981, pp. 271-284.

[R5.55] Triantaphyllon, E. et. al., "The Problem of Determining Membership Values in Fuzzy Sets in Real World Situations," in D.E. Brown et. al. (eds), *Operations Research and Artificial Intelligence: The Integration of Problem-solving Strategies*, Boston, Mass., Kluwer, 1990, pp. 197-214.

[R5.56] Tsichritzis, D., "Participation Measures," *Jour. Math. Anal. and Appln.*, Vol. 36, 1971, pp. 60-72.

[R5.57] Turksens, I.B., "Four Methods of Approximate Reasoning with Interval-Valued Fuzzy Sets," *Intern. Journ. of Approximate Reasoning*, Vol. 3, #2, 1989, pp. 121-142.

[R5.58] Turksen, I.B., "Measurement of Membership Functions and Their Acquisition," *Fuzzy Sets and Systems*, Vol. 40, #1, 1991, pp. 5-38.

[R5.59] Wang, P.P. (ed.), *Advances in Fuzzy Sets, Possibility Theory, and Applications*, New York, Plenum Press, 1983.

[R5.60] Wang, Zhenyuan, and George Klir, *Fuzzy Measure Theory,* New York, Plenum Press, 1992.

[R5.61] Wang, P.Z. et. al. (eds.), *Between Mind and Computer: Fuzzy Science and Engineering*, Singapore, World Scientific Press, 1993.

[R5.62] Wang, S., "Generating Fuzzy Membership Functions: A Monotonic Neural Network Model," *Fuzzy Sets and Systems*, Vol. 61, #1, 1994, pp. 71-82.

[R5.63] Wong, C.K., "Fuzzy Points and Local Properties of Fuzzy Topology," *Jour. Math. Anal. and Appln.*, Vol. 46, 19874, pp. 316-328.

[R5.64] Wong, C.K., "Categories of Fuzzy Sets and Fuzzy Topological Spaces," *Jour. Math. Anal. and Appln.*, Vol. 53, 1976, pp. 704-714.

[R5.65] Yager, R.R. et. al., (Eds.), *Fuzzy Sets, Neural Networks, and Soft Computing*, New York, Nostrand Reinhold, 1994.

[R5.66] Zadeh, L.A., "A Computational Theory of Decompositions," *Intern. Jour. of Intelligent Systems*, Vol. 2, #1, 1987, pp. 39-63.

[R5.67] Zimmerman, H.J., *Fuzzy Set Theory and Its Applications,* Boston, Mass, Kluwer, 1985.

R6. FUZZY OPTIMIZATION, INFORMATION, DECISION-CHOICE THEORY AND THE SCIENCE OF KNOWING

[R6.1] Bose, R.K. and Sahani D, "Fuzzy Mappings and Fixed Point Theorems," *Fuzzy Sets and Systems*, Vol. 21, 1987, pp. 53-58.

[R6.2] Butnariu D. "Fixed Points for Fuzzy Mappings," *Fuzzy Sets and Systems*, Vol. 7, 1982, pp.191-207.

[R6.3] Dompere, Kofi K., "Fuzziness, Rationality, Optimality and Equilibrium in Decision and Economic Theories" in Weldon A. Lodwick and Janusz Kacprzyk (Eds.), *Fuzzy Optimization: Recent Advances and Applications* (Series: Studies in Fuzziness and Soft Computing, Vol. 254), Berlin, Heidelberg, Springer, 2010.

[R6.4] Eaves, B.C., "Computing Kakutani Fixed Points," *Journal of Applied Mathematics*, Vol. 21, 1971, pp. 236-244.

[R6.5] Heilpern, S. "Fuzzy Mappings and Fixed Point Theorem," *Journal of Mathematical Analysis and Applications*, Vol. 83, 1981, pp.566-569.

[R6.6] Kacprzyk, J. et. al., (eds.), *Optimization Models Using Fuzzy Sets and Possibility Theory*, Boston, Mass., D. Reidel, 1987.

[R6.7] Kaleva, O. "A Note on Fixed Points for Fuzzy Mappings", *Fuzzy Sets and Systems,* Vol. 15, 1985, pp. 99-100.

[R6.8] Lodwick, Weldon A and Janusz Kacprzyk (eds.), *Fuzzy Optimization: Recent Advances and Applications,* (Studies in Fuzziness and Soft Computing, Vol. 254), Berlin Heidelberg, Springer, 2010.

[R6.9] Negoita, C.V., "The Current Interest in Fuzzy Optimization," *Fuzzy Sets and Systems*, Vol. 6, #3, 1981, pp. 261-270.

[R6.10] Negoita, C.V., et. al., "On Fuzzy Environment in Optimization Problems," in J. Rose et. al., (eds.), *Modern Trends in Cybernetics and Systems,* Springer, Berlin, 1977, pp.13-24.

[R6.11] Zimmerman, H.-J., "Description and Optimization of Fuzzy Systems," *Intern. Jour. Gen. Syst.* Vol. 2, #4, 1975, pp. 209-215.

R7. IDEOLOGY, DISINFORMATION, MISINFORMATION AND PROPAGANDA IN INTRA- AND INTER-EPISTEMOLOGICAL MASSEGING SYSTEMS

[R7.1] Abercrombie, Nicholas et al., *The Dominant Ideology Thesis*, London, Allen and Unwin, 1980.

[R7.2] Abercrombie, Nicholas, *Class, Structure, and Knowledge: Problems in the Sociology of Knowledge*, New York, New York University Press, 1980.

[R7.3] Aron, Raymond, *The Opium of the Intellectuals*, Lanham, MD, University Press of America, 1985.

[R7.4] Aronowitz, Stanley, *Science as Power: Discourse and Ideology in Modern Society,* Minneapolis, University of Minnesota Press, 1988.

[R7.5] Barinaga, M. and E. Marshall, *Confusion on the Cutting Edge*, Science, Vol. 257, July 1992, pp. 616-625.

[R7.6] Barnett, Ronald, *Beyond All Reason: Living with Ideology in the University,* Philadelphia, PA., Society for Research into Higher Education and Open University Press, 2003.

[R7.7] Barth, Hans, *Truth and Ideology*, Berkeley, University of California Press,1976.

[R7.8] Basin, Alberto, and Thierry Verdie, "The Economics of Cultural Transmission and the Dynamics of Preferences," *Journal of Economic Theory*, Vol. 97, 2001, pp. 298-319.

[R7.9] Bikhchandani, Sushil et al., "A Theory of Fads, Fashion, Custom, and Cultural Change," *Journal of political Economy*, Vol. 100 1992, pp 992-1026.

[R7.10] Boyd Robert and Peter J Richerson, *Culture and Evolutionary Process*, Chicago, University of Chicago Press, 1985.

[R7.11] Buczkowski, Piotr and Andrzej Klawiter, *Theories of Ideology and Ideology of Theories*, Amsterdam, Rodopi, 1986.

[R7.12] Chomsky, Norm, *Manufacturing Consent*, New York, Pantheo Pess, 1988.

[R7.13] Chomsky, N., *Problem of Knowledge and Freedom*, Glasgow, Collins, 1972.

[R7.14] Cole, Jonathan, R., "Patterns of Intellectual influence in Scientific Research," *Sociology of Education*, Vol. 43, 1968, pp.377-403.

[R7.15] Cole Jonathan, R. and Stephen Cole, *Social Stratification in Science*, Chicago, University of Chicago Press, 1973.

[R7.16] Debackere, Koenraad and Michael A. Rappa, "Institutioal Variations in Problem Choice and Persistence among Scientists in an Emerging Fields," *Research Policy*, Vol. 23, 1994, pp425- 441.

[R7.17] Dompere, K. K., "Ideology and Decision-Choice Rationalities" in Dompere, Kofi. Kissi., *Fuzzy Rationality: Methodological Critique and Unity of Classical, Bounded and Other Rationalities*, (Studies in Fuzziness and Soft Computing, vol.235) New York, Springer, 2009, pp.143-165.

[R7.18A] Dompere, K. K., "Culture, Ideology and Categorial Conversion of Social Polarities in Systems Dynamics," in Dompere, K. K., *The Theory of Philosophical Consciencism*; Practice Foundations of Nkrumaism in Social Systemicity, London, Adonis & Abbey Pub. 2017, pp.124 –158.

[R7.18B] Dompere, K. K., "Abstract Ideas and Practice of Ideas in Social Settings: Extensions and Reflections on Nkrumah and Africa under Systems Thinking," in Dompere, K. K., *The Theory of Categorial Conversion: Rational Foundations of Nkrumaism in Socio-natural Systemicity and Complexity*; London, Adonis & Abbey Pub. 2017, pp.61 85 .

[R7.19] Dompere, K. K., "Zones of Thought: Reflections on the theories of Thought," in Dompere, K. K, *Fuzziness and Foundations of Exact and Inexact Sciences*, (Studies in Fuzziness and Soft Computing, vol.290) New York, Springer, 2013, pp.103-116.

[R7.20] Fraser, Colin and George Gaskell (eds.), *The Social Psychological Study of Widespread Beliefs*, Oxford, Clarendon Press, 1990.

[R7.21] Gieryn, Thomas, F. "Problem Retention and Problem Change in Science," *Sociological Inquiry*, Vol. 48, 1978, pp96-115.

[R7.22] Harrington, Joseph E. Jr, "The Rigidity of social Systems," *Journal of Political Economy*, Vol. 107, pp. 40-64.

[R7.23] Hinich, Melvin and Michael Munger, *Ideology and the Theory of Political Choice*, Ann Arbor University of Michigan Press, 1994.

[R7.24] Hull, D. L., *Science as a Process: An Evolutionary Account of the Social and Conceptual Development of Science,* Chicago, University of Chicago Press, 1988.

[R7.25] Marx, Karl and Friedrich Engels, *The German Ideology*, New York, International Pub, 1970.

[R7.26] Mészáros, István , *Philosophy, Ideology and Social Science*: Essay in Negation and Affirmation, Brighton, Sussex, Wheatsheaf, 1986.

[R7.27] Mészáros, István *The Power of Ideology*, New York, New York University Press, 1989.

[R7.28] Newcomb, Theodore M. et. al., *Persistence and Change*, New York, John Wiley, 1967.

[R7.29] Pickering, Andrew, *Science as Practice and Culture,* Chicago, University of Chicago Press, 1992.

[R7.30] Therborn, Göran, *The Ideology of Power and the Power of Ideology*, London, NLB Publications, 1980.

[R7.31] Thompson, Kenneth, *Beliefs and Ideology*, New York, Tavistock Publication,1986.

[R7.32] Trine, Ralph W., *In Tune with the Infinite*: A Modern Classic of Spiritual Inspiration, New York, The Bobbs Merrill Company. 1970.

[R7.33] Ziman, John, "The Problem of 'Problem Choice'," *Minerva*, Vol. 25, 1987, pp. 92-105.

[R7.34] Ziman, John, *Public Knowledge: An Essay Concerning the Social Dimension of Science*, Cambridge, Cambridge University Press, 1968.

[R7.35] Zuckerman, Harriet, "Theory Choice and Problem Choice in Science," *Sociological Inquiry*, Vol.48, 1978, pp. 65-95.

R 8. INFORMATION, THOUGHT, KNOWING AND KNOWLEDGE

[R8.1] Aczel, J. and Z. Daroczy, *On Measures of Information and their Characterizations*, New York, Academic Press, 1975

[R8.2] Afanasyev, *Social Information and Regulation of Social Development*, Moscow, Progress,

1978

[R8.3] Anderson, J. R., *The Architecture of Cognition*, Cambridge, Mass., Harvard University Press, 1983.

[R8.4] Angelov, Stefan and Dimitr Georgiev, "The Problem of Human Being in Contemporary Scientic Knowledge," *Soviet Studies in Philosophy*, Summer, 1974, pp. 49-66.

[R8.5] Ash, Robert, *Information Theory*, New York, John Wiley and Sons, 1965.

[R8.6] Bergin, J., "Common Knowledge with Monotone Statistics," *Econometrica*, Vol. 69, 2001, pp. 1315-1332.

[R8.7] Bestougeff, Hélène and Gerard Ligozat, *Logical Tools for Temporal Knowledge Representation*, New York, Ellis Horwood, 1992.

[R8.8] Brillouin, L., *Science and information Theory,* New York, Academic Press, 1962.

[R8.9] Bruner, J. S., et. al., *A Study of Thinking*, New York, Wiley, 1956.

[R8.10] Brunner, K. and A. H. Meltzer (eds.), *Three Aspects of Policy and Policy Making: Knowledge, Data and Institutions*, Carnegie-Rochester Conference Series, Vol. 10, Amsterdam, North-Holland, 1979.

[R8.11] Burks, A. W., *Chance, Cause, Reason: An Inquiry into the Nature of Scientific Evidence*, Chicago, University of Chicago Press, 1977.

[R8.12] Calvert, Randall, *Models of Imperfect Information in Politics*, New York, Hardwood Academic Publishers, 1986.

[R8.13] Cornforth, Maurice, *The Theory of Knowledge*, New York, International Pub. 1972

[R8.14] Cornforth, Maurice, *The Open Philosophy and the Open Society,* New York, International Pub. 1970

[R8.15] Coombs, C. H., *A Theory of Data*, New York, Wiley, 1964.

[R8.16a] Dompere, Kofi K., The Theory of Problem-Solution Dualities and Polarities: Information-Decision-Choice Foundations of Unity of Knowing and Unity of Science, Springer, New York, 2022.

[R8.16b] Dompere, Kofi K., *The Theory of Organizational Analytics in the Functor Space and the Space of Categorial Converters,* (Research Institute of Information, Decision and Economic Sciences (RIIDES) Unpublished Monograph, Washington, DC.USA, 2022.

[R8.17] Dretske, Fred. I., *Knowledge and the Flow of Information*, Cambridge, Mass., MIT Press 1981.

[R8.18] Dreyfus, Hubert L., "A Framework for Misrepresenting Knowledge," in Martin Ringle (ed.) *Philosophical Perspectives in Artificial Intelligence,* Atlantic Highlands, N.J., Humanities press, 1979.

[R8.19] Fagin R. et al., *Reasoning About Knowledge,* Cambridge, Mass, MIT Press, 1995.

[R8.20] George, F. H., *Models of Thinking*, London, Allen and Unwin, 1970.

[R8.21] George, F. H., "Epistemology and the problem of perception," *Mind*, Vol.66, 1957, pp.491-506.

[R8.22] Harwood, E. C., *Reconstruction of Economics,* Great Barrington, Mass, American Institute for Economic Research, 1955.

[R8.23] Hintikka, J., *Knowledge and Belief*, Ithaca, N. Y., Cornell University Press, 1962.

[R8.24] Hirshleifer, Jack., "The Private and Social Value of Information and Reward to inventive activity," *American Economic Review,* Vol. 61, 1971, pp.561-574.

[R8.25] Kapitsa, P. L., "The Influence of Scientific Ideas on Society," *Soviet Studies in Philosophy*, Fall, 1979, pp.52-71.

[R8.26] Kedrov, B. M., "The Road to Truth," *Soviet Studies in Philosophy*, Vol. 4, 1965, pp 3 – 53.

[R8.27a] Klatzky, R. L., *Human Memory: Structure and Processes*, San Francisco, Ca., W. H. Freeman Pub., 1975.

[R8.27b] Klix, Friedhart,(ed.) Human and Artfificial Intelligence, (Fundamental Studies in Computer Science, Vol. 8),New York, North-Holland, 1979

[R8.28] Kreps, David and Robert Wilson, "Reputation and Imperfect Information," *Journal of Economic Theory,* Vol. 27. 1982, pp253-279.

[R8.29] Kubát , Libor and J. Zeman (eds.), *Entropy and Information*, Amsterdam, Elsevier, 1975.

[R8.30] Kurcz, G. and W. Shugar et al (eds.), *Knowledge and Language*, Amsterdom, North-Holland, 1986.

[R8.31] Lakemeyer, Gerhard,and Bernhard Nobel (eds.), *Foundations of Knowledge*

Representation and Reasoning, Berlin, Springer-Verlag, 1994.

[R8.32] Lektorskii, V. A., "Principles involved in the Reproduction of Objective in Knowledge,", *Soviet Studies in Philosophy*, Vol. 4, #4, 1967, pp. 11-21.

[R8.33] Levi, I., *The Enterprise of Knowledge*, Cambridge, Mass. MIT Press1980.

[R8.34] Levi, Isaac, "Ignorance, Probability and Rational Choice", *Synthese* Vol. 53, 1982, pp. 387-417.

[R8.35] Levi, Isaac, "Four Types of Ignorance," *Social Science*, Vol. 44, pp745-756.

[R8.36] Marschak, Jacob, *Economic Information, Decision and Prediction: Selected Essays*, Vol. II, Part II, Boston, Mass. Dordrecnt-Holland, 1974.

[R8.37] Menges, G. (ed.), *Information, Inference and Decision*, Dordrecht, Holland, D. Reidel Pub.,1974.

[R8.38] Michael Masuch and Lászlò Pólos (eds.), *Knowledge Representation and Reasoning Under Uncertainty*, New York, Springer-Verlag, 1994.

[R8.39] Moses, Y. (ed.), *Proceedings of the Fourth Conference of Theoretical Aspects of Reasoning about Knowledge*, San Mateo, Morgan Kaufmann, 1992.

[R8.40] Nielsen, L.T. et al., "Common Knowledge of Aggregation Expectations," *Econometrica*, Vol. 58, 1990, pp. 1235-1239.

[R8.41] Newell, A., *Unified Theories of Cognition*, Cambridge, Mass. Harvard University Press, 1990.

[R8.42] Newell, A and Herbert A. Simon., *Human Problem Solving*, Englewood Cliff, N.J. Prentice- Hall, 1972.

[R8.43] Ogden, G. K. and I. A., *The Meaning of Meaning*, New York, Harcourt-Brace Jovanovich, 1923.

[R8.44] Planck, Max, Scientific Autobiography and Other Papers, Westport, Conn., Greenwood, 1968.

[R8.45] Pollock, J., *Knowledge and Justification*, Princeton, Princeton University Press, 1974.

[R8.46] Polanyi, M., *Personal Knowledge*, London, Routledge and Kegan Paul, 1958.

[R8.47] Popper, K. R., *Objective Knowledge*, London, Macmillan, 1949.

[R8.48] Popper, K. R., *Open Society and it Enemies, Vols. 1 and 2* Princeton, Princeton Univ. Press, 2013.

[R8.49] Popper, K. R., *The Poverty of Historicism* New York, Taylor and Francis, 2002.

[R8.50] Price, H. H., *Thinking and Experience*, London, Hutchinson, 1953.

[R8.51] Putman, H., *Reason, Truth and History*, Cambridge, Cambridge University Press, 1981.

[R8.52] Putman. H., *Realism and Reason*, Cambridge, Cambridge University Press, 1983.

[R8.53] Putman, H., *The Many Faces of Realism*, La Salle, Open Court Publishing Co., 1987.

[R8.54] Russell, B., *Human Knowledge, its Scope and Limits*, London, Allen and Unwin, 1948.

[R8.55] Russell, B., *Our Knowledge of the External World,* New York, Norton, 1929.

[R8.56] Samet, D., "Ignoring Ignorance and Agreeing to Disagree," *Journal of Economic Theory*, Vol. 52, 1990, pp. 190-207.

[R8.57] Schroder, Harold, M. and Peter Suedfeld (eds.), *Personality Theory and Information Processing*, New York, Ronald Pub. 1971.

[R8.58] Searle J., *Minds, Brains and Science*, Cambridge, Mass., Harvard University Press, 1985.

[R8.59] Shin, H., "Logical Structure of Common Knowledge," *Journal of Economic Theory*, Vol. 60, 1993, pp. 1-13.

[R8.60] Simon, H. A., *Models of Thought*, New Haven, Conn., Yale University Press,1979.

[R8.61] Smithson, M., *Ignorance and Uncertainty, Emerging Paradigms*, Springer-Verlag, New York 1989.

[R8.62] Sowa, John F., *Knowledge Representation: Logical, Philosophical, and Computational Foundations*, Pacific Grove, Brooks Pub., 2000.

[R8.63] Stigler, G. J., The Economics of Information," *Journal of Political Economy*, Vol.69, 1961, pp.213-225.

[R8.64] Tiukhtin, V. S., "How Reality Can be Reflected in Cognition: Reflection as a Property of All Matter," *Soviet Studies in Philosophy*, Vol.3 #1, 1964, pp.3-12.

[R8.65] Tsypkin, Ya Z., *Foundations of the Theory of Learning Systems*, New York, Academic Press, 1973.

[R8.66] Ursul, A. D., "The Problem of the Objectivity of Information," in Libor Kubát , and J. Zeman (eds.), *Entropy and Information*, Amsterdam, Elsevier, 1975, pp. 187 – 230.

[R8.67] Vardi, M. (ed.), *Proceedings of Second Conference on Theoretical Aspects of Reasoning about Knowledge*, Asiloman, Ca., Los Altos, Ca, Morgan Kaufman1988.

[R8.68] Vazquez, Mararita, et al., "Knowledge and Reality: Some Conceptual Issues in System Dynamics Modeling," *Systems Dynamics Review,* Vol. 12, 1996, pp. 21-37.

[R8.69] Zadeh, L. A., "A Theory of Commonsense Knowledge," in Skala, Heinz J. et al., (eds.), *Aspects of Vagueness*, Dordrecht, D. Reidel Co. 1984. pp 257 – 295.

[R8.70] Zadeh, L. A., "The Concept of Linguistic Variable and its Application to Approximate reasoning," *Information Science*, Vol. 8, 1975, pp. 199 – 249 (Also in Vol. 9, pp. 40 – 80).

R9. LANGUAGES AND INFORMATION

[R9.1] Agha, Agha, *Language and Social Relations,* Cambridge, Cambridge University Press 2006.

[R9.2]Aitchison, Jean (ed.), *Language Change: Progress or Decay?* Cambridge, New York, Melbourne: Cambridge University Press, 2001.

[R9.3] Anderson, Stephen, *Languages: A Very Short Introduction.* Oxford: Oxford University Press (2012).

[R9.4] Aronoff, Mark and Fudeman, Kirsten, *What is Morphology. New York, John Wiley & Sons,* 2011.

[R9.5] Bauer, Laurie (ed.), *Introducing linguistic morphology* Washington, D.C.: Georgetown University Press, 2003.

[R9.6] Barber Alex and Robert J Stainton *(eds.). Concise Encyclopedia of Philosophy of Language and Linguistics.* New York, Elsevier, 2010.

[R9.7] Brown, Keith; Ogilvie, Sarah, (eds.), *Concise Encyclopedia of Languages of the World,* New York Elsevier Science 2008.

[R9.8] Campbell, Lyle_ (ed.), *Historical Linguistics: an Introduction Cambridge,* MASS, MIT Press, 2004.

[R9.9] Chomsky, Noam, *Syntactic Structures.* The Hague: Mouton,1957.

[R9.10] Chomsky, Noam, *The Architecture of Language*. Oxford, Oxford University Press, 2000.

[R9.11] Clarke, David S. *Sources of semiotic: readings with commentary from antiquity to the present* Carbondale: Southern Illinois University Press, 1990.

[R9.12] Collinge, N.E.(ed.), *An Encyclopedia of Language.* London: New York: Routledge (1989).

[R9.13] Comrie, Bernard (ed.), *Language universals and linguistic typology: Syntax and morphology,* Oxford, Blackwell,1989.

[R9.14] Comrie, Bernard (ed.), *The World's Major Languages.* New York: Routledge, 2009.

[R9.15] Coulmas, Florian, *Writing Systems: An Introduction to Their Linguistic Analysis.* Cambridge, Cambridge University Press, 2002.

[R9.16] Croft, William, Cruse, D. Alan, *Cognitive Linguistics.* Cambridge, Cambridge University Press, 2004.

[R9.17] Croft, William, "Typology". In Mark Aronoff; Janie Rees-Miller(ed.), *The Handbook of Linguistics,* Oxford, Blackwell, 2001. pp. 81–105.

[R9.18] Crystal, David (ed.) *The Cambridge Encyclopedia of Language. Cambridge: Cambridge University Press,* 1997.

[R9.19] Deacon, Terrence , *The Symbolic Species: The Co-evolution of Language and the Brain,* New York: W.W. Norton & Company, 1997.

[R9.20] Deming W. Edwards, "On Errors in Surveys," *American Sociological Rev*., Vol. IX, 1944, pp. 359-369.

[R9.21] Devitt, Michael and Sterelny, Kim, *Language and Reality: An Introduction to the Philosophy of Language. Boston:* MIT Press 1999.

[R9.22] Duranti, Alessandro "Language as Culture in U.S. Anthropology: Three Paradigms" *Current Anthropology* Vol. 44 (3), pp. 323–348, 2003.

[R9.23] Evans, Nicholas and Levinson, Stephen C, "The myth of language universals: Language diversity and its importance for cognitive science," Vol. 32 *(5). Behavioral and Brain Sciences.* pp. 429–492, 2009.

[R9.24] Fitch, W. Tecumseh, *The Evolution of Language,* Cambridge, Cambridge University Press, 2010.

[R9.25] Foley, William A., *Anthropological Linguistics: An Introduction,* Oxford, Blackwell, 1997.

[R9.26] Ginsburg, Seymour, *Algebraic and Automata-Theoretic Properties of Formal Languages*, New York, North-Holland, 1973.

[R9.27] Goldsmith, John A, *The Handbook of Phonological Theory: Blackwell Handbooks in Linguistics.* Oxford, Blackwell Publishers, 1995.

[R9.28] Greenberg, Joseph, *Language Universals: With Special Reference to Feature Hierarchies.* The Hague, Mouton & Co., 1966.

[R9.29] Hauser, Marc D.; Chomsky, Noam; Fitch, W. Tecumseh, "The Faculty of Language: What Is It, Who Has It, and How Did It Evolve?". *Science,* 22 **298** (5598), 2002, pp. 1569–1579.

[R9.30] Hörz, Herbert, "Information, Sign, Image," in Libor Kubát and Jiří Zeman (eds.*) Entropy and Information in Science and Philosophy*, New York, Elsevier, 1975.

[R9.31] International Phonetic Association, *Handbook of the International Phonetic Association: A guide to the use of the International Phonetic Alphabet. Cambridge,* Cambridge University Press (1999).

[R9.32] Katzner, Kenneth, *The Languages of the World,* New York: Routledge, 1999.

[R9.33] Labov, William, *Principles of Linguistic Change vol.I Internal Factors,* Oxford, Blackwell, 1994.

[R9.34] Labov, William, *Principles of Linguistic Change vol.II Social Factors,* Oxford, Blackwell, 2001.

[R9.35] Levinson, Stephen C. *Pragmatics.* Cambridge: Cambridge University Press, 1983.

[R9.36] Lewis, M. Paul (ed.), *Ethnologue: Languages of the World,* Dallas, Tex.: SIL International 2009.

[R9.37] Lyons, John, *Language and Linguistics,* Cambridge, Cambridge University Press, 1981.

[R9.38] MacMahon, *April M.S., Understanding Language Change,* Cambridge, Cambridge University Press, 1994.

[R9.39] Matras, Yaron; Bakker, Peter, (eds.). *The Mixed Language Debate: Theoretical and Empirical Advances.* Berlin: Walter de Gruyter, 2003.

[R9.40] Moseley, Christopher (ed.), *Atlas of the World's Languages in Danger, Paris: UNESCO Publishing* (2010).

[R9.41] Nerlich, B. "History of pragmatics". In L. Cummings (ed.), *The Pragmatics Encyclopedia.* New York: Routledge. pp. 192–93, 2010.

[R9.42] Newmeyer, Frederick J., *The History of Linguistics*. Linguistic Society of America 2005.

[R9.43] Newmeyer, Frederick J., *Language Form and Language Function* (PDF). Cambridge, MA: MIT Press 1998.

[R9.44] Nichols, Johanna, *Linguistic diversity in space and time.* Chicago, University of Chicago Press,1992.

[R9.45] *Nichols, Johanna. "Functional Theories of Grammar". Annual Review of Anthropology* Vol. **13**: 1949, pp. 7–117.

[R9.46] *Sandler, Wendy; Lillo-Martin, Diane,* "Natural Sign Languages", *In Mark Aronoff; Janie Rees-Miller (eds.). The Handbook of Linguistics.* Oxford, Blackwell. pp. 533–563, 2001.

[R9.47] Swadesh, Morris, *"The phonemic principle", Language,* Vol. 10 (2): 1934, pp.117–129.

[R9.48] Tomasello, Michael, "The Cultural Roots of Language". In B. Velichkovsky and D. Rumbaugh 9EDS.), *Communicating Meaning: The Evolution and Development of Language.* New York, Psychology Press. 1996, pp. 275–308.

[R9.49] Tomasello, Michael, *Origin of Human Communication.* Cambridge Mass., MIT Press 2008.

[R9.50] Thomason, Sarah G, *Language Contact – An Introduction,* Edinburgh, Edinburgh University Press,2001.

[R9.51] Ulbaek, Ib, "The Origin of Language and Cognition", In J. R. Hurford and C. Knight (eds.). *Approaches to the evolution of language.* Cambridge, Cambridge University Press. *1998,* pp. 30–43.

[R9.52] Van Valin, jr. Robert D., "Functional Linguistics," In Mark Aronoff; Janie Rees-Miller (eds.), *The Handbook of Linguistics.* Oxford, Blackwell 2001, pp. 319–337.

R10. LANGUAGE, KNOWLEDGE-PRODUCTION PROCESS AND EPISTEMICS

[R10.1] Aho, A. V. "Indexed Grammar - An Extension of Context-Free Grammars" *Journal of the Association for Computing Machinery*, Vol. 15, 1968, pp. 647-671.

[R10.2] Black, Max (ed.), *The Importance of Language*, Englewood Cliffs, N.J, Prentice-Hall, 1962.

[R10.3] Carnap, Rudolff, Meaning and Necessity: A Study in Semantics and Modal Logic, Chicago, University of Chicago Press, 1956.

[R10.4] Chomsky, Norm, "Linguistics and Philosophy" in S. Hook (ed.) *Language and Philosophy*, New York, New York University Press, 1968, pp. 51-94.

[R10.5] Chomsky, Norm, *Language and Mind*, New York, Harcourt Brace Jovanovich, 1972.

[R10.6] Cooper, William S., *Foundations of Logico-Linguistics: A Unified Theory of Information, Language and Logic*, Dordrecht, D. Reidel, 1978

[R10.7] Cresswell, M.J.., *Logics and Languages*, London, Methuen Pub. 1973.

[R10.8] Dilman, Ilham, *Studies in Language and Reason*, Totowa, N.J., Barnes and Nobles, Books, 1981.

[R10.9] Fodor, Jerry A., *The Language and Thought*, New York, Thom as Y. Crowell Co, 1975.

[R10.10] Givon, Talmy, *On Understanding Grammar*, New York, Academic Press, 1979.

[R10.11] Gorsky, D.R., *Definition,* Moscow, Progress Publishers, 1974.

[R10.12] Hintikka, Jaakko, The Game of Language, Dordrecht, D. Reidel Pub. 1983.

[R10.13] Johnson-Lair, Philip N. *Mental Models: Toward Cognitive Science of Language, Inference and Consciousness*, Cambridge, Mass, Harvard University Pres, 1983.

[R10.14] Kandel, A., "Codes Over Languages," *IEEE Transactions on Systems Man and Cybernetics*, Vol. 4, 1975, pp. 135-138.

[R10.15] Keenan, Edward L. and Leonard M. Faltz, *Boolean Semantics for Natural Languages*, Dordrecht, D. Reidel Pub.,1985.

[R10.16] Lakoff, G. Linguistics and Natural Logic, *Synthese*, Vol. 22, 1970, pp. 151-271.

[R10.17] Lee, E.T., et. al., "Notes On Fuzzy Languages," *Information Science*, Vol. 1, 1969, pp. 421-434.

[R10.18] Mackey, A. and D. Merrill (eds.) *Issues in the Philosophy of Language,* New Haven, CT. Yale University Press, 1976.

[R10.19] Nagel, T., "Linguistics and Epistemology" in S. Hook(ed.) *Language and Philosophy*, New York, New York University Press, 1969, pp. 180-184.

[R10.20] Pike, Kenneth, *Language in Relation to a Unified Theory of Structure of Human Behavior*, The Hague, Mouton Pub., 1969.

[R10.21] Quine, W.V. O. *Word and object,* Cambridge, Mass, MIT Press, 1960.

[R10.22] Russell, Bernard, *An Inquiry into Meaning and Truth,* Penguin Books, 1970.

[R10.23] Tarski, Alfred, *Logic, Semantics and Metamathematics,* Oxford, Clarendon Press, 1956.

[R10.24] Whorf, B.L. (ed.), *Language, Thought and Reality*, New York, Humanities Press, 1956.

R11 PHILOSOPHY OF ECONOMICS AND PRODUCTION-CONSUMPTION DUALITIES

[R11.1] Alchian, A., "Uncertainty, Evolution and Economic Theory", *Journal of Political Economy*, Vol. 57,1950: pp.211–21.

R11.2] Angyrous, G., "Refutation or Comparison?" *British Journal for the Philosophy of Science*, Vol. 17, 1967, pp.279–96.

[R11.3] Arrow, K., "Economic Equilibrium", *International Encyclopedia of the Social Sciences*, New York: Macmillan 1968, pp. 376–89.

[R11.3] Arrow, K, *The Limits of Organization*, New York: Norton, 1974.

[R11.4] Aruka, Y., *Evolutionary Foundations of Economic Science: How Can Scientists Study Evolving Economic Doctrines from the Last Centuries*, New York: Springer, 2015.

[R11.5] Ayer, A., 1936. *Language, Truth and Logic*, 2nd edition, reprinted New York: Dover, 1946.

[R11.6] Backhouse, R., *Truth and Progress in Economic Knowledge*, Cheltenham: Edward Elgar 1997.

[R11.7] Bavetta, Sebastiano, Pietro Navarra, and Dario Maimone, , *Freedom and the Pursuit of Happiness: An Economic and Political Perspective*, Cambridge: Cambridge University Press 2014.

[R11.8] Becker, G., *The Economic Approach to Human Behavior*, Chicago: University of Chicago Press, 1976.

[R11.9] Berger, L., "Economics and Hermeneutics", *Economics and Philosophy*, Vol. 5: 1989, pp. 209–34.

[R11.10] Boehm, S. *et al* (eds.), *Is There Progress in Economics? Knowledge, Truth and the*

History of Economic Thought, Cheltenham: Edward Elgar, 2002.

[R11.11] Boylan, T. and P. O'Gorman, *Beyond Rhetoric and Realism in Economics: Towards a Reformulation of Economic Methodology*, London: Routledge,1995.

[R11.12] Brousseau, E. and J. Glachant (eds.), *New Institutional Economics: A Guidebook*, Cambridge: Cambridge University Press, 2008.

[R11.13] Brunner, K., "'Assumptions' and the Cognitive Quality of Theories", *Synthese*, Vol. 20, 1969. pp.501–25.

[R11.14] Brzezinski, J., F. Coniglione, R. Kuipers, and L. Nowak (eds.), *Idealization I: General Problems* (Poznan Studies in the Philosophy of the Sciences and Humanities, Vol, 16), Amsterdam: Rodopi, 1990.

[R11.15] Camerer, C., *Behavioral Game Theory: Experiments in Strategic Interaction*, Princeton: Princeton University Press, 2003.

[R11.16] Camerer, C, "Neuroeconomics: Using Neuroscience to Make Economic Predictions", *Economic Journal*, Vol. 117, 2007. pp. C26–C42.

[R11.17] Camerer, C., G. Loewenstein, and M. Rabin (eds.), *Advances in Behavioral Economics*, Princeton: Princeton University Press, 2003.

[R11.18] Camerer, C., G. Loewenstein and D. Prelec, "Neuroeconomics: How Neuroscience Can Inform Economics", *Journal of Economic Literature*, Vol.43, 2005, pp.9–64.

[R11.19] Clark, Andrew and Sarah Flèche, *The Origins of Happiness: The Science of Well-Being over the Life Course*, Princeton: Princeton University Press. 2018.

[R11.20] Cohen, I., N. Daniels, and N.Eyal (eds.), ,*Identified Versus Statistical Lives: An Interdisciplinary Perspective*, New York: Oxford University Press, 2015.

[R11.21] Coleman, J., "Economics and the Law: A Critical Review of the Foundations of the Economic Approach to Law", *Ethics*, Vol. 94, 1984, pp. 649–79.

[R11.22] Davis, J. (ed.),. *Recent Developments in Economic Methodology*, Cheltenham: Edward Edgar 2006.

[R11.23] De Alessi, L., "Reversals of Assumptions and Implications", *Journal of Political Economy*, Vol. 79 1971, pp.867–77.

[R11.24] Deaton, A "Instruments, Randomization, and Learning about Development", *Journal of Economic Literature*, Vol. 48, 2010, pp. 424 - 55.

[R11.25] De Marchi, D. and M. Blaug (eds.),. *Appraising Modern Economics: Studies in the Methodology of Scientific Research Programs*, Cheltenham: Edward Elgar,1991.

[R11.26] Diesing, P., *Science and Ideology in the Policy Sciences*, New York: Aldine, 1982.

[R11.27] Dolan, Paul and Daniel Kahneman , *Happiness by Design: Change What You Do, Not How You Think* New York: Avery Press, 2014.

[R11.28] Engelskirchen, H., *Capital as a Social Kind: Definitions and Transformations in the Critique of Political Economy*, New York: Routledge, 2011.

[R11.29] Epstein, Brian, *The Ant Trap: Rebuilding the Foundations of the Social Sciences*, New York: Oxford University Press, 2015.

[R11.30] Fox, G., *Reason and Reality in the Methodologies of Economics*, Cheltenham: Edward Elgar 1997.

[R11.31] Frey, B., *Economics as a Science of Human Behaviour: Towards a New Social Science Paradigm*, 2nd edition, Dordrecht: Kluwer, 1999.

[R11.32] Gaus, G., *On Philosophy, Politics, and Economics*, Belmont, CA: Wadsworth, 2008.

[R11.33] George, D. (ed.), *Issues in Heterodox Economics*, *Journal of Economic Surveys*, Volume 21, No. 3, Malden, MA and Oxford: Blackwell, 2007.

[R11.34 Glimcher, P., *Foundations of Neuroeconomic Analysis.* New York: Oxford University Press, 2010.

[R11.35] Glimcher, P., C. Camerer, R. Poldrack, and E. Fehr (eds.), *Neuroeconomics: Decision Making and the Brain*, Amsterdam: Elsevier, 2009.

[R11.36] Gustafsson, B., C. Knudsen, and U. Mäki (eds.), *Rationality, Institutions and Economic Methodology*, London: Routledge, 1993.

[R11.37] Henderson, W., T. Dudley-Evans, and R. Backhouse (eds.), *Economics and Language*, London: Routledge, 1993.

[R11.38] Hodgson, G., "What Is the Essence of Institutional Economics?" *Journal of Economic Issues*, Vol. 34, 2000, pp.317–29.

[R11.39] Hutchison, T., *Knowledge and Ignorance in Economics*, Chicago: University of Chicago Press1977.

[R11.40] Jalladeau, J., "Research Program versus Paradigm in the Development of Economics",*Journal of Economic Issues*, Vol. 12, 1978, pp. 583–608.

[R11.41] Katouzian, H., *Ideology and Method in Economics*, New York: New York University Press,1980.

[R11.42] Kincaid, H., *Philosophical Foundations of the Social Sciences: Analyzing Controversies in Social Research*, Cambridge: Cambridge University Press,1996.

[R11.43] Knight, F., 1935. "Economics and Human Action", from Knight 1935b; reprinted in Hausman, ed. 2008b, pp. 111–118.

[R11.44] Koopmans, T., *Three Essays on the State of Economic Science*, New York: McGraw-Hill,1957.

[R11.45] Krupp, S. (ed.), *The Structure of Economic Science*, Englewood Cliffs: Prentice-Hall 1966.

[R11.46] Kuipers, T. (ed.), *What is Closer-to-the Truth? A Parade of Approaches to Truthlikeness*(Poznan Studies in the Philosophy of the Sciences and Humanities, Volume 10), Amsterdam: Rodopi, 1987.

[R11.47] Lavoie, D. (ed.), *Economics and Hermeneutics*, London: Routledge, 1990.

[R11.48] Lawson, T., *Economics and Reality*, London: Routledge, 1997.

[R11.49] Lee, F., and B. Cronin (eds.), *Handbook of Research Methods and Applications in Heterodox Economics*, Cheltenham, UK: Elgar 2016.

[R11.50] Leontief, W., "Theoretical Assumptions and Non-observed Facts", *American Economic Review*, Vol. 61,1971. Pp. 1–7.

[R11.51] Loasby, B., *Choice, Complexity and Ignorance*, Cambridge: Cambridge University Press, 1976.

[R11.52] Loewenstein, G., S. Rick, and J.D. Cohen, "Neuroeconomics", *Annual Review of Psychology*, Vol. 59, 2008, pp. 647–672.

[R11.53] McCloskey, D., *The Rhetoric of Economics*, Madison: University of Wisconsin Press, 1985.

[R11.54] Marchionni, C. and J. Vromen (eds.), *Neuroeconomics: Hype or Hope?*, London, Routledge, 2014.

[R11.55] Mayer, T., *Truth Versus Precision in Economics*, Cheltenham: Edward Elgar, 1993.

[R11.56] Mises, L. von, *The Ultimate Foundation of Economic Science: An Essay on Method*,

2nd edition, Kansas City: Sheed Andrews, 1978.

[R11.57] Mishan, E., *Cost Benefit Analysis: An Introduction*, New York: Praeger, 1971.

[R11.58] Mishan, E, "Imagination and Imaging in Model Building", *Philosophy of Science*, Vol. 71, 2004, pp. 753–66.

[R11.59] Nagel, E., "Assumptions in Economic Theory", *American Economic Review Papers and Proceedings*, Vol. 53,1963, pp. 211–19.

[R11.60] Neuberg, L., *Conceptual Anomalies in Economics*, Cambridge: Cambridge University Press, 1988.

[R11.61] Oakley, A., *Reconstructing Economic Theory: The Problem of Human Agency*, Cheltenham, Edward Elgar, 2002.

[R11.62] Ochangco, A., *Rationality in Economic Thought.: Methodological Ideas on the History of Political Economy*, Cheltenham: Edward Elgar, 1999.

[R11.63] O'Sullivan, P., *Economic Methodology and Freedom to Choose*, London: Allen & Unwin, 1987.

[R11.64] Pratten, S. (ed.), *Social Ontology and Modern Economics*, London: Routledge, 2015.

[R11.65] Rabin, M., "Psychology and Economics", *Journal of Economic Literature*, Vol. 36,1998, pp. 11–46.

[R11.66] Radcliff, Benjamin, *The Political Economy of Human Happiness*, Cambridge: Cambridge University Press 2013.

[R11.67] Redman, D., *Economic Methodology: A Bibliography with References to Works in the Philosophy of Science, 1860–1988*, New York: Greenwood Press, 1989.

[R11.68] Reuter, M. and C. Montag (eds.), *Studies in Neuroscience, Psychology and Behavioral*

Economics, New York: Springer, 2016.

[R11.69] Richter, R. (ed.), *Essays on New Institutional Economics.* New York: Springer, 2015.

[R11.70] Ross, D., *Economic Theory and Cognitive Science: Microexplanation*, Cambridge, MA: MIT Press, 2005.

[R11.71] Schlefer, J., *The Assumptions Economists Make*, Cambridge, MA: Harvard University Press, 2012.

[R11.72] Sen, A., and B. Williams (eds.), *Utilitarianism and Beyond*, Cambridge: Cambridge University Press1982.

[R11.73] Shrader-Frechette, K., *Science Policy, Ethics, and Economic Methodology: Some Problems of Technology Assessment and Environmental-Impact Analysis*, Dordrect: D. Reidel,1984.

[R11.74] Simon, H., "Theories of Decision-Making in Economics and Behavioral Science", *American Economic Review*, 49: 1959. 253–83.

[R11.75] Sowell, T., *Knowledge and Decisions*, New York: Basic Books,1980.

[R11.76] Vromen, J., *Economic Evolution: An Inquiry into the Foundations of Institutional Economics*, London: Routledge.1995.

[R11.77] Vromen, J. and C. Marchionni (eds.), *Neuroeconomics*, London: Routledge, 2018.

[R11.78] Weimann, Joachim and Andreas Knabe, *Measuring Happiness: The Economics of Well-Being*, Cambridge, MA: MIT Press,2015.

[R11.79] Wilson, D., and A. Kirman (eds.), *Complexity and Evolution: Toward a New Synthesis for Economics*, Cambridge, MA: MIT Press,2016.

R12. PHILOSOPHY OF INFORMATION AND SEMANTIC INFORMATION

[R12.1] Aisbett, J., Gibbon, G.: "A practical measure of the information in a logical theory" *Journal of Experimental and Theoretical Artificial Intelligence* Vol. 11(2), 1999, pp. 201–218.

[R12.2] Badino, M.: "An Application of Information Theory to the Problem of the Scientific Experiment" *Synthese* Vol. 140, 2004, pp.355–389.

[R12.3] Bar-Hillel, Y. (ed.), *Language and Information: Selected Essays on Their Theory and Application,* Reading, Addison-Wesley, 1964.

[R12.4] Bar-Hillel, Y., Carnap, R. "An Outline of a Theory of Semantic Information," (1953); in Bar-Hillel, Y. (ed.), *Language and Information: Selected Essays on Their Theory and Application,* Reading, Addison-Wesley, 1964, pp. 221–274.

[R12.5] Barwise, J., Seligman, J.: *Information Flow: The Logic of Distributed Systems*, Cambridge, University Press, Cambridge 1997.

[R12.6] Braman, S.: "Defining Information," *Telecommunications Policy*, Vol. 13, 1989, pp. 233–242.

[R12.7] Bremer, M.E.: "Do Logical Truths Carry Information?" *Minds and Machines* Vol. 13(4), 2003, pp. 567–575.

[R12.8] Bremer, M. and Cohnitz, D., *Information and Information Flow: An Introduction*, Ontos Verlag, Frankfurt, Lancaster, 2004.

[R12.9] Chaitin, G.J., *Algorithmic Information Theory*, Cambridge, Cambridge University Press, 1987.

[R12.10] Chalmers, D.J.: *The Conscious Mind: In Search of a Fundamental Theory*, New York, Oxford Univ. Press, (1996).

[R12.11] Cherry, C.: *On Human Communication: A Review, a Survey, and a Criticism*, Cambridge, MIT Press, 1978.

[R12.12] Colburn, T.R., *Philosophy and Computer Science*, Armonk, M.E. Sharpe, 2000.

[R12.13] Cover, T.M., Thomas, J.A.: *Elements of Information Theory*, New York, Wiley, 1991.

[R12.14] Dennett, D.C.: "Intentional Systems," *The Journal of Philosophy,* Vol. 68, 1971, pp. 87–106.

[R12.15] Deutsch, D., *The Fabric of Reality*, London Penguin, 1997.

[R12.16] Devlin, K.J., *Logic and Information.* Cambridge, Cambridge University Press, 1991.

[R12.17] Fetzer, J.H., "Information, Misinformation, and Disinformation" *Minds and Machines*, Vol. 14(2), 2004, pp. 223–229.

[R12.18] Florida, L., *Philosophy and Computing: An Introduction*, London, Routledge, (1999)

[R12.19] Floridi, L., "What Is the Philosophy of Information?" *Metaphilosophy*, Vol. 33(1-2) (2002), pp. 123–145,

[R12.20] Floridi, L., "Two Approaches to the Philosophy of Information," *Minds and Machines,* Vol. 13(4), (2003), pp. 459–469.

[R12.21] Floridi, L.: "Open Problems in the Philosophy of Information," *Metaphilosophy* Vol. 35(4), 2004, pp. 554–582.

[R12.22] Floridi, L., "Outline of a Theory of Strongly Semantic Information," *Minds and Machines,* Vol. 14(2), 2004, pp. 197–222.

[R12.23] Floridi, L., "Is Information Meaningful Data?," *Philosophy and Phenomenological Research,* Vol. 70(2), 2005, pp. 351–370.

[R12.24] Fox, C.J.: *Information and Misinformation: An Investigation of the Notions of Information, Misinformation, Informing, and Misinforming,* Westport Greenwood

Press, 1983.

[R12.25] Frieden, B.R., *Science from Fisher Information: A Unification*, Cambridge University Press, Cambridge, 2004.

[R12.26] Golan, A., "Information and Entropy Econometrics - Editor's View". Journal of Econometrics, Vol. 107(1-2), 2002, pp.1–15.

[R12.27] Graham, G., *The Internet: A Philosophical Inquiry.* , London, Routledge, 1999.

[R12.28] Grice, H.P. *Studies in the Way of Words,* Cambridge, Harvard University Press, 1989.

[R12.29] Hanson, P.P. (ed.), *Information, language, and cognition*, Vancouver, University of British Columbia Press, 1990.

[R12.30] Harms, W.F., "The Use of Information Theory in Epistemology," *Philosophy of Science*, Vol. 65(3), 472–501 (1998).

[R12.31] Heil, J., "Levels of Reality," *Ratio* Vol. 16(3), 2003, pp. 205–221.

[R12.32] Hintikka, J., Suppes, P. (eds.), *Information and Inference*, Reidel, Dordrecht, 1970.

[R12.33] Kemeny, J., "A Logical Measure Function," *Journal of Symbolic Logic* Vol. 18, 1953, pp. 289–308.

[R12.34] Kolin K.K. "The Nature of Information and Philosophical Foundations of Informatics". *Open Education,* Vol.*2*, (2005) pp. 43-51.

[R12.35] Kolin K.K.). "The Evolution of Informatics," *Information Technologies,* Vol.1, 2005, pp. 2-16.

[R12.36] Kolin K.K., "The Formation of Informatics as Basic Science and Complex Scientific Problems," In K.Kolin (Ed.), *Systems and Means of Informatics. Special Issue. Scientific and Methodological Problems of Informatics.* Moscow: IPI RAS, 2006, Pp. 7-57.

[R12.37] Kolin K.K., Fundamental Studies in Informatics: A General Analysis, Trends and Prospects. *Scientific and Technical Information,* Vol.1, (7), 2007, pp. 5-11.

[R12.38] Kolin K.K. "Structure of Reality and the Phenomenon of Information," *Open Education,* Vol. 5, 2008, pp. 56-61.

[R12.39] Losee, R.M., "A Discipline Independent Definition of Information," *Journal of the American Society for Information Science* Vol. 48(3), 1997, pp. 254–269.

[R12.40] Lozinskii, E. "Information and evidence in logic systems," *Journal of Experimental and Theoretical Artificial Intelligence*, Vol. 6, 1994, pp. 163–193.

[R12.41] Machlup, F., Mansfield, U. (eds.), *The Study of Information: Interdisciplinary Messages*, New York, Wiley, 1983.

[R12.42] MacKay, D.M., *Information, Mechanism and Meaning*, Cambridge MIT Press, 1969.

[R12.43] Marr, D., Vision: *A Computational Investigation into the Human Representation and Processing of Visual Information*. San Francisco, W.H. Freeman, 1982.

[R12.44] Mingers, J., "The Nature of Information and Its Relationship to Meaning," In: Winder, R.L., et al. (eds.) *Philosophical Aspects of Information Systems*, London, Taylor and Francis, 1997, pp. 73–84.

[R12.45] Nauta, D., *The Meaning of Information*, The Hague, Mouton, 1972.

[R12.46] Newell, A. "The Knowledge Level". *Artificial Intelligence* Vol. 18, 1982, pp. 87–127.

[R12.47] Newell, A., Simon, H.A., "Computer Science as Empirical Inquiry: Symbols and Search," *Communications of the ACM*, Vol. 19, 1976 pp.113–126.

[R12.48] Pierce, J.R., *An Introduction to Information Theory: Symbols, Signals and Noise*, New York, Dover Publications, 1980.

[R12.49] Poli, R., "The Basic Problem of the Theory of Levels of Reality" *Axiomathes*,

Vol. 12, 2001, pp. 261–283.

[R12.50] Sayre, K.M.: *Cybernetics and the Philosophy of Mind*. London, Routledge and Kegan Paul, (1976).

[R12.51] Simon, H.A., *The Sciences of the Artificial*, Cambridge, MIT Press, 1996.

[R12.52] Smokler, H., "Informational Content: A Problem of Definition," *The Journal of Philosophy,* Vol. 63(8), 1966, pp. 201–211.

[R12.53] Ursul A.D., *The Nature of the Information. Philosophical Essay*. Moscow: Politizdat, (1968).

[R12.54] Ursul A.D., *Information. Methodological Aspects*. Moscow: Nauka, 1971.

[R12.55] Ursul A.D., *Reflection and Information.* Moscow: Nauka (1973).

[R12.56] Ursul A.D. *The Problem of Information in Modern Science: Philosophical Essays.* Moscow: Nauka. (1975).

[R12.57] Ursul, A. D. "The Problem of the Objectivity of Information" in Libor Kubát íř, and Jiří Zeman (eds.), *Entropy and Information in Science and Philosophy,* New York, Elsevier, 1975.

[R12.58] Weaver, W., "The Mathematics of Communication," *Scientific American* Vol. 181(1), 1949, pp. 11–15.

[R12.59] Winder, R.L., Probert, S.K., Beeson, I.A.: *Philosophical Aspects of Information Systems*, London, Taylor & Francis 1997.

R13 PLANNING, PRESCRIPTIVE SCIENCE AND INFORMATION IN COST-BENEFIT ANALYTICS

[R13.1] Alexander, Ernest R., *Approaches to Planning*, Philadelphia, Pa. Gordon and Breach, 1992.

[R13.2] Bailey, J., *Social Theory for Planning*, London, Routledge and Kegan Paul, 1975.

[R13.3] Burchell, R.W. and G. Sternlieb (eds.), *Planning Theory in the 1980's: A Search for Future Directions*, New Brunswick, N. J., Rutgers University Center for Urban and Policy Research, 1978.

[R13.4] Cambis, Marios, *Planning Theory and Philosophy*, London, Tavistock Publicationa, 1979.

[R13.5] Chadwick, G., *A Systems View of Planning,* Oxford, Pergamon, 1971.

[R13.6] Cooke, P., *Theories of Planning and Special Development*, London, Hutchinson,1983

[R13.7] Dompere, Kofi K., and Taresa Lawrence, "Planning," in Syed B Hussain, *Encyclopedia of Capitalism,* Vol. II, New York, Facts on File, Inc., 2004, pp. 649-653.

[R13.8] Dompere, Kofi K., *Social Goal-Objective Formation, Democracy and National Interest: A Theory of Political Economy under Fuzzy Rationality*, (Studies in Systems, Decision and Control, Vol. 4), New York, Springer, 2014.

[R13.9] Dompere, Kofi K., *Fuzziness, Democracy, Control and Collective Decision-Choice System: A Theory on Political Economy of Rent-Seeking and Profit-Harvesting,* (Studies In Systems, Decision and Control, Vol. 5), New York, Springer, 2014.

[R13.10] Dompere, Kofi K., *The Theory of Aggregate Investment in Closed Economic Systems*, Westport, CT, Greenwood Press, 1999.

[R13.11] Dompere, Kofi K., *The Theory of Aggregate Investment and Output Dynamics in Open Economic Systems*, Westport, CT, Greenwood Press, 1999.

[R13.12] Faludi, A., *Planning Theory*, Oxford, Pergamon, 1973.

[R13.13] Faludi, A.(ed.), *A Reader in Planning Theory,* Oxford, Pergamon, 1973.

[R13.14] Harwood, E.C. (ed.), Reconstruction of Economics, American Institute For Economic Research, Great Barrington, Mass, 1955., Also in John Dewey and Arthur Bently, 'Knowing and the known', Boston, Beacon Press, 1949, p. 269.

[R13.15] Kickert, W.J.M., *Organization of Decision-Making A Systems-Theoretic Approach*, New York, North-Holland, 1980.

[R13.16] Knight, Frank H. *Risk, Uncertainty and Profit,* Chicago, University of Chicago Press, 1971.

[R13.17] Knight, Frank H. *On History and Method of Economics,* Chicago, University of Chicago Press, 1971.

R.14. POSSIBLE-ACTUAL WORLDS AND INFORMATION ANALYTICS

[R14.1] Adams, Robert M., "Theories of Actuality," *Noûs*, Vol. 8, 1974, pp211-231.

[R14.2] Allen, Sture (ed.) *Possible Worlds in Humanities, Arts and Sciences*, Proceedings of Nobel Symposium, Vol. 65, New York, Walter de Gruyter Pub., 1989.

[R14.3] Armstrong, D. M., *A Combinatorial Theory of Possibility.* Cambridge University Press, 1989.

[R14.4] Armstrong, D.M *A World of States of Affairs*, Cambridge, Cambridge University Press 1997.

[R14.5] Bell, J.S., "Six Possible Worlds of Quantum Mechanics" in Allen, Sture (Ed.) *Possible Worlds in Humanities, Arts and Sciences*, Proceedings of Nobel Symposium, Vol. 65, New York, Walter de Gruyter Pub., 1989.pp. 359-373.

[R14.6] Bigelow, John. "Possible Worlds Foundations for Probability", *Journal of Philosophical Logic,* 5 (1976), pp. 299-320.

[R14.7] Bradley, Reymond and Norman Swartz, *Possible World: An Introduction to Logic and its Philosophy,* Oxford, Bail Blackwell, 1997.

[R14.8] Castañeda, H.-N. "Thinking and the Structure of the World", *Philosophia*, 4 (1974), pp. 3-40.

[R14.9] Chihara, Charles S. *The Worlds of Possibility: Modal Realism and the Semantics of Modal Logic*, Clarendon, 1998.

[R14.10] Chisholm, Roderick. "Identity through Possible Worlds: Some Questions", *Noûs,* 1 1967, pp. 1-8; reprinted in Loux, *The Possible and the Actual.*

[R14.11] Divers, John, *Possible Worlds*, London: Routledge, 2002.

[R14.12] Forrest, Peter. "Occam's Razor and Possible Worlds", *Monist,* 65 (1982), pp. 456-64.

[R14.13] Forrest, Peter, and Armstrong, D. M. "An Argument Against David Lewis' Theory of Possible Worlds", *Australasian Journal of Philosophy,* 62 (1984), pp. 164-168.

[R14.14] Grim, Patrick, "There is No Set of All Truths", *Analysis*, Vol. 46, 1986, pp. 186-191.

[R14.15] Heller, Mark. "Five Layers of Interpretation for Possible Worlds", *Philosophical Studies*, 90 (1998), pp. 205-214.

[R14.16] Herrick, Paul, *The Many Worlds of Logic,* Oxford: Oxford University Press1999.

[R14.17] Krips, H. "Irreducible Probabilities and Indeterminism", *Journal of Philosophical Logic*, Vol. 18, 1989, pp. 155-172.

[R14.18] Kuhn, Thomas S., "Possible Worlds in History of Science" in Allen, Sture (ed.) *Possible Worlds in Humanities, Arts and Sciences,* Proceedings of Nobel Symposium, Vol. 65, New York, Walter de Gruyter Pub., 1989 pp. 9-41.

[R14.19] Kuratowski, K. and Mostowski, A. *Set Theory: With an Introduction to Descriptive Set Theory*, New York: North-Holland, 1976.

[R14.20] Lewis, David, *On the Plurality of Worlds*, Oxford, Basil Blackwell, 1986.

[R14.21] Loux, Michael J. (ed.) *The Possible and the Actual*: *Readings in the Metaphysics of Modality,* Ithaca & London: Cornell University Press,1979.

[R14.22] Parsons, Terence, *Nonexistent Objects,* New Haven, Yale University Press, 1980.

[R14.23] Perry, John, "From Worlds to Situations", *Journal of Philosophical Logic*, Vol. 15, 1986, pp. 83-107.

[R14.24] Rescher, Nicholas and Brandom, Robert. *The Logic of Inconsistency: A Study in Non-Standard Possible-World Semantics and Ontology*, Rowman and Littlefield, 1979.

[R14.25] Skyrms, Brian. "Possible Worlds, Physics and Metaphysics", *Philosophical Studies,* Vol. 30, 1976, pp. 323-32.

[R14.26] Stalmaker, Robert C. "Possible World", *Noûs*, Vol. 10, 1976, pp. 65-75

[R14.27] Quine, W.V.O. *Word and Object*, M.I.T. Press, 1960.

[R14.28] Quine, W.V.O "Ontological Relativity", *Journal of Philosophy*, Vol. 65, 1968, pp. 185-212.

R15. RATIONALITY, INFORMATION, GAMES, CONFLICTS AND EXACT REASONING

[R15.1] Border, Kim, *Fixed Point Theorems with Applications to Economics and Game Theory,* Cambridge, Cambridge University Press 1985.

[R15.2] Brandenburger, Adam, "Knowledge and Equilibrium Games," *Journal of Economic Perspectives*, Vol.6, 1992, pp. 83-102.

[R15.3] Campbell, Richmond and Lanning Sowden, *Paradoxes of Rationality and Cooperation: Prisoner's Dilemma and Newcomb's Problem*, Vancouver, University of British Columbia Press, 1985.

[R15.4] Gates Scott and Brian Humes, *Games, Information, and Politics: Applying Game Theoretic Models to Political Science*, Ann Arbor, University of Michigan Press, 1996.

[R15.5] Gjesdal, Froystein, "Information and Incentives: The Agency Information Problem," *Review of Economic Studies*, Vol.49, 1982, pp373-390.

[R15.6] Harsanyi, John, "Games with Incomplete Information Played by 'Bayesian' Players I: The Basic Model," *Management Science,* Vol.14, 1967, pp.159-182.

[R15.7] Harsanyi, John, "Games with Incomplete Information Played by 'Bayesian' Players II: Bayesian Equilibrium Points," *Management Science*, Vol.14, 1968, pp.320-334.

[R15.8] Harsanyi, John, "Games with Incomplete Information Played by 'Bayesian' Players III: The Basic Probability Distribution of the Game," *Management Science*, Vol.14, 1968, pp.486-502.

[R15.9] Harsanyi, John, *Rational Behavior and Bargaining Equilibrium in Games and Social Situations*, New York Cambridge University Press, 1977.

[R15.10] Krasovskii, N.N. and A.I. Subbotin, *Game-theoretical Control Problems*, New York, Springer-Verlag, 1988.

[R15.11] Lagunov, V. N., *Introduction to Differential Games and Control Theory*, Berlin, Heldermann Verlag, 1985.

[R15.12] Maynard Smith, John, *Evolution and the Theory of Games,* Cambridge, Cambridge University Press, 1982.

[R15.13] Myerson, Roger, *Game Theory: Analysis of Conflict*, Cambridge, Mass. Harvard University Press, 1991.

[R15.14] Rapoport, Anatol and Albert Chammah, *Prisoner's Dilemma: A Study in Conflict and Cooperation*, Ann Arbor, University of Michigan Press, 1965.

[R15.15] Roth, Alvin E., "The Economist as Engineer: Game Theory, Experimentation, and Computation as Tools for Design Economics," *Econometrica,* Vol.70, 2002, pp1341-1378.

[R15.16] Shubik, Martin, *Game Theory in the Social Sciences: Concepts and Solutions*, Cambridge, Mass., MIT Press, 1982.

R16. SOCIAL SCIENCES, MATHEMATICS AND THE PROBLEMS OF EXACT AND INEXACT INFORMATION

[R16.1] Ackoff, R.L., *Scientific Methods: Optimizing Applied Research Decisions*, New York, John Wiley, 1962.

[R16.2] Angyal, A. "The Structure of Wholes," *Philosophy of Sciences*, Vol.6, #1, 1939, pp 23-37.

[R16.3] Bahm, A.J., "Organicism: The Philosophy of Interdependence" *International Philosophical Quarterly*, Vol. VII # 2, 1967.

[R16.4] Bealer, George, *Quality and Concept*, Oxford, Clarendon Press, 1982.

[R16.5] Black, Max, *Critical Thinking*, Englewood Cliffs, N.J., Prentice-Hall, 1952.

[R16.6] Brewer, Marilynn B., and Barry E Collins (eds.) *Scientific Inquiry and Social Sciences,* San Francisco, Ca, Jossey-Bass Pub., 1981.

[R16.7] Campbell, D.T., "On the Conflicts Between Biological and Social Evolution and Between Psychology and Moral Tradition", *American Psychologist*, Vol. 30, 1975, pp1103-1126.

[R16.8] Churchman, C. W. and P. Ratoosh (eds.) *Measurement: Definitions and Theories*, New York, John Wiley, 1959.

[R16.9] Foley, Duncan, "Problems versus Conflicts Economic Theory and Ideology" American Economic Association Papers and Proceedings, Vol. 65, May 1975, pp. 231-237.

[R16.10] Garfinkel, Alan, *Forms of Explanation: Structures of Inquiry in Social Science,* New Haven, Conn., Yale University Press, 1981.

[R16.11] Georgescu-Roegen, Nicholas, *Analytical Economics*, Cambridge, Harvard University Press, 1967.

[R16.12] Gillespie, C., *The Edge of Objectivity*, Princeton, Princeton University press, 1960.

[R16.13] Hayek, F.A., *The Counter-Revolution of Science*, New York, Free Press of Glencoe Inc, 1952.

[R16.14] Laudan, L., *Progress and Its Problems: Towards a Theory of Scientific Growth,* Berkeley, CA, University of California Press, 1961.

[R16.15] Marx, Karl, *The Poverty of Philosophy,* New York, International Pub. 1971.

[R16.16] Miller, John W., *The Philosophy of History with Reflection and Aphorisms,* Norton, New York, 1951.

[R16.17] Phillips, Denis C., *Holistic Thought in Social Sciences*, Stanford, CA, Stanford University Press, 1976.

[R16.18] Popper, K., *Objective Knowledge*, Oxford, Oxford University Press, 1972.

[R16.19] Rashevsky, N. "Organismic Sets: Outline of a General Theory of Biological and Social Organism," *General Systems*, Vol XII, 1967, pp. 21-28.

[R16.20] Roberts, Blaine, and Bob Holdren, *Theory of Social Process*, Ames, Iowa University Press, 1972.

[R16.21] Rudner, Richard S., *Philosophy of Social Sciences*, Englewood Cliff, N.J., Prentice Hall, 1966.

[R16.22] Simon, H. A., "The Structure of Ill-Structured Problems," *Artificial Intelligence,* Vol. 4, 1973, pp. 181-201.

[R16.23] Toulmin, S., *Foresight and understanding: An Enquiry into the Aims of Science,* New York, Harper and Row, 1961.

[R16.24] Winch, Peter, *The Idea of a Social Science*, New York, Humanities Press, 1958.

17 TRANFORMATIONS, DECISIONS, POLARITY, DUALITY AND CONFLICTS

[R17.1a] Abraham, W.E., *The Mind of Africa,* Chicago, The University of Chicago Press, 1962.

[R17.1b] Amen, Ra Un Nefer, *Metu Neter Vol.1*, Khamit Corp., Bronx, NY, 1990.

[R17.2a] Anovsky, 0mely M.E., *Linin and Modern Natural Science, Moscow*, Progress Pub. 1978.

[R17.2b] Arrow, Kenneth J., "Limited Knowledge and Economic Analysis", American Economic Review, Vol. 64, 1974, pp. 1-10.

[R17.3a] Asante, Molefi K, *Revolutionary Pedagogy,* Chicago African American Image Pub.,

[R17.3b] Asante, Molefi K, *Afrocentricity:The Theory of Social Change*, African American Image. Pub. Chicago, 2003.

[R17.4a] Ben-Jochannan, Yosef A, A., *Cultural Genocide in the Black and African Studies Curriculum,* Baltimore, U.S.A..,Black Classic Press,2004.

[R17.4b] Ben-Jochannan, Yosef A, A., *Africa: Mother of Western Civilization,* Baltimore, MD, U.S.A., Black Classic Press, 1997.

[R17.5a] Berkeley, George, *Treatise Concerning the Principles of Human Knowledge*, *Works*, Vol. I (edited by A. Fraser), Oxford, Oxford University Press, 1871-1814.

[R17.5b] Berkeley, George, "Material Things are Experiences of Men or God" in [R1.5], 1967, pp. 658-668.

[R17.6a] Boulding, Kenneth E., *A New Theory of Societal Evolution,* Sage Pub. Beverly Hills Ca, 1978.

[R17.6b] Brody, Baruch A. (ed.), *Readings in the Philosophy of Science*, Englewood Cliffs, NJ., Prentice-Hall Inc., 1970.

[R17.7a] Brouwer, L.E.J., "Consciousness, Philosophy, and Mathematics," in Benecerraf, P. and H. Putnam (eds.), *Philosophy of Mathematics: Selected Readings* , Cambridge, Cambridge University Press, 1983, pp. 90-96.

[R17.7b] Brown, B. and J Woods (eds.), *Logical Consequence; Rival Approaches and New Studies in exact Philosophy: Logic, Mathematics and Science*, Vol. II Oxford, Hermes, 2000.

[R17.8] Cornforth, Maurice, *Dialectical Materialism and Science*, New York, International Pub.1960.

[R17.9] Cornforth, Maurice, *Materialism and Dialectical Method*, New York, International Pub.1953.

[R17.10] Cornforth, Maurice, *Science and Idealism: An Examination of "Pure Empiricism"*, New York International Pub.1947.

[R17.11a] Cornforth, Maurice, *The Open Philosophy and the Open Society: A Reply to Dr. Karl Popper's Refutations of Marxism* New York, International Pub.1968.

[R17.11b] Cornforth, Maurice, *The Theory of Knowledge*, New York, International Pub.1960.

[R17.12a] Diop, Cheikh A., *The African Origins of Civilization: Myth or Reality*, Brooklyn, New York, Lawrence Hill, 1974.

[R17.12b] Diop, Cheikh A., *Civilization or Barbarism*, Brooklyn, New York, Lawrence Hill, 1991.

[R17.13] Dompere, Kofi K., "On Epistemology and Decision-Choice Rationality" in R. Trapple (ed.), *Cybernetics and System Research*, New York, North Holland, 1982, pp. 219-228.

[R17.14] Dompere, Kofi K. and M. Ejaz, *Epistemics of Development Economics: Toward a Methodological Critique and Unity*, Westport, CT, Greenwood Press, 1995.

[R17.15] Dompere, Kofi K., *The Theory of Categorial Conversion: Rational Foundations of Nkrumaism in socio-natural Systemicity and Complexity,* London, Adonis-Abbey Pubs. 2016-2017.

[R17.16] Dompere, Kofi K., *The Theory of Philosophical Consciencism: Practice Foundations of Nkrumaism in Social Systemicity,,* London, Adonis-Abbey Pubs. 2016-2017.

[R17.17] Dompere, Kofi K., *The Theory of Info-Statics: Conceptual Foundations of Information and Knowledge*, New York, Springer (Series: Studies in Systems, Decision and Control, Vol. 112), 2017.

[R17.18] Dompere, Kofi K., *The Theory of Info-Dynamics: Rational Foundations of Information-Knowledge Dynamics,* New York, Springer (Series: Studies in Systems, Decision and Control, Vol. 114), 2017.

[R17.19a] Dompere, Kofi K., *Polyrhythmicity: Foundations of African Philosophy*, London, Adonis, and Abbey Pub, 2006.

[R17.19b] Dubois, W. E. B., *Dubois Speaks: Speeches and Addresses 1890-1919*, New York, Pathfinder,1970.

[R17.20] Engels, Frederick, *Dialectics of Nature*, New York, International Pub.,1971

[R17.21] Engels, Frederick, *Origin of the Family, Private Property and State*, New York, International Pub.,1971.

[R17.22] Ewing, A.C., "A Reaffirmation of Dualism" in [R1.5], pp. 454-461.

[R17.23] Fedoseyer, P.N. *et al., Philosophy in USSR: Problems of Dialectical Materialism*, Moscow, Progress Pub., 1977.

[R17.24] Kedrov, B. M., "On the Dialectics of Scientific Discovery," *Soviet Studies in Philosophy*, Vol. 6 1967, pp16 – 27.

[R17.25] Lenin, V. I. *Materialism and Empirio-Criticism: Critical Comments on Reactionary Philosophy*, New York, International Pub., 1970.

[R17.26] Lenin, V. I. *Collected Works Vol. 38: Philosophical Notebooks*, New York, International Pub., 1978.

[R17.27] Lenin, V. I., *On the National Liberation Movement*, Peking, Foreign Language Press, 1960.

[R17.28] Hegel, George, *Collected Works*, Berlin, Duncher und Humblot, 1832 – 1845 [also

Science of Logic, translated by W. H. Johnston and L. G. Struther , London, 1951].

[R17.29] Hempel, Carl G., and P. Oppenheim, "Studies in the Logic of Explanation," in [R15.5], pp. 8 – 27.

[R17.30] Ilyenkov, E.V., *Dialectical Logic: Essays on its History and Theory*, Moscow, Progress Pub. 1977.

[R17.31] Keirstead, B.S., "The Conditions of Survival," American Economic Review, Vol. 40, #2, pp.435- 445.

[R17.32] Kühne, Karl, *Economics and Marxism, Vol.I: The Renaissance of the Marxian System*, New York, St Martin's Press, 1979.

[R17.33] Kühne, Karl, *Economics and Marxism, Vol.II: The Dynamics of the Marxian System*, New York, St Martin's Press, 1979.

[R17.34] March, J. C., "Bounded Rationality, Ambiguity and Engineering of Choice," *The Bell Journal of Economics*, Vol. 9 (2), 1978.

[R17.35] Marx, Karl, *Contribution to the Critique of Political Economy*, Chicago, Charles H. Kerr and Co. 1904.

[R17.36] Marx, Karl, *Economic and Philosophic Manuscripts of 1884,* Moscow, Progress Pub., 1967.

[R17.37] Marx, Karl, *The Poverty of Philosophy*, New York, International Publishers, 1963.

[R17.38] Marx, Karl, Economic and Philosophic Manuscripts of 1844, Moscow, Progress Pub, 1963.

[R17.39] Massey, Gerald, *A Book of the Beginnings*, Vols. 1-2, London, William and Norgate, 1881.

[R17.40] Massey, Gerald, *Pyramid Text*, Vols. 1-4, New York, Longmans Green, 1952.

[R17.41] Massey, Gerald, *The Natural Genesis*, Vols. 1 and 2, Baltimore, MD, Black Classic Press, 1998 (First published 1883).

[R17.42] Massey, Gerald, *Ancient Egypt: The Light of the World*, Baltimore, MD, Black Classic Press, 1992.

[R17.43] Niebyl, Karl, H., "Modern Mathematics and Some Problems of Quantity, Quality and Motion in Economic Analysis," *Philosophy of Science*, Vol 7, # 1, January,1940, pp. 103 – 120.

[R17.44] Nkrumah, Kwame, *Consciencism*, London, Heinemann, 1964.

[R17.45] Obenga, Theophile T, *African Philosophy During The Period Of The Pharaohs, 2800-300BC*, Popenguine, Senegal, W.A., Per Ankh Pub., 2006.

[R17.46] Obenga, Theophile T, *African Philosophy in World History,* Popenguine, Senegal, W.A., Per Ankh Pub., 1998.

[R17.47] Price, H. H., *Thinking and Experience*, London, Hutchinson, 1953.

[R17.48] Putman, H., *Reason, Truth and History*, Cambridge, Cambridge University Press, 1981.

[R17.49] Putman. H., *Realism and Reason*, Cambridge, Cambridge University Press, 1983.

[R17.50] Quiggin, John, *Zombie Economics: How Dead Ideas Still Walk Among Us, Princeton* University Press, Prince ton, NJ. 2010.

[R17.51] Robinson, Joan, *Economic Philosophy.* New York, Anchor Books, 1962.

[R17.52] Robinson, Joan, *Freedom and Necessity: An Introduction to the Study of Society,* New York, Vintage Books, 1971.

[R17.53] Robinson, Joan, *Economic Heresies: Some Old-Fashioned Questions in Economic Theory,* New York, Basic Books, 1973.

[R17.54] Schumpeter, Joseph A., *The Theory of Economic Development*, Cambridge, Mass. Harvard University Press, 1934.

[R17.55] Schumpeter, Joseph A., *Capitalism, Socialism and Democracy*, New York, Harper & Row, 1950.

[R17.56] Schumpeter, Joseph A., "March to Socialism," *American Economic Review,* Vol. 40 May 1950, pp. 446 456.

[R17.57] Schumpeter, Joseph A., "Theoretical Problems of Economic Growth" *Journal of Economic History* Vol. 8, Supplement 1947, pp. 1-9.

[R17.58] Schumpeter, Joseph A., "The Analysis of Economic Change," *Review of Economic Statistics*, Vol. 17, 1935, pp 2-10.

[R17.59] Schwaller de Lubicz, R.A., *The Egyptian Miracle: An Introduction to the Wisdom of the Temple*, Rochester (Vermont) Inner Traditions International,1985.

[R17.60] Schwaller de Lubicz, R.A., *A Study of Numbers: A Guide to The constant creation of The Universe*, Rochester, (Vermont) Inner Traditions International, 1986.

[R17.61] Schwallier de Lubicz, R.A., *The Temple in Man: The Secrets of Ancient Egypt*, Brookline,

(Massachusetts), Autumn Press, 1977.

[R17.62] Schaller de Lubicz, R.A., *The Temple In Man: Sacred Architecture and The Perfect Man.*, Rochester, (Vermont), Inner Traditions Published, 1981.

[R17.63] Schwaller de Lubicz, R.A., *Symbol and the Symbolic: Egypt, Science, and The Evolution of Consciousness*, Brookline, (Massachusetts) Autumn Press, 1978.

[R17.64] Schwaller de Lubicz, R.A., *The Temple of Man: Apet of The South at Luxor*, Rochester, (Vermont) Inner Traditions, 1998.

[R17.65] Wright, Richard A. (ed.), *African Philosophy: An Introduction*, New York, University Press of America, 1984.

[R17.66] Weber, Luc E., and James J. Duderswtadt (Eds.), Reinventing the Research University, Economica, London, 2004.

[R17.67] Welsing, Frances, C., *The Isis Papers The Key to the Colors*, Chicago, Third World Pres, 1991.

[R17.68] Woodson, Carter G, *Miseducation of the Negro,* Washington, D.C, The Associated Pub.1969.

[R17.69] Yesufu, T. M. (ed.) *Creating the African University: Emerging Issues of the 1970s*, Oxford, Oxford University Press, 1973.

R18. VAGUENESS, THINKING, APPROXIMATION AND REASONING IN THE INFORMATION-KNOWLEDGE PROCESS

[R18.1] Adams, E. w., and H. F. Levine, "On the Uncertainties Transmitted from Premises to Conclusions in deductive Inferences," *Synthese*, Vol. 30, 1975, pp. 429 – 460.

[R18.2] Arbib, M. A., *The Metaphorical Brain*, New York, McGraw-Hill, 1971.

[R18.3] Bečvář, Jiří , "Notes on Vagueness and Mathematics," in Skala, Heinz J. et al., (eds.), *Aspects of Vagueness*, Dordrecht, D. Reidel Co. 1984, pp.1-11.

[R18.4] Black, M, "Vagueness: An Exercise in Logical Analysis," *Philosophy of Science*, Vol. 17, 1970, pp141-164.

[R18.5] Black, M. "Reasoning with Loose Concepts," *Dialogue*, Vol. 2, 1973, pp. 1-12.

[R18.6] Black, Max, *Language and Philosophy*, Ithaca, N.Y.: Cornell University Press, 1949.

[R18.7] Black, Max, *The Analysis of Rules*, in Black, Max [R18.8] *Models and Metaphors: Studies in Language and Philosophy*, Ithaca, New York: Cornell University Press 1962 pp. 95-139.

[R18.8] Black, Max, *Models and Metaphors: Studies in Language and Philosophy*, Ithaca, NewYork: Cornell University Press. 1962.

[R18.9] Black, Max, *Margins of Precision*, Ithaca: Cornell University Press,1970.

[R18.10] Boolos, G. S. and R. C. Jeffrey, *Computability and Logic,* New York, Combridge University Press, 1989.

[R18.11] Cohen, P. R., *Heuristic Reasoning about uncertainty: An Artificial Intelligent Approach*, Boston, Pitman, 1985.

[R18.12] Darmstadter, H., "Better Theories," *Philosophy of Science*, Vol. 42, 1972, pp. 20 – 27.

[R18.13] Davis, M., *Computability and Unsolvability*, New York, McGraw-Hill, 1958.

[R18.14] Dummett, M., "Wang's Paradox," *Synthese,* Vol. 30, 1975, pp301 – 324.

[R18.15] Dummett, M., *Truth and Other Enigmas*, Cambridge, Mass. Harvard University Press, 1978.

[R18.16] Endicott, Timothy, *Vagueness in the Law*, Oxford, Oxford University Press, 2000.

[R18.17] Evans, Gareth, "Can there be Vague Objects?" *Analysis,* Vol. 38, 1978, p. 208.

[R18.18] Fine, Kit, "Vagueness, Truth and Logic," *Synthese,* Vol.54, 1975, pp. 235-259.

[R18.19] Gale, S., "Inexactness, Fuzzy Sets and the Foundation of Behavioral Geography," *Geographical Analysis*, Vol.4, #4, 1972, pp.337-349.

[R18.20] Ginsberg, M. L. (ed.), *Readings in Non-monotonic Reason*, Los Altos, Ca., Morgan Kaufman, 1987.

[R18.21] Goguen, J. A., "The Logic of Inexact Concepts," *Synthese*, Vol. 19, 1968/69, pp. 325 – 373.

[R18.22] Grafe, W., "Differences in Individuation and Vagueness," in A. Hartkamper and H. –J. Schmidt, *Structure and Approximation in Physical Theories*, New York, Plenum Press, 1981. pp.113–122.

[R18.23] Goguen, J. A, "The Logic of Inexact Concepts" *Synthese*, Vol.19, 1968-1969.

[R18.24] Graff, Delia and Timothy (eds.), *Vagueness*, Aldershot, Ashgate Publishing, 2002.

[R18.25] HartkäAmper and H.J. Schmidt (eds.), *Structure and Approximation in Physical Theories*, New York, Plenum Press, 1981.

[R18.26] Hersh, H.M. et. al., "A Fuzzy Set Approach to Modifiers and Vagueness in Natural Language," *J. Experimental*, Vol. 105, 1976, pp. 254-276.

[R18.27] Hilpinen, R., "Approximate Truth and Truthlikeness," in M. Pprelecki et al. (eds.) *Formal Methods in the Methodology of Empirical Sciences*, Wroclaw, Reidel, Dordrecht and Ossolineum, 1976 pp. 19 – 42.

[R18.28] Hockney D. et al. (eds.), *Contemporary Research in Philosophical Logic and Linguistic Semantics*, Dordrecht-Holland, Reidel Pub. Co. 1975.

[R18.29] Höhle Ulrich et al (eds.), *Non-Clasical Logics and their Applications to Fuzzy Subsets: A Handbook of the Mathematical Foundations of Fuzzy Set Theory*, Boston, Mass. Kluwer, 1995.

[R18.30] Katz, M., "Inexact Geometry," *Notre-Dame Journal of Formal Logic*, Vol.21, 1980, pp. 521-535.

[R18.31] Katz, M., "Measures of Proximity and Dominance," *Proceedings of the Second World Conference on Mathematics at the Service of Man*, Universidad Politecnica de Las Palmas, 1982, pp. 370 – 377.

[R18.32] Katz, M., "The Logic of Approximation in Quantum Theory," *Journal of Philosophical Logic*, Vol. 11, 1982, pp. 215 – 228.

[R18.33] Keefe, Rosanna, *Theories of Vagueness*, Cambridge, Cambridge University Press, 2000.

[R18.34] Keefe, Rosanna and Peter Smith (eds.) *Vagueness: A Reader*, Cambridge, MIT Press, 1996.

[R18.35] Kling, R., "Fuzzy Planner: Reasoning with Inexact Concepts in a Procedural Problem-solving Language," *Jour. Cybernetics*, Vol. 3, 1973, pp. 1-16.

[R18.36] Kruse, R.E. et. al., *Uncertainty and Vagueness in Knowledge Based Systems: Numerical Methods*, New York, Springer-Verlag 1991.

[R18.37] Ludwig, G., "Imprecision in Physics," in A. Hartk ämper and H.J. Schmidt (eds.), *Structure and Approximation in Physical Theories*, New York, Plenum Press, 1981, pp. 7 – 19.

[R18.38] Kullback, S. and R. A. Leibler, "Information and Sufficiency," *Annals of Math.*

Statistics, Vol. 22, 1951, pp. 79 – 86.

[R18.39] Lakoff, George, "Hedges: A Study in Meaning Criteria and Logic of Fuzzy Concepts," in, Hockney D. et al. (eds.), *Contemporary Research in Philosophical Logic and Linguistic Semantics*, Dordrecht-Holland, Reidel Pub. Co. 1975, pp.221-271.

[R18.40] Lakoff, G., "Hedges: A Study in Meaning Criteria and the Logic of Fuzzy Concepts," *Jour. Philos. Logic,* Vol. 2, 1973, pp. 458-508.

[R18.41] Levi, I., *The Enterprise of Knowledge*, Cambridge, Mass. MIT Press 1980.

[R18.42] Łucasiewicz, J., *Selected Works: Studies in the Logical Foundations of Mathematics,* Amsterdam, North-Holland, 1970.

[R18.43] Machina, K.F., "Truth, Belief and Vagueness," *Jour. Philos. Logic*, Vol. 5, 1976, pp. 47-77.

[R18.44] Menges, G., et. al., "On the Problem of Vagueness in the Social Sciences," in Menges, G. (ed.), *Information, Inference and Decision*, D. Reidel Pub., Dordrecht, Holland, 1974, pp. 51-61.

[R18.45] Merricks, Trenton, "Varieties of Vagueness," *Philosophy and Phenomenological Research*, Vol.53, 2001, pp. 145-157.

[R18.46] Mycielski, J., "On the Axiom of Determinateness," *Fund. Mathematics*, Vol. 53, 1964, pp. 205 – 224.

[R18.47] Mycielski, J., "On the Axiom of Determinateness II," *Fund. Mathematics,* Vol. 59, 1966, pp. 203 – 212.

[R18.48] Naess, A., "Towards a Theory of Interpretation and Preciseness," in L. Linsky (ed.) *Semantics and the Philosophy of Language*, Urbana, Ill. Univ. of Illinois Press, 1951.

[R18.49] Narens, Louis, "The Theory of Belief," *Journal of Mathematical Psychology*, Vol. 49, 2003, pp1-31.

[R18.50] Narens, Louis, "A Theory of Belief for Scientific Refutations," *Synthese,* Vol.145, 2005, pp397-423.

[R18.51] Netto, A. B., "Fuzzy Classes," *Notices, Amar, Math. Society*, Vol. 68T- H28, 1968, pp.945.

[R18.52] Neurath, Otto et al. (eds.), *International Encyclopedia of Unified Science, Vol.*

1 – 10, Chicago, University of Chicago Press, 1955.

[R18.53] Niebyl, Karl, H., "Modern Mathematics and Some Problems of Quantity, Quality and Motion in Economic Analysis," *Science*, Vol 7, # 1, January,1940, pp. 103 – 120.

[R18.54] Orlowska, E., "Representation of Vague Information," *Information Systems*, Vol. 13, #2, 1988, pp. 167-174.

[R18.55] Parrat, L. G., *Probability and Experimental Errors in Science*, New York, John Wiley and Sons, 1961.

[R18.56] Raffman. D., "Vagueness and Context-sensitivity," *Philosophical Studies*, Vol. 81, 1996, pp.175-192.

[R18.57] Reiss, S., *The Universe of Meaning*, New York, The Philosophical Library, 1953.

[R18.58] Russell, B., "Vagueness," *Australian Journal of Philosophy*, Vol.1, 1923 pp. 84-92.

[R18.59] Russell, B., *An Inquiry into Meaning and Truth*, New York, Norton, 1940.

[R18.60] Shapiro, Stewart, *Vagueness in Context*, Oxford, Oxford University Press, 2006.

[R18.61] Skala, H. J. "Modelling Vagueness," in M. M. Gupta and E. Sanchez, *Fuzzy Information and Decision Processes*, Amsterdam North-Holland, 1982, pp 101 – 109.

[R18.62] Skala, Heinz J. et al., (eds.), *Aspects of Vagueness*, Dordrecht, D. Reidel Co. 1984.

[R18.63] Sorensen, Roy, *Vagueness and Contradiction*, Oxford, Oxford University Press, 2001.

[R18.64] Tamburrini, G. and S. Termini, "Some Foundational Problems in Formalization of Vagueness," in M. M. Gupta et al (eds.), *Fuzzy Information and Decision Processes*, Amsterdam, North Holland, 1982, pp. 161-166.

[R18.65] Termini, S. "Aspects of Vagueness and Some Epistemological Problems Related to their Formalization," in Skala, Heinz J. et al., (eds.), *Aspects of Vagueness,* Dordrecht, D. Reidel Co. 1984, pp.205 – 230.

[R18.66] Tikhonov, Andrey N. and Vasily Y. Arsenin., *Solutions of Ill-Posed Problems*, New York, John Wiley and Sons, 1977.

[R18.67] Tversky, A. and D. Kahneman, "Judgments under Uncertainty: Heuristics and Biases," *Science,* Vol. 185, September 1974, pp. 1124-1131.

[R18.68] Ursul, A. D., "The Problem of the Objectivity of Information," in Kuba't,

Libor and J. Zeman (eds.), *Entropy and Information*, Amsterdam, Elsevier, 1975, pp. 187-230.

[R18.69] Vardi, M. (ed.), *Proceedings of Second Conference on Theoretical Aspects of Reasoning about Knowledge*, Asiloman, Ca, Los Altos, Ca, Morgan Kaufman, 1988.

[R18.70] Verma, R.R., "Vagueness and the Principle of the Excluded Middle," *Mind*, Vol. 79, 1970, pp. 66-77.

[R18.71] Vetrov, A. A., "Mathematical Logic and Modern Formal Logic," *Soviet Studies in Philosophy*, Vol. 3, #1, 1964 pp. 24 – 33.

[R18.72] von Mises, Richard, *Probability, Statistics and Truth,* New York, Dover Pub. 1981.

[R18.73] Williamson, Timothy, *Vagueness,* London, Routledge, 1994.

[R18.74] Wiredu, J.E., "Truth as a Logical Constant with an Application to the Principle of the Excluded Middle," *Philos. Quart.*, Vol. 25, 1975, pp. 305-317.

[R18.75] Wright, C., "On Coherence of Vague Predicates," *Synthese*, Vol. 30, 1975. pp. 325 – 365.

[R18.76] Wright, Crispin, "The Epistemic Conception of Vagueness," *Southern Journal of Philosophy,* Vol. 33, Supplement, 1995, pp. 133-159.

[R18.77] Zadeh, L. A., A Theory of Commonsense Knowledge," in Skala, Heinz J. et al., (eds.), *Aspects of Vagueness*, Dordrecht, D. Reidel Co. 1984., pp 257 – 295.

[R18.78] Zadeh, L. A., "The Concept of Linguistic Variable and its Application to Approximate reasoning," *Information Science*, Vol. 8, 1975, pp. 199 – 249 (Also, in Vol. 9, pp. 40 – 80).

R19 VAGUENESS, DISINFORMATION, MISINFORMATIOMN AND FUZZY-GAME THEORY IN SOCIO-NATURAL TRANSFORMATIONS

[R19.1] Aubin, J.P. "Cooperative Fuzzy Games", Mathematics of Operations Research, Vol.6, 1981, pp1 – 13.

[R19.2] Aubin, J.P. Mathematical Methods of Game and Economics Theory, New York, North Holland. 1979.

[R19.3] Barone, J.M. "Morphisms in Set H and Fuzzy Move Games," *Annual meeting of the North American Fuzzy Information Processing Society, NAFIPS* , 2006, pp.290-295.

[R19.4] Butnaria, D., "Fuzzy Games: A description pf the concepts," *Fuzzy sets and systems* Vol.1, 1978, pp. 181 – 192.

[R19.5] Butnaria, D., "Stability and shapely value for a n – persons Fuzzy Games," *Fuzzy sets and systems,* Vol. 4, #1, 1980, pp 63 – 72.

[R19.6] Chakeri, A., Dariani, A.N., Lucas, C. "How can fuzzy logic determine game equilibriums better?" *Intelligent Systems (IS'08), 4th International IEEE Conference,* Volume 1, 2008, pp.2-56.

[R19.7] Ereshko F.I. "Reflexive games with uncertain and fuzzy interests," *Proceedings of the IEEE International Conference on Fuzzy Systems, FUZZIEEE'02,* Volume 2, 2002, pp.1608-1612.

[R19.8] Li, K.W., Karray, F., Hipel, K.W. and Kilgour, D.M.). "Fuzzy approaches to the game of Chicken," IEEE Transactions on Fuzzy Systems, Volume 9, Issue 4, 2001, pp.608-623.

[R19.9] Michael Aristidou, Sudipta Sarangi. Games in fuzzy environments. Southern Economic Journal, Vol. 72, #3, 2006, pp.645-659.

[R19.10] P.K. Nayak, M. Pal. Bi-matrix games with intuitionistic fuzzy goals. Iranian Journal of Fuzzy Systems. Vol. 7, No. 1, 2010. pp. 65-79.

[R19.11] Nurmi, H.., "A Fuzzy Solution to a Majority Voting Game, "Fuzzy sets and systems, Vol.5, 1981 pp187-198.

[R19.12] Al-Mutairi, M.S. "Two-decision-maker cooperative games with fuzzy preferences," *IEEE International Conference on Industrial Engineering and Engineering Management (IEEM)*, (2010), pp.6-12.

[R19.13] Qian Song, and Kandel, A. A fuzzy approach to strategic games. *IEEE Transactions on Fuzzy Systems*, vol.7, no.6, (1999), pp.634-642.

[R19.14] Regade., R. K., "Fuzzy Games in the Analysis of Options," jour. Of Cybernetics, Vol. 6, 1976, pp 213 – 221.

[R19.15] Rodica Branzei, Dinko Dimitrov, Stef Tijs. Models in Cooperative Game Theory: Crisp, Fuzzy and Multichoice Games. Springer, 2005.

[R19.16] Russell, S.,Lodwick, W.A. Fuzzy game theory and Internet commerce: e-strategy and metarationality. *Proceedings of the Annual Meeting of the North American Fuzzy Information Processing Society*, 2002 NAFIPS, 2002, pp. 93- 98.

[R19.17] Sharifian, S., Chakeri, A., Sheikholeslam, F. "Linguistic representation of Nash equilibriums in fuzzy games." *Annual Meeting of the North American Fuzzy Information Processing Society (NAFIPS)*, (2010). pp.1-6.

[R19.18] Sharma, R., Gopal, M. "Hybrid Game Strategy in Fuzzy Markov-Game Based Control." *IEEE Transactions on Fuzzy Systems*, Volume 16, Issue.5, (2008), pp.1315- 1327.

[R19.19] Shih-Hung Wu, Von Wun-Soo. A Fuzzy Game Theoretic Approach to Multi-Agent Coordination. Multiagent Platforms. Lecture Notes in Computer Science, Volume 1599, 1998. pp.76-87.

[R19.20] Spillman, B. et al., "Coalition Analysis with Fuzzy Sets," *Kybernetes*, Vol. 8, 1979, pp. 203-211.

[R19.21] Wernerfelt, B., "Semifuzzy Games" *Fuzzy sets and systems*, Vol. 19, 1986, pp. 21 – 28

.

[R19.22] Xiaohui Yu, Qiang Zhang. (2010). Fuzzy Nash equilibrium of fuzzy n-person non-cooperative game. Journal of Systems Engineering and Electronics, Vol. 21, No. 1, February 2010, pp.47–56.

[R19.23] Yan-Qing Zhang, Kandel, A., Henne, M. "On normal fuzzy reasoning and fuzzy moves" *Proceedings of the Sixth IEEE International Conference on Fuzzy Systems*, Volume 1, (1997). pp.549-553.

[R19.24] Zhaoming Wang, Yonghui Ling.. "A Static Negotiation Model of Ekectronic Commerce," *International Journal of Nonlinear Science*,Volume 5, #1, 2008, pp.43-50.

R20 WEAPON FOUNDATIONS FOR INFORMATION SYSTEM

[R20.1] Forte, B., "On a System of Functional Equation in Information Theory," *Aequationes Math.* Vol5, 1970, pp. 202-211.

[R20.2] Gallick, James, *The Information: A History, a Theory, a Flood.* Pantheon, New York, NY, 2011.

[R20.3] Hopcroft, John, E. Rajeev Motwani and D. Ljeffrey, Ullman, *Introduction to automata Theory, Languages, and Computation*, Pearson Education, 2000.

[R20.4] Howard, N., *Paradoxes of Rationality*, Cambridge, Mass., MIT Press, 1972.

[R20.5] Ingarden, R.S, "A Simplified Axiomatic Definition of Information," *Bull. Acad. Polo Sci. Ser, Sci. Math Astronomy Phys.*, Vol.11, 1963, pp. 209-212.

[R20.6] Lee, P. M., "On the Axioms of Information Theory," *Annals of Math. Statistics*, vol.35, 1964, pp 415-418.

[R20.7] Luce, R. D. (ed.) , *Development in Mathematical Psychology*, Westport, Greenwood Press, 1960.

[R20.8] Luciano Floridi, "Is Information Meaningful Data?," *Philosophy and Phenomenological Research*, 70 (2), 2005 pp. 351 – 370.

[R20.9] Meyer, L, "Meaning in Music and Information Theory," *Journal of Aesthetics and Art Criticism,* Vol. 15, 1957, pp412 – 424.

[R20.10] Rich, Elaine, *Automata, Computability, and Complexity: Theory and Applications*, Pearson, 2008.

[R20.11] Shannon, Claude E., "The Mathematical Theory of Communication," *Bell System Technical Journal,* Vol. 27, #3, 1945, pp.379-423 and Vol. 27, #4, 1948, pp623-666.

[R20.12] *Shannon, Claude E.* and Warren Weaver, *The Mathematical Theory of Communication.* University of Illinois Press,1949.

[R20.13] Theil, Henri, *Statistical Decomposition Analysis*, Amsterdam, North-Holland, 1974.

[R20.14] Vigo, R. "Representational information: a new general notion and measure of

Information". *Information Sciences.* Vol.18, *(*2011*)*, pp. *4847–4859.*

[R20.15] Vigo, R "Complexity over Uncertainty in Generalized Representational Information Theory (GRIT): A Structure-Sensitive General Theory of Information". *Information.* Vol. 4. (1), 2013, pp. 1–30.

[R20.16] Vigo, R. *Mathematical Principles of Human Conceptual Behavior: The Structural Nature of Conceptual Representation and Processing*, Routledge, New York and London, 2014.

[R20.17] von Mises, R., *Probability, Statistics and Truth*, New York, Dover, 1981.

[R20.18] Wicker Stephen B. and Saejoon Kim*, Fundamentals of Codes, Graphs, and Iterative Decoding. Springer*, New York, 2003.

[R20.19] Wiener, N., Cybernetics, New York, John Wiley and Sons, 1948.

[R20.20] Wiener, N., The Human use of Human Beings, Boston, Mass, Houghton, 1950.

[R20.21] Young, Paul. *The Nature of Information,* Greenwood Publishing Group, Westport, Ct, 1987.

21. WEAPON FOUNDATIONS FOR FUZZY INFORMATION AND ENTROPY

[R21.1] Belis, M. and S. Guiasu, "A Quantitative–Qualitative Measure of Information in Cybernetic Systems," *IEEE Trans. Inform. Theory* Vol.14, 1968, pp. 593–594.

[R21.2] Burillo, P. and H. Bustince, "Entropy on Intuitionistic Fuzzy Sets and on Interval-valued Fuzzy Sets," *Fuzzy Sets and Systems*, vol. 78, no. 3, pp. 305–316, 1996.

[R21.3] Ceng, H.D., Chen, Y.H. & Sun, Y., "A Novel Fuzzy Entropy Approach to Image Enhancement and Thresholding," *Signal Processing, 75,* 1999, pp.277-301.

[R21.4] De Luca, A. S. Termini, "A Definition of Non-probabilistic Entropy in Setting of Fuzzy Set Theory", *Inform. Control.,* Vol. 20, 1972, pp. 301–312.

[R21.5] Dompere, Kofi K., *A General Theory of Entropy: Rational Foundations of Information-Knowledge Dynamics,* New York, Springer (Series: Studies in Systems, Decision and Control, Vol. 114), 2019.

[R21.6] Dumitrescu, D., "Fuzzy Measures and the Entropy of Fuzzy Partitions," *J. Math. Anal. Appl.*, Vol.176, 1993, pp. 359–373.

[R21.7] Dumitrescu, D., "Entropy of Fuzzy Process," *Fuzzy Sets Syst.*, Vol.55, 1993, pp. 169–177.

[R21.8] Dumitrescu, D., "Entropy of a Fuzzy Dynamical System," *Fuzzy Sets Syst.*, Vol. 70, 1995, pp.45–57.

[R21.9] Garbaczewski, P., "Differential Entropy and Dynamics of Uncertainty," *J. Stat. Phys.* Vol.123, 2006, pp. 315–355.

[R21.10] Hu, Q. and Yu, D., "Entropies of Fuzzy Indiscernibility Relation and its Operations," *International Journal of Uncertainty, Fuzziness and Knowledge-Based Systems* Vol.12, 2004, pp.575-589.

[R21.11] Hu, Q., Yu, D., Xie, Z. and Liu, J. "Fuzzy Probabilistic Approximation Spaces and Their Information Measures," *IEEE Trans. Fuzzy Syst.*, Vol. 14, 2006, pp.191–201.

[R21.12] Hudetz, T. "Space-time Dynamical Entropy for Quantum Systems," *Lett. Math. Phys.*, Vol.16, 1988, pp.151–161.

[R21.14] Hung, W. L. and M. S. Yang, "Fuzzy Entropy on Intuitionistic Fuzzy Sets," *International Journal of Intelligent Systems,* vol. 21, no. 4, 2006, pp. 443–451.

[R21.15] Kapur, J. N., "Four Families of Measures of Entropy," *Indian J. Pure Appl. Math.* Vol.17, 1986, pp. 429-449.

[R21.16] Kapur, J.N. *Measures of Fuzzy Information*. Mathematical Sciences Trust Society, New Delhi 1997.

[R21.17] Kasko, B. "Fuzzy Entropy and Conditioning," *Information Sciences,* Vol.40, 1986, pp. 165-174.

[R.21.18] Klir, G.J. "Generalized Information Theory: Aims, Results and Open Problems," *Reliab. Eng. Syst. Safety*, Vol. 85 (1–3), 2004, pp. 21–38.

[R21.19] Kolmogorov, A.N. *Foundations of the Theory of Probability*; Chelsea Publishing Company: New York, NY, 1950.

[R21.20] Liu, B. and Y.K. Liu, "Expected Value of Fuzzy Variable and Fuzzy Expected Value Models," *IEEE Transactions on Fuzzy Systems,* Vol.10, #4, 2002, pp. 445–450.

[R21.21] S.G. Loo, "Measures of Fuzziness," *Cybernetica,* Vol. 20, 1977, pp. 201–210.

[R21.22] Markechová, D. "The Entropy of Fuzzy Dynamical Systems and Generators," *Fuzzy Sets Syst.*, Vol. 48, 1992, pp. 351–363.

[R21.23] Mesiar, R. and Rybárik, J. "Entropy of Fuzzy Partitions: A General Model," *Fuzzy Sets Syst.*, 99, 1998, pp. 73–79.

[R21.24] Mesiar, R. "The Bayes Principle and the Entropy on Fuzzy Probability Spaces," Int. J. Gen. Syst., Vol. 20, 1991, pp. 67–72.

[R21.25] Parkash, O. "A New Parametric Measure of Fuzzy Entropy," *Inform. Process. Manage. Uncertain.* Vol.2, 1998, pp. 1732–1737.

[R21.26] Parkash, O. and Sharma, P.K. "Measures of Fuzzy Entropy and their Relations," *International Journal of Management & Systems,* Vol. 20, 2004, pp. 65-72.

[R21.27] Rahimi, M.and Riazi, A., "On Local Entropy of Fuzzy Partitions," *Fuzzy Sets Syst.*, Vol. 234, 2014, pp. 97–108.

[R21.28] Rieˇcan, B. "An Entropy Construction Inspired by Fuzzy Sets," Soft Comput., Vol.7, 2003, pp. 486–488.

[R21.29] Szmidt, E. and J. Kacprzyk, "Entropy for Intuitionistic Fuzzy Fets," *Fuzzy Sets and Systems,* vol. 118, no. 3, 2001, pp. 467–477.

[R21.30] Verma R. and B. D. Sharma, "On Generalized Exponential Fuzzy Entropy," *Engineering and Technology,* vol. 5, 2011, pp. 956–959.

[R21.31] Zeng W. and H. Li, "Relationship between Similarity Measure and Entropy of Interval Valued Fuzzy Sets," *Fuzzy Sets and Systems*, vol. 157, #11, 2006 pp. 1477–1484.

R22 WEAPON FOUNDATIONS FOR KNOWLEDGE BY ACQUAINTANCE AND KNOWLEDGE BY DESCRIPTION

[R22.1] Balog, K. "Acquaintance and the Mind-Body Problem." In Simone Gozzano and Christopher Hill (eds.), *New Perspectives on Type Identity: The Mental and the Physical* Cambridge: Cambridge University Press 2012, pp. 16-43.

[R22.2] BonJour, L., "Toward a Defense of Empirical Foundationalism." In Michael Raymond DePaul (ed.), *Resurrecting Old-Fashioned Foundationalism,* Lanham: Rowman and Littlefield, 2001, pp. 21-38.

[R22.3] Brewer, B. *Perception and its Objects*. Oxford: Oxford University Press. 2011.

[R22.4] Chalmers, D. *The Conscious Mind.* Oxford: Oxford University Press,1996.

[R22.5] Chisolm, Roderick, *Acquaintance and the Mind-Body Problem*, Oxford, Oxford University Press, (2008).

[R22.6] Churchman, C. W., and Ratoosh, P., *Measurement, Definitions and Theories*, Wiley and Sons, 1959.

[R22.7] Fales, E. *A Defense of the Given.* Lanham: Rowman and Littlefield, 1996.

[R22.8] Gertler, B. "A Defense of the Knowledge Argument," *Philosophical Studies,* Vol. 93, 1999., pp. 317-336.

[R22.9] Gertler, B. "Introspecting Phenomenal States." *Philosophy and Phenomenological Research* Vol. 63, 2001, (pp. 305–328).

[R22.10a] Gertler, B. "Renewed Acquaintance." In Declan Smithies and Daniel Stoljar (eds.), *Introspection and Consciousness*, Oxford, Oxford University Press, 2012, pp. 93-128.

[R22.10b] Harrison, Glenn W., and John A List, "*Field Experiments,*" *Journal of Economic Literature,* Vol. 42, Dec. 2004.

[R22.11] Hasan, Ali and Fumerton, Richard, "Knowledge by Acquaintance vs. Description," *The Stanford Encyclopedia of Philosophy*, Edward N. Zalta (ed.), (Spring 2014 Edition).

[R22.12] Lazerowitz, M., "Knowledge by Description", *The Philosophical Review*, Vol.46, No.4, July 1937, pp. 402–415.

[R22.13] Morgenstern, O., *On the Accuracy of Economic Observation*, Princeton, Princeton University Press, 1973.

[R22.14] Parker, D.H., "Knowledge by Acquaintance", *The Philosophical Review*, Vol.54, No.1, January 1945A, (pp. 1–18).

[R22.15] Parker, D.H., "Knowledge by Description", *The Philosophical Review*, Vol.54, No.5, September1945B, pp. 458–488.

[R22.16] Poston, T., "Similarity and Acquaintance." *Philosophical Studies* Vol. 147, 2010, pp. 369-378.

[R22.17] Russell, B. "On Denoting." *Mind*, Vol.14, 1905, pp. 479-493.

[R22.18] Russell, B. "Knowledge by Acquaintance and Knowledge by Description." *Proceedings of the Aristotelian Society* Vol.11, 1910-1911, pp.108-128.

[R22.19] Russell, B., "On the Nature of Acquaintance," in *Bertrand Russell: Logic and Knowledge: Essays 1901-1950*, R. C. Marsh (ed.) New York, Capricorn Books,1971, pp.,127-174.

[R22.20] Russell, B., *Logic and Knowledge: Essays 1901-1950*, R. C. Marsh (ed.) New York, Capricorn Books,1971.

R23 WRITTEN AND AUDIO LANGUAGES AND INFORMATION

[R23.1] Agha, Agha, *Language and Social Relations,* Cambridge, Cambridge University Press, 2006.

[R23.2] Aitchison, Jean (ed.), *Language Change: Progress or Decay?* Cambridge, New York, Melbourne: Cambridge University Press 2001.

[R23.3] *Allerton, D. J.* "Language as Form and Pattern: Grammar and its Categories". In Collinge, N.E. (ed.) *An Encyclopedia of Language.* London and New York: Routledge, 1989.

[R23.4] Anderson, Stephen, *Languages: A Very Short Introduction.* Oxford: Oxford University Press, 2012.

[R23.5] Aronoff, Mark; Fudeman, Kirsten, *What is Morphology.* New York, John Wiley & Sons, 2011.

[R23.6] Barber Alex & Robert J Stainton (eds.), *Concise Encyclopedia of Philosophy of Language and Linguistics.* New York, Elsevier, 2010.

[R23.7] Bauer, Laurie (ed.), *Introducing linguistic morphology* Washington, D.C., Georgetown University Press, 2003.

[R23.8] Brown, Keith; Ogilvie, Sarah, (eds.), *Concise Encyclopedia of Languages of the World,* New York Elsevier Science, 2008.

[R23.9] Campbell, Lyle (ed.), *Historical Linguistics: an Introduction,* Cambridge, MA: MIT Press, 2004.

[R23.10] Chandler, W.L. "Abstract Families of Deterministic Languages," *Proceedings of the First ACM Symposium on Theory of Computing,* 1969, pp21-30.

[R23.11] Chao, Yuen Ren, *Language and Symbolic Systems,* Cambridge, Cambridge University Press. 1968.

[R23.12] *Chomsky, Noam, Syntactic Structures. The Hague: Mouton,* 1957.

[R23.13] Chomsky, Noam, *The Architecture of Language.* Oxford: Oxford University Press 2000.

[R23.14] Clarke, David S., *Sources of semiotic: readings with commentary from antiquity to the present.* Carbondale: Southern Illinois University Press, 1990.

[R23.15] Comrie, Bernard (ed.), *Language universals and linguistic typology: Syntax and morphology.* Oxford: Blackwell, 1989.

[R23.16] Comrie, Bernard, (ed.), *The World's Major Languages.* New York: Routledge, 2009.

[R23.17] Coulmas, Florian, *Writing Systems: An Introduction to Their Linguistic Analysis.* Cambridge University Press, 2002.

[R23.18] Croft, William, Cruse, D. Alan, *Cognitive Linguistics. Cambridge, Cambridge University Press,* 2004.

[R23.19] Crystal, David, *The Cambridge Encyclopedia of Language.* Cambridge: Cambridge University Press, 1997.

[R23.19] Culik, K. H., "Abstract Families of Relations," *Journal. of Computer and System Science,* Vol.5, 1971, PP596-606.

[R23.20] Deacon, Terrence , *The Symbolic Species: The Co-evolution of Language and the Brain,* New York: W.W. Norton & Company, 1997.

[R23.21] Devitt, Michael and Sterelny, Kim *Language and Reality: An Introduction to the Philosophy of Language. Boston:* MIT Press, 1999.

[R23.22] Evans, Nicholas; Levinson, Stephen C.. "The myth of language universals: Language diversity and its importance for cognitive science". *Behavioral and Brain Sciences*32 (5), 2009, pp. 429–492.

[R23.23] Fitch, W. Tecumseh, *The Evolution of Language. Cambridge: Cambridge University Press,* 2010.

[R23.24] Foley, William A., *Anthropological Linguistics: An Introduction,* Oxford, Blackwell, 1997.

[R23.25] Ginsburg, S, *Algebraic and Automata-Theoretic Properties of Formal Languages,* (Fundamental Studies in Computer Science, Vol. 2), New Amsterdam, North- Holland, 1975.

[R23.26] Greenberg, Joseph, *Language Universals: With Special Reference to Feature Hierarchies. The Hague: Mouton & Co.* 1966.

[R23.27] Hauser, Marc D.; Chomsky, Noam; Fitch, W. Tecumseh, "The Faculty of Language: What Is It, Who Has It, and How Did It Evolve," *Science 22, **298,*** (5598) , 2002, pp. 1569–1579.

[R23.28] Hopcroft, J, E. and J. D. Ullman, *Formal Languages and their relation to Automata*, Reading, Mass, Addison-Wesley, 1959.

[R23.29] International Phonetic Association, *Handbook of the International Phonetic Association: A guide to the use of the International Phonetic Alphabet.* Cambridge: Cambridge University Press, 1999.

[R23.30] Katzner, Kenneth, *The Languages of the World,* New York: Routledge, 1999.

[R23.31] Labov, William, *Principles of Linguistic Change vol. I: Internal Factors,* Oxford, Blackwell, 1994.

[R23.32] Labov, William, *Principles of Linguistic Change vol. II: Social Factors,* Oxford, Blackwell 2001.

[R23.33] Ladefoged, Peter, Maddieson, Ian, *The sounds of the world's languages*, Oxford: Blackwell., 1996, pp. 329–330.

[R23.34] Levinson, Stephen C. *Pragmatics.* Cambridge: Cambridge University Press 1983.

[R23.35] Lewis, M. Paul (ed.), *Ethnologue: Languages of the World*, Dallas, Tex.: SIL International 2009.

[R23.36] Lyons, John, *Language and Linguistics,* Cambridge, Cambridge University Press 1981.

[R23.37] MacMahon, *April M.S., Understanding Language Change,* Cambridge, Cambridge University Press, 1994.

[R23.38] Matras, Yaron; Bakker, Peter, (eds). *The Mixed Language Debate: Theoretical and Empirical Advances.* Berlin: Walter de Gruyter 2003.

[R23.39] Moseley, Christopher (ed.), *Atlas of the World's Languages in Danger,* Paris: UNESCO Publishing, 2010.

[R23.40] Nerlich, B. "History of pragmatics". In L. Cummings (ed.), *The Pragmatics Encyclopedia.* New York: Routledge. 2010, pp. 192–93.

[R23.41] Newmeyer, Frederick J., *The History of Linguistics*., Linguistic Society of America, 2005.

[R23.42] Newmeyer, Frederick J., *Language Form and Language Function,* Cambridge, MA: MIT Press, 1998.

[R23.43] Nichols, Johanna *Linguistic diversity in space and time.* Chicago: University of Chicago Press,1992.

[R23.44] Nichols, Johanna "Functional Theories of Grammar". *Annual Review of Anthropology,* Vol. 13, 1984, pp.97–117.

[R23.45] Saloman, A., Formal Languages, New York, Academic Press, 1973.

[R23.46] Senft, Gunter, *Systems of Nominal Classification.* Cambridge University Press. (ed.), 2008.

[R23.47] Swadesh, Morris, "The phonemic principle", *Language,* Vol. **10** (2): (1934), 117–129.

[R23.48] Tomasello, Michael "The Cultural Roots of Language". In B. Velichkovsky and D. Rumbaugh (Eds.), *Communicating Meaning: The Evolution and Development of Language.* Psychology Press, 1996, pp. 275–308.

[R23.49] Tomasello, Michael, *Origin of Human Communication.* MIT Press, 2008.

[R23.50] Ulbaek, Ib, "The Origin of Language and Cognition", In J. R. Hurford & C. Knight (eds.). *Approaches to the evolution of language.* Cambridge University Press. 1998, pp. 30–43.

[R 24] ZONES OF EPISTEMIC CONFLICTS ON POSSIBILITY AND PROBABILITY

.

[R24.1] Alberoni, F. Contribution to the study of subjective probability. Part I. *Journal of General Psychology,* 1962, 66, 241-264.

[R24.2] Berger, James O., *Statistical Decision Theory and Bayesian Analysis* (Springer Series in Statistics). New York, Springer-Verlag 1985.

[R24.3] Benacerraf, P.: "Mathematical Truth," *Journal of Philosophy* Vol. 70, 1973, pp.661-79.

[R24.4] de Finetti, Bruno, *Theory of Probability. A Critical Introductory Treatment,* (translation by A.Machi and AFM Smith of 1970 book) 2 volumes. New York, Wiley, 1974 –75.

[R.24.5] Dompere, Kofi K., "Savage Axioms, Ellsberg's Paradox and Fuzzy Optimal Decision-Choice Rationality, *Fuzzy Economic Review,* Vol 17, #1, 2012, pp. 15 – 51.

[R24.6] DeGroot, Morris, *Optimal Statistical Decisions.* Wiley Classics Library, New York, Wiley, 2004.

[R24.7] Edwards, W. "Conservatism in human information processing," In B. Kleinmuntz (ed.), *Formal representation of human judgment.* New York: Wiley, 1968. pp. 17-52.

[R24.8] Eells, E. and E. Sober, "Probabilistic Causality and the Question of Transitivity," *Philosophy of Science* Vol.50, 1983, 35-57.

[R24.9] Feller, W., *An introduction to probability theory and its applications.* Vol.1 New York: Wiley, 1968.

[R24.10] Fetzer, J. H., "Dispositional Probabilities," in R. C. Buck and R. S. Cohen (eds.), *Boston Studies in the Philosophy of Science VIII* 1971, Reidel, Dordrecht, pp. 473-482.

[R24.11] Fetzer, J. H., "Statistical Probabilities: Single Case Propensities vs. Long-Run Frequencies," in W. Leinfellner and E. Kohler (eds.), *Developments in the Methodology of Social Science,* Reidel, Dordrecht,1974, pp. 387-397.

[R24.12] Fetzer, J. H. and D. E. Nute: 1979, "Syntax, Semantics, and Ontology: A Probabilistic Causal Calculus," *Synthese* Vol. 40, 1979, 453-95.

[R24.13a] Gärdenfors, P. and Sahlin, Nils-Eric. *Decision, Probability, and Utility: Selected Readings*. Cambridge, Cambridge University Press,1988.

[R24.13b] Good, Irving, J. Good, *Thinking: Foundations of Probability and its Applications*, New York, Dover, 1983.

[R24.14] Hacking, Ian "Slightly More Realistic Personal Probability*". Philosophy of Science.* Vol. **34** (4): (December 1967). 311–325.

[R24.15] James H. Fetzer (ed.), *Probability and Causality,* Reidel, Dordrecht, 1988.

[R24.16] Jazwinski, A. H., *Stochastic Processes and Filtering Theory,* New York, Dover, 1998

[R24.17] Jeffrey, R. C., *The Logic* of *Decision,* University of Chicago Press, Chicago 1990.

[R24.18] Kolmogorov, A.N. *Foundations of the Theory of Probability*; Chelsea Publishing Company: New York, NY, 1950.

[R24.19] Kyburg, H. E., Jr. "Propensities and Probabilities," *British Journal for the Philosophy of Science,* Vol. 25, 1974, pp359-75.

[R24.20a] Lindley, Dennis, V. *Introduction to Probability and Statistics: From a Bayesian Viewpoint Part I, Probability*, Cambridge, Cambridge Univ. Press, 1965.

[R24.20b] Lindley, Dennis, V. *Introduction to Probability and Statistics: From a Bayesian Viewpoint Part 2, Inference,* Cambridge, Cambridge Univ. Press, 1965.

[R24.21] Morgenstern, Oskar *(1978). "Utility". In Andrew Schotter. Selected Economic Writings of Oskar Morgenstern.* New York, New York University Press.1978, pp. 65–70.

[R24.22] Ozga, Andrew S. *Expectations in Economic Theory,* Chicago, Aldine Pub., 1965.

[R24.23] Pfanzagl, J., "Subjective Probability Derived from the Morgenstern-von Neumann Utility Theory". In Martin Shubik.(ed.) *Essays in Mathematical Economics In Honor of Oskar Morgenstern.* Princeton, N.J., Princeton University Press. 1967, pp. 237–251.

[R24.24] Popper, K. R., "The Propensity Interpretation of the Calculus of Probability, and the Quantum Theory", in S. Korner (ed.), *Observation and Interpretation in the Philosophy* of *Physics,* New York, Dover Publications, 1957, pp. 65-70.

[R24.25] Popper, K. R., 'The Propensity Interpretation of Probability', *British Journal for the Philosophy of Science* Vol. 10, 1959, pp. 25-42.

[R24.26] Salmon, W. C., *The Foundations of Scientific Inference,* Pittsburg, University of Pittsburgh Press1967.

[R24.27] Savage, L. J., *The Foundations of Statistics*, New York, Dover Pub, 1972.

[R24.28] Savage, L. J., "Subjective Probability and Statistical Practice," in Robert E Machol and Paul Gray (eds.), *Recent Developments in Decision and Information Processes*, New York, Macmillan, 1962.

[R24.29] Savage, L. J., "Difficulties in the theory of Personal Probability," *Philosophy of Science*, Vol.34, 1967, pp. 305 – 310.

[R24.30] Sklar, L.: "Is Probability a Dispositional Property?," *Journal of Philosophy* Vol.67, 1970, pp.355-66.

[R24.31] Skyrms, B.: *Causal Necessity,* New Haven, Yale University Press, 1980.

[R24.32] Stigler, SM. *The History of Statistics: The Measurement of Uncertainty before 1900. Cambridge, Mass.,* Belknap Press/Harvard University *Press* 1990.

[R24.33] Suppes, P., "New Foundations of Objective Probability: Axioms for Propensities," in P. Suppe, L. Henkin, A. Joja, and G. C. Moisil (eds.), *Logic, Methodology and Philosophy of Science IV,* North-Holland, Amsterdam, London, 1973, pp. 515-29.

[R24.34] Tversky, A., & Kahneman, D. The belief in the law of small numbers. Psychological Bulletin, 1971, 76, 105-110.

[R24.35] Von Mises, R., *Probability, Statistics and Truth,* Macmillan, New York 1957.

[R24.36] von Neumann, John and Oskar Morgenstern, *Theory of Games and Economic Behavior,* Princeton University Press, 1943, 1947.

[R24.37] Wagenaar, W. A. "Subjective randomness and the capacity to generate information,"

In A. F. Sanders (ed.) *Attention and performance III*, Acta Psychologica, Vol. 33,1970, pp. 233-242.

R25. ZONES OF EPISTEMIC CONFLICTS IN UNITY OF SCIENCE AND KNOWING.

[R25.1] Andler, Daniel, "Unity Without Myths" In John Symons, Juan Manuel Torres & Olga Plomb (eds.), *New approaches to the Unity of Science, vol. 1: Otto Neurath and the Unity of Science.* New York, Springer. 201.

[R25.2] Anton, J. P., "The Unity of Scientific Inquiry and Categorial Theory in Aristotle," *Boston Studies in the Philosophy of Science* 1990, pp. 121:29-43.

[R25.3] Barrow, John D. *New Theories of Everything: The Quest for Ultimate Explanation,* Oxford, Oxford University Press, 2007.

[R25.4] Bechtel, William P. & Andrew Hamilton "Reduction, Integration, and the Unity of Science: Natural, Behavioral, and Social Sciences and the Humanities," In T. Kuipers (ed.), *Philosophy of Science: Focal Issues (Volume 1 of the Handbook of the Philosophy of Science).* New York, Elsevier, 2007.

[R25.5] Boersema, David , "Metaphysics, Mind, and the Unity of Science," *Behavioral and Brain Sciences* 27 (5), 2004 pp. 627-628.

[R25.6] Briskman, Larry, "Three Views Concerning the Unity of Science," In Gerard Radnitzky (ed.), *Centripetal Forces in the Sciences*, New York, Paragon House Publishers, 1987. pp. 1--105.

[R25.7] Bunge Mario Augusto (ed.) *The Methodological Unity of Science.*, Boston, Reidel1973.

[R25.8] Burian, Richard M. "Conceptual Change, Cross-Theoretical Explanation, and the Unity of Science" *Synthese* 32 (#1-2), 1975, pp. 1 - 28.

[R25.9]Byrne, Lee "An Educational Application of Resources of the Unity of Science Movement," - *Philosophy of Science,* Vol. 7 (#2): 1940, pp. 241-262.

[R25.10] Carnap, Rudolf, *The Unity of Science*, London: K. Paul, Trench, Trubner & Co., 1934.

[R25.11] Carrier, Martin, "The Unity of Science," *International Studies in the Philosophy of Science* Vol. 4 (1): 1990, pp. 17-31.

[R25.12] Causey, Robert L., "Unity of Science," *Philosophical Review* Vol. 90 (1) 1977, pp. 150-153.

[R25.13] Crave, Carl F. r, "Beyond Reduction: Mechanisms, Multifield Integration and the Unity of Neuroscience," *Studies in History and Philosophy of Science Part C* ,Vol. 36 (#2), 2005 . pp. 373-395.

[R25.14] Damasio, Antonio R. (ed.) *Unity of Knowledge: The Convergence of Natural and Human Science.* New York, New York Academy of Sciences, 2001.

[R25.15] Davies, David "Explanatory Disunities and the Unity of Science," *International Studies in the Philosophy of Science* Vol. 10 (1): 1996, pp.5 – 21.

[R25.16] de Santillana, George & Edgar Zilsel "Foundations of the Unity of Science. II 8: The Development of Rationalism and Empiricism". *Philosophical Review*, 1943 Vol. 52 (1):87-87.

[R25.17] Dewey, John "Foundations of the Unity of Science," Vol. II, No. 4: Theory of Valuation. *Philosophical Review* Vol. 50 (4): 1941, pp. 443-446.

[R25.18] Donovan, Rickard, "Science Without Unity," *International Philosophical Quarterly* Vol. 30 (1): 1990, pp. 122-125.

[R25.19] Dupre J. "Against Scientific Imperialism," - 1994 - *PSA: Proceedings of the Biennial Meeting of the Philosophy of Science Association* 1994, pp. 374 - 381.

[R25.20] Dupré, John "The Disunity of Science," *Mind* Vol. 92 (367): 1983, pp. 321-346.

[R25.21] Finch III, Charles, S, *The Star of Deep Beginnings : The Genesis of African sciences and Technology,Decatur,* GA, USA, Khenti INC, 2001.

[R25.22] Galison, Peter & David J. Stump (eds.), *The Disunity of Science: Boundaries, Contexts, and Power.* Stanford, Ca. Stanford University Press. 1996.

[R25.23] Neurath, Otto et al. (eds.), *International Encyclopedia of Unified Science, Vol. 1 – 10,* Chicago, University of Chicago Press, 1955.

[R25.24] Neurath, Otto, "Unified Science as Encyclopedic," in Neurath, Otto et al. (eds.), *International Encyclopedia of Unified Science, Vol. 1 – 10,* Chicago, University of Chicago Press, 1955, pp.1-27.

[R25.25] **Sertima**, Ivan Van (ed.) *Blacks in Science: ancient and modern,* New Brunswick, NJ. USA, Transaction Pub., 1998.

[R26] ZONES OF EPISTEMIC CONFLICTS ON DECIDABILITY, COMPUTABILITY AND PREDICTABILITY IN THE SPACE OF THE PROBLEM-SOLUTION DUALITIES AND POLARITIES

[R26.1] Davis, Martin, *The Undecidable: Basic Papers on Undecidable Propositions, Unsolvable Problems and Computable Functions*, Dover, Dover Books, 2004.

[R26.2] Gödel, Kurt, *On formally undecidable Propositions of Principia Mathematica and Related Systems*, Dover, Dover Pub.,1992.

[R26.3] Gödel, K. Kurt, *Gödel, Collected Works, Vols.1- 3.* Oxford University Press, Oxford, 1995.

[R26.4] Hartshorne, C. *The Logic of Perfection and Other Essays in Neoclassical Metaphysics*. Open Court Press, Lasalle, Illinois,1962.

[R26.5] Tarski, Alfred, A Mostowski et al., *Undecidable Theories: Studies in Logic and Foundation of Mathematics*, Dover, Dover Books, 2010.

[R26.6] Turing, A. M., "Solvable and Unsolvable Problems"., *UroBooks,* pp 576-595, 1954.

[R26.7] Turing, A. M., "Intelligent Machinery, A Heretical Theory", in *The Essential Turing*, 2004, pp465-475.

[R26.6] Wang, Hao, *Reflections on Kurt* Gödel, Cambridge, Mass. MIT Press,1987.

[R27] ZONE OF SYMMETRY, ADINKRA, ADINKRAMATICS, ADINKRA GRAPHS AND POLYRHYTHMS

[R27.1] Douglas B. L.,[1]S. James Gates,[2]B. L. Segler,[1]and J. B. Wang *Automorphism Properties and Classification of Adinkras,* Vol. 2015| Article IID584542 htt://doi.org/10.1155/2016/584542, 2016.

,

[R27.2] Doran C. F., M. G. Faux, S. J. Gates, Jr., T. Hübsch, K. M. Iga and G. D. Landweber,"Relating doubly-even error-correcting codes, graphs, and irreducible representations of *N*-supersymmetry," in: *Discrete and Computational Mathematics,* pp. 53–71 (Nova Sci. Publ., 2008.

[R27.3] Doran C. F., M. G. Faux, S. J. Gates, Jr., T. Hübsch, K. M. Iga and G. D. Landweber, "On Graph-Theoretic Identifications of Adinkras, Supersymmetry Representations and Superfields," *Int. J. Mod. Phys.* Vol. A 22, pp.869–930, 2012.

[R27.4] Faux, M. and S. J. Gates, Jr., "Adinkras: a graphical technology for supersymmetric representation theory," *Phys. Rev.* D **71**(3), 065002, 2005.

[R27.5] . Gates, S, K. Iga,, K. Stiffler *Generating All 36, 864 Four-Color Adinkras via Signed Permutations and Organizing into ℓ- and ℓ ˜ -Equivalence Classes,* Published 21 December 2017.

[R27.6] Hrnciar P. and A. Haviar, "All trees of diameter five are graceful," ˇ Discrete Mathematics, vol. 233, no. 1–3, 2001, pp. 133–150.

[R27.7] Huang, C., A. Kotzig, and A. Rosa, "Further results on tree Labellings," Utilitas Mathematica, vol. 21, pp. 31–48, 1982.

[R27.8] Iga, Kevin and Yan Zhang, "Structural Theory and Classification of 2D Adinkras," *High Energy Physics,* Vol, 2016, 2016 12pages, (SCOAP3).

[R27.9] Korankye, Charles M., *Adinkra Alphabets: The Adinkra Symbols as Alphabets and Meanings*, Washington, USA, Korankye 2017.

[R27.10] Marcolli, M., Zolman, N. Adinkras, Dessins, Origami, and Supersymmetry Spectral Triples. *P-Adic Num Ultrametr Anal Appl* **11**, 223–247 (2019).

[R27.11] Ofori-Ansah, Kwaku, P, *Symbols of Adinkra Cloth,* Hyattsville, MD, USA Sankofa Pub. 1995.

[R27.12] Ofori-Ansah, Kwaku, P, *Weights of Wisdom: History,Uses and Sybolism of Akan Goldweights*, Hyattsville, MD, USA, Sankofa Pub. 1997.

[R27.13] Owusu-Ansah, J.V., Nana. *New Versions of the Traditional Motifs*, Kumasi, Ghana, Degraft Graphics ang Pub. 1992.

[R27.14] A. Panpa and T. Poomsaard "On Graceful Spider Graphs with at Most Four Legs of Lengths Greater than One" , *Hindawi Journal of Applied Mathematics Volume 2016*, Article ID 5327026, 5 pages.

[R27.15] Rosa, A. "On certain valuations of the vertices of a graph," in *Theory of Graphs, International Symposium, Rome, July 1966*, New York, NY, Gordon and Breach 1967, pp. 349–355.

[R27.16] Sarpong, P., *The Sacred Stools of the Akan*, Tema, Ghana, Ghana Pub. Co, 1971.

[R27.17] Willis, W. Bruce, *The Adinkra dictionary: A visual primer on the language of Adinkra,* Washington, D. C., Pyramid Complex, 1998.

[R27.18] Zhang Yan X *Adinkras for Mathematicians,* Massachusetts Institute of Technology

November28, 201.

[R28] ZONE OF SYMMETRY, FRATAL GRAPH, MATHEMATICS, ART, RHYTHMS AND GEOMETRY

[R28.1] Bangura, Abdul K., *African Mathematics: From Bones to Computers*, Lanham, MD, University Press of America, 2012.

[R28.2] Burkle-Elizondo, Gerardo; Valdéz-Cepeda, Ricardo David "Fractal analysis of Mesoamerican pyramids". *Nonlinear Dynamics, Psychology, and Life Sciences.* Vol. 10 (1): (2006). pp,105–122.

[R28.3] Cannon, W. J. Floyd, W. R. Parry. "*Finite subdivision rules*", *Conformal Geometry and Dynamics*, vol. 5, 2001, pp. 153–196.

[R28.4] Cheng, Qiuming *"Multifractal Modeling and Lacunarity Analysis". Mathematical Geology.* Vol. 29 (7): (1997), pp. 919–932.

[R28.5] Cross, Simon S. "Fractals in Pathology". *The Journal of Pathology. 182 (1):* 1997, pp.1–8.

[R28.6] Edgar, Gerald, *Measure, Topology, and Fractal Geometry,* New York, *Springer Science & Business Media., 2007.*

[R28.7] Eglash, Ron, *African Fractals: Modern Computing and Indigenous Design.,* New Brunswick: Rutgers University Press. (Archived from the original on January 3, 2018).

[R28.8] Edgar, Gerald *Measure, topology, and fractal geometry. New York: Springer-Verlag.* 2008.

[R28.9] Enright, Matthew B.; Leitner, David M. *"Mass fractal dimension and the compactness of proteins". Physical Review E. 71 (1)*, January 27, 2005.

[R28.10] Falconer, Kenneth *Fractals, A Very Short Introduction,* New York, Oxford

University Press2013.

[R28.11] Falconer, Kenneth *Fractal Geometry: Mathematical Foundations and Applications.* New York, John Wiley & Sons, 2003.

[R28.12] Falconer, Kenneth *Fractals, A Very Short Introduction,* New York, Oxford University Press2013.

[R28.13] Gouyet, Jean-François *physics and fractal structures. new* york: masson springer, 1996.

[R28.14] Hahn, Horst K.; Georg, Manfred; Peitgen, Heinz-Otto "Fractal aspects of three-dimensional vascular constructive optimization". In Losa, Gabriele A. and Nonnenmacher, Theo F. (eds.). *Fractals in biology and medicine,* New York, Springer. (2005), pp. 55–66.

[R28.15] Hu, Shougeng; Cheng, Qiuming; Wang, Le; Xie, Shuyun "Multifractal characterization of urban residential land price in space and time". *Applied Geography.* Vol34:, 2012, pp. 161–170.

[R28.16] Jelinek, Herbert F; Fernandez, Eduardo "Neurons and fractals: how reliable and useful are calculations of fractal dimensions?". *Journal of Neuroscience Methods. Vol 81 (1–2):* (June 1998). pp. 9–18.

[R28.17] Krapivsky, P. L.; Ben-Naim, E., "Multiscaling in Stochastic Fractals". *Physics Letters A. 196 ,1994 ,* pp*3–4: 168.*

[R28.18] Karperien, Audrey L.; Jelinek, Herbert F.; Buchan, Alastair M. "Box-Counting Analysis of Microglia Form in Schizophrenia, Alzheimer's Disease and Affective Disorder". *Fractals.* Vol.16 (2): 2008. p.103.

[R28.19] Landini, Gabriel *"Fractals in microscopy". Journal of Microscopy.* Vol.241 (#1*):* 2011 pp. 1–8.

[R28.20] Lee, S.; Olsen, S.; Gooch, B. "Simulating and Analyzing Jackson Pollock's Paintings*". Journal of Mathematics and the Arts.* Vol.1 (2): (2007). Pp.73–83.

[R28.21] Liu, Jing Z.; Zhang, Lu D.; Yue, Guang H. "Fractal Dimension in Human Cerebellum Measured by Magnetic Resonance Imaging". *Biophysical Journal. 85 (6):,* 2003. Pp. 4041–4046.

[R28.22] Losa, Gabriele A., Nonnenmacher, Theo F. *Fractals in biology and medicine.* New York, *Springer.* 2005.

[R28.23] Mandelbrot, Benoît B. *Fractals and Chaos. Berlin: Springer.* 2004.

[R28.24] Mandelbrot, B., and Michael Frame, *Fractals, Graphics. And Mathematics Education,* Mathematical Association of America, Notes Series Number 58, *(Translation, Mathematical Monograph)* 2002.

[R28.25] Mandelbrot, Benoît B., *The fractal geometry of nature.* New York, *Macmillan. 1983.*

[R28.26] Mandelbrot, B. B.: *The Fractal Geometry of Nature*. W. H. Freeman and Company, New York 1982.

[R28.27] Nelson, Bryn; "Sophisticated Mathematics Behind African Village Designs Fractal patterns use repetition on large, small scale," *San Francisco Chronicle*, February 23, 2009.

[R28.28] Ofori-Ansah, Kwaku, P, *Kente is More than A Cloth: History and Significance of Ghana's Kente Cloth,* Hyattsville, MD, USA Sankofa Pub. 1995.

[R28.29] Ostwald, Michael J., and Vaughan, Josephine, *The Fractal Dimension of Architecture*. Birhauser, Basel. 2016.

[R28.30] Ozhovan M. I., Dmitriev I. E., Batyukhnova O. G. "Fractal structure of pores of clay soil". *Atomic Energy*, Vol. 74, 1993, pp.241–243.

[R28.31] Peters, Edgar, *Chaos and order in the capital markets: a new view of cycles, prices, and market volatility.* New York, Wiley. *(1996).*

[R28.32] Tan, Can Ozan; Cohen, Michael A.; Eckberg, Dwain L.; Taylor, J. Andrew "Fractal properties of human heart period variability: Physiological and methodological implications". *The Journal of Physiology. 587* (15), 2009, pp, 3929–41.

[R28.33] Russ, John C. *Fractal surfaces. Vol. 1.* New York, *Springer.* (1994).

[R28.34] D. Seekell, M. L. Pace, L. J. Tranvik, C. Verpoorter, "A fractal-based approach to lake size-distributions" *Geophysical Research Letters,* Vol 40 (3), 2013 pp. 517–521.

- **[R28.35]** Shamar, L "What Makes a Pollock Pollock: A Machine Vision Approach," *International Journal of Arts and Technology.* Vol 8: , 2015 . pp.1–10.

[R28.36] Sornette, Didier, *Critical phenomena in natural sciences: chaos, fractals, selforganization, and disorder: concepts and tools.* New York, *Springer,* 2004. *pp. 128–140.*

[R28.37] Spencer, John; Thomas, Michael S. C.; McClelland, James L. *Toward a unified theory of development: connectionism and dynamic systems theory re-considered.* New York, Oxford University Press, 2009.

[R28.38] Sreenivasan, K. R.; Meneveau, C. "The Fractal Facets of Turbulence". *Journal of Fluid Mechanics. 173:* 1986. *357–386.*

[R28.39] Stumpff, Andrew "The Law is a Fractal: The Attempt to Anticipate Everything". *Loyola University Chicago Law Journal Vol. 44.*, 2013, p. 649.

[R28.40] Takayasu, H. *Fractals in the physical sciences,* Manchester, Manchester University Press. 1990.

[R28.41] Tan, Can Ozan; Cohen, Michael A.; Eckberg, Dwain L.; Taylor, J. Andrew "Fractal properties of human heart period variability: Physiological and methodological implications". *The Journal of Physiology. 587 (15),* 2009, pp,3929–41.

[R28.42] Taylor, Richard P. "Fractal Fluency: An Intimate Relationship Between the Brain and Processing of Fractal Stimuli". In Di Ieva, Antonio (ed.). *The Fractal Geometry of the Brain,* Springer Series in Computational Neuroscience. Springer. *(2016).* pp. 485–496.

[R28.43] Taylor, R. P.; Spehar, B.; Van Donkelaar, P.; Hagerhall, C. M. "Perceptual and Physiological Responses to Jackson Pollock's Fractals". *Frontiers in Human Neuroscience.*Vol. *5:* (2011). 1–13.

[R28.44] Zhiwei Huang; Yunho Hwang; Vikrant Aute; Reinhard Radermacher "Review of Fractal Heat Exchangers" *International Refrigeration and Air Conditioning Conference. Paper 1725,* 2016.

[R29] ALGEBRAIC-GEEOMETRIC THEORY OF SPIDER GRAPHS, SPIDERMATICS AND ANANSIMATICS

[R29.1] Barwise, J., Etchemendy, J.: "Visual information and valid reasoning." In: Allwein, G., Barwise, J. (eds.) *Logical Reasoning with Diagrams*, pp. 3–25. Oxford University Press, New York (1996).

[R29.2] Bottoni, P., Fish, A.: "Coloured Euler diagrams: a tool for visualizing dynamic systems and structured information." In: *Proceeding Diagrams 2010, LNCS*, (Low Noise Cable Sensors) Vol. 6170, pp. 39–53 (2010).

[R29.3] Bottoni, P., Andrew Fish & Francesco Parisi Presicce Spider Graphs: a graph transformation system for spider diagrams *Software & Systems Modeling* **volume 14**, pages1421–1453 (2015).

[R29.4] Bottoni, P., Koch, M., Parisi-Presicce, F., Taentzer, G.: "Consistency checking and visualization of OCL constraints." In: *Proceedings of the UML 2000, LNCS,* (Low Noise Cable Sensors), vol. 1939, pp. 294–308 (2000).

[R29.5] Bottoni, P., Koch, M., Parisi-Presicce, F., Taentzer, G.: "A visualization of OCL using collaborations." In: Gogolla, M., Kobryn, C. (eds.) *Proceedings of the UML (unified Model Language) 2001, LNCS,* Vol. 2185 (2001) , pp. 257–271.

[R29.6] Bottoni, P., Schürr, A., Taentzer, G.: "Efficient parsing of visual languages based on critical pair analysis (and contextual layered graph transformation." In: *Proceedings of the IEEE-VL'00,* IEEE CS Press (2000), pp. 59-61.

[R29.7] Cabot, J., Clariso, R., Riera, D.: "Verification of UML/OCL class diagrams using constraint programming." *In: Proceedings of the ICSTW '08,* IEEE CS Press (2008) pp. 73–80.

[R29.8] Cordasco, G., De Chiara, R., Fish, A.: "Interactive visual classification with Euler diagrams." In*: Proceedings of the VL/HCC 2009*, IEEE CS Press (2009) pp. 185–192.

[R29.9] Ehrig, H., Ehrig, K., Habel, A., Pennemann, K.H.: "Theory of constraints and application conditions: from graphs to high-level structures." *Fundam. Inform.* **74**(1), (2006) pp.135–166.

[R29.10] Ehrig, H., Ehrig, K., Prange, U., Taentzer, G.: "Fundamental theory for typed attributed graphs and graph transformation based on adhesive HLR categories." *Fundam. Inform.* **74**(1), 31–61 (2006).

[R29.11] Ehrig, H., Ehrig, K., Prange, U., Taentzer, G.: *Fundamentals of Algebraic Graph Transformation*. Berlin, Springer, (2006).

[R29.12] Ehrig, H., Heckel, R., Korff, M., Löwe, M., Ribeiro, L., Wagner, A., Corradini, A.: "Algebraic approaches to graph transformation—part II: single pushout approach and comparison with double pushout approach." In: Rozenberg, G. (ed.) *Handbook of Graph Grammars and Computing by Graph Transformations, Volume 1: Foundations*, World Scientific, Singapore (1997) pp. 247–312.

[R29.13] Fish, A., Flower, J., Howse, J.: The semantics of augmented constraint diagrams. (JVLC) *Journal of Visual Language and Computing* Vol.**16** (2005) , pp.541–573

[R29.14] Fish, A., John, C., Taylor, J.: "A normal form for Euler diagrams with shading." *In: Proceedings of the diagrams 2008, LNCS, vol. 5223,*. Berlin, Springer, (2008) pp. 206–221

[R29.15] Flower, J., Masthoff, J., Stapleton, G.: "Generating proofs with spider diagrams using heuristics." *In: Proceedings of DMS-VLC,* Knowledge Systems Institute (2004), pp. 279–285.

[R29.16] Flower, J., Masthoff, J., Stapleton, G.: "Generating readable proofs: a heuristic approach to theorem proving with spider diagrams." In*: Proceedings of Diagrams 2004, LNAI, vol. 2980*, Berlin, Springer, (2004) pp. 166–181.

[R29.17] Flower, J., Stapleton, G.: "Automated theorem proving with spider diagrams." In: *Proceedings of Computing: The Australasian Theory Symposium, ENTCS,* vol. 91, Elsevier (2004) pp. 116–132

[R29.18] Goedicke, M., Meyer, T., Taentzer, G.: "Viewpoint-oriented software development by distributed graph transformation: towards a basis for living with inconsistencies." In: *Proceedings of IEEE RE'99*, (1999), pp. 92–99

[R29.18] Habel, A., Heckel, R., Taentzer, G.: "Graph grammars with negative application conditions." *Fundam. Inform.* **26**(3,4), (1996), pp. 287–313

[R29.19] Habel, A., Pennemann, K.H.: "Correctness of high-level transformation systems relative to nested conditions." *Math. Structructures in. Computer. Sci.* **19**(2), (2009), pp. 245–296

[R29.20] Hammer, E., Shin, S.J.: "Euler's visual logic." *History and Philosophy of Logic,* (1998) pp. 1–29

[29.21] Howse, J., Molina, F., Taylor, J., Kent, S., Gil, J.: Spider diagrams: a diagrammatic reasoning system. JVLC (Journal of Visual Language and Computing)**, Vol. 12**(3), 299–324 (2001)

[R29.22] Howse, J., Stapleton, G., Taylor, J.: "Spider diagrams." *LMS J. Comput. Math.* Vol. **8**, (2005), 145–194

[R29.23] Kreowski, H.J., Kuske, S., Schürr, A.: "Nested graph transformation units." (Int. J. SEKE) *International Journal of Software Engineering and Knowledge Engineering*, Vol. 7(4), (1997) pp.479–502

[R29.24] Shin, S.J., *The Logical Status of Diagrams*., Cambridge, Cambridge University Press (1994)

[R29.25] Sowa, J.: *Conceptual Structures: Information Processing in Mind and Machine.* London, Addison-Wesley, (1984)

[R29.26] Stapleton, G., Howse, J., Taylor, J., "A decidable constraint diagram reasoning system." *J. Logic and Computaion.* Vol. **15**(6), (2005), pp. 975–1008.

[R29.27] Stapleton, G., Masthoff, J., Flower, J., Fish, A., Southern, J.: "Automated Theorem Proving in Euler Diagram systems." *J. Automated. Reason.* Vol. **39**(4), (2007), 431–470.

[R29.28] Stapleton, G., Thompson, S., Howse, J., Taylor, J.: "The expressiveness of spider diagrams." *J. Logic and. Computation*. Vol. **14**(6), (2004), pp. 857–880.

[R29.29] Urbas, M., Jamnik, M., Stapleton, G., Flower, J., "Speedith: a diagrammatic reasoner for spider diagrams." In: Proceedings of Diagrams 2012, LNCS, (Low Noise Cable Sensors). Vol. 7352, pp. 163–177. Springer, Berlin (2012)

[R29.30] Warmer, J., Kleppe, A.: *The Object Constraint Language: Precise Modeling with UML.* London Addison-Wesley, (1999)

[R29.31] Zykov, A.: Hypergraphs. Russian. Math. Survs. Vol. **29**(6), (1974) pp. 89–156

Index

K. K. Dompere, *The Theory of Epistemic Fields*, Studies in Systems, Decision and Control 498, https://doi.org/10.1007/978-3-031-42470-0

D

J

K

Q

R

S

T

U

The manufacturer's authorised representative in the EU is Springer Nature Customer Service Centre GmbH, Europaplatz 3, 69115 Heidelberg, Germany. If you have any concerns regarding our products, please contact ProductSafety@springernature.com

Printed and bound by CPI Group (UK) Ltd, Croydon, CR0 4YY
23/09/2026
02221534-0005